ROTATION-VIBRATION OF POLYATOMIC MOLECULES

Higher Order Energies and Frequencies of Spectral Transitions

G. AMAT
Faculté des Sciences de Paris
Paris, France

H. H. NIELSEN
Department of Physics
The Ohio State University
Columbus, Ohio

G. TARRAGO
Faculté des Sciences de Paris
Paris, France

with the collaboration of

M. Goldsmith, M. L. Grenier-Besson, S. K. Kurtz, P. Kupecek

MARCEL DEKKER, INC., NEW YORK 1971

MARCEL DEKKER, INC.

95 Madison Avenue, New York, New York 10016

LIBRARY OF CONGRESS CATALOG CARD NUMBER: 71-152569

ISBN NO.: 0-8247-1007-X

Printed in the United States of America

FOREWORD

The field of high resolution molecular spectroscopy has shown during the past few years many impressive developments, thanks to the improvement of the instrumentation resulting in particular from the availability of sensitive detectors, the progress of interferometric methods, the use of laser sources and the appearance of new techniques, such as non-linear absorption. High quality rotation and rotation-vibration spectra are being obtained for many polyatomic molecules, both in the infrared and the microwave regions. The interpretation of such spectra requires the use of very accurate formulae for the frequencies of the spectral lines expressed in terms of quantum numbers and molecular parameters. For this purpose it is necessary to compute rotation-vibration energies to a high order of accuracy (i.e. including fourth order or even sixth order perturbations).

The first half of this monograph deals with the theoretical material upon which these high order energy calculations are based : the expansion of the Hamiltonian and the perturbation computation are performed to fourth order for any polyatomic molecule, and the calculation of non-vanishing matrix elements is discussed for linear, axially symmetric and asymmetric top molecules.

The second half of the book gives explicit formulae for rotation-vibration energies and for frequencies of spectral lines, namely rotational lines (up to sixth order of approximation) in the ground state and in the low excited vibrational states, as well as rotation-vibration lines (up to fourth order) in the fundamentals and in the lowest overtones and combination bands. These formulae are given for linear molecules and for axially symmetric molecules with a threefold or a fourfold axis.

In an endeavour of this type it is difficult to acknowledge the support and assistance given by many individuals and organizations. It should be mentioned that the close cooperation in research activities between The Ohio State University and the Faculté des Sciences de Paris played an important role.

We thank Dr. M. Goldsmith, Dr. M.-L. Grenier-Besson, Dr. S.K. Kurtz and Dr. P. Kupecek who have done part of the computations presented in this book. We are also grateful to Dr. H. Hanson for helpful discussions of some of the material included herein, to Dr. S. Maes, who has verified many of the calculations, to Dr. M.H. Andrade e Silva for her critical reading of the manuscript and also to Mrs. C. Aubert-Arnardi who has drawn the tables and the mathematical symbols and to Mrs. D. Seletti who has typed the text.

Our gratitude goes furthermore to Dr. J.W. Stout, editor of the Journal of Chemical Physics and Dr. Hugh Wolfe of the American Institute of Physics who gave their permission to the republication of some of the material included in this work. The kind interest shown by Professor K. Narahari Rao of The Ohio State University is also gratefully acknowledged.

G. AMAT

H.H. NIELSEN

G. TARRAGO

CONTENTS

Page

Foreword III

Introduction 1

Part 1 : Expansion of the Hamiltonian and Contact Transformations

Chapter I : Expansion of the Hamiltonian 5

Chapter II : First Contact Transformation (to third order) 15

Chapter III : Second Contact Transformation (to third order) 37

Chapter IV : First and Second Contact Transformations (fourth order terms) 51

Chapter V : Contact Transformations in the case of a Fermi Resonance 67

Part 2 : Matrix Elements of the Transformed Hamiltonian

Chapter VI : Non-vanishing Matrix Elements for a Molecule of Given Symmetry 85

Chapter VII : Orders of Magnitude of Matrix Elements and of their Contributions to the Energy 91

Chapter VIII : Matrix Elements of the Transformed Hamiltonian in Terms of Quantum Numbers (to third order) 111

Chapter IX : Matrix Elements of the Transformed Hamiltonian in Terms of Quantum Numbers (fourth order) 119

Part 3 : Energy Levels, Frequencies of Rotation and Rotation-Vibration Transitions

Chapter X : Molecules with a threefold Symmetry 203

Chapter XI : Molecules with a fourfold Symmetry 325

Chapter XII : Linear Molecules 391

References 439

INTRODUCTION

This Monograph is divided into three Parts. In Part 1, the rotation-vibration Hamiltonian is first expanded in a power series with respect to normal coordinates. This is done in Chapter I to fourth order :

$$H = H_0 + H_1 + H_2 + H_3 + H_4$$

Then two contact transformations are performed :

$$H' = THT^{-1} = H_0 + h'_1 + h'_2 + h'_3 + h'_4$$

$$H^{\dagger} = \mathcal{T} H' \mathcal{T}^{-1} = H_0 + h'_1 + h^{\dagger}_2 + h^{\dagger}_3 + h^{\dagger}_4$$

in order to partially diagonalize the Hamiltonian matrix, namely with respect to the vibrational quantum numbers v_s . The first three terms in the expansion of $H^{\dagger}$ are diagonal with respect to the quantum numbers v_s . When no accidental resonance occurs, this partial diagonalization proves to be sufficient in order to compute the rotation-vibration energy to fifth order, since the matrix elements of $h^{\dagger}_3$, nondiagonal in v_s , will contribute to the energy only in the sixth order. Operators h'_1 , h'_2, h'_3, operators $h^{\dagger}_2$, $h^{\dagger}_3$ and operators h'_4 , $h^{\dagger}_4$ are computed in Chapters II, III and IV respectively. The modifications to the contact transformations required in the event of an accidental Fermi type resonance are then described in Chapter V.

The rotation-vibration energy will be obtained by solving for the roots of a secular equation in order to diagonalize the Hamiltonian matrix with respect to the quantum numbers ℓ_s, m_s ... associated with degenerate vibrations and with respect to the rotational quantum number K . Part 2 deals with the computation of the matrix elements of the transformed Hamiltonian $H^{\dagger}$ appearing in this secular equation. In Chapter VI, the non-vanishing matrix elements of $H^{\dagger}$ are determined for a molecule of a given symmetry. In Chapter VII, the orders of magnitude of these matrix elements are discussed together with the orders of magnitude of their contributions to the rotation-vibration energy. The matrix elements of the transformed Hamiltonian $H^{\dagger}$ are then given to third order in Chapter VIII in terms of quantum numbers. The matrix elements of $h_4^{\dagger}$ are subsequently computed in Chapter IX.

In Part 3,the rotation-vibration energy is computed for the simplest vibration levels. Frequencies of rotational transitions are then given for these various vibrational levels up to the sixth order of approximation and frequencies of rotation-vibration transitions are derived up to the fourth order of approximation. This is done for axially symmetric molecules with a threefold symmetry in Chapter X, for axially symmetric molecules with a fourfold symmetry in Chapter XI, and for linear molecules in Chapter XII.

The topics presented in Chapters I to V and in Chapter VII originate from a series of six papers published in the Journal of Chemical Physics by G.Amat, H.H.Nielsen, M.Goldsmith and M.L.Grenier-Besson,under the Title : "Higher Order Rotation-Vibration Energies of Polyatomic Molecules" [1 to 6] ; the materials presented in Chapters IX, X, XI and XII originate from theses written by S.K.Kurtz [7], G.Tarrago [8] and P.Kupecek [9].

Throughout the present Monograph we shall make use of the notation developed by H.H.Nielsen in reference [10].

PART 1

EXPANSION OF THE HAMILTONIAN AND CONTACT TRANSFORMATIONS

CHAPTER I

EXPANSION OF THE HAMILTONIAN

INTRODUCTION

In the present Chapter we expand the rotation-vibration Hamiltonian of a polyatomic molecule in orders of magnitude as[1] :

$$H = H_0 + \lambda H_1 + \lambda^2 H_2 + \lambda^3 H_3 + \lambda^4 H_4 + \cdots \tag{I,1}$$

and give the expressions of H_0 , H_1 , H_2 , H_3 and H_4 in terms of molecular parameters.

THE SCHROEDINGER EQUATION

The wave equation

$$(H - E)\Psi = 0 \tag{I,2}$$

of a rotating vibrating polyatomic molecule was discussed first by E.B.Wilson and J.B.Howard [11] and then by B.T.Darling and D.M.Dennison[2] [12] whose formulation, which is now known to be equivalent [10] to that of the former authors, proves more convenient in the present work. We therefore take :

$$H = \tfrac{1}{2}\Big[\mu^{\frac{1}{4}} \sum_{\alpha\beta} (P_\alpha - p_\alpha)\mu_{\alpha\beta}\,\mu^{-\frac{1}{2}}(P_\beta - p_\beta)\mu^{\frac{1}{4}}$$

$$+ \mu^{\frac{1}{4}} \sum_{s\sigma} p^*_{s\sigma}\,\mu^{-\frac{1}{2}}\,p^*_{s\sigma}\,\mu^{\frac{1}{4}}\Big] + V \tag{I,3}$$

where, in accord with previous notation [10], α and β have the range x, y, z (the principal axes of inertia of the molecule). P_α and p_α are components of total and internal angular momentum respectively, $p^*_{s\sigma}$ is the momentum conjugate to the normal coordinate $Q_{s\sigma}$ (i.e. $p^*_{s\sigma} = -i\hbar\,(\partial/\partial Q_{s\sigma})$), and V is the potential function of the molecular force field. The inertial parameters are given by

$$\mu_{\alpha\alpha} = (I'_{\beta\beta} I'_{\gamma\gamma} - I'^{2}_{\beta\gamma}) \mu \qquad \text{(I,4a)}$$

$$\mu_{\alpha\beta} = (I'_{\gamma\gamma} I'_{\alpha\beta} + I'_{\alpha\gamma} I'_{\gamma\beta}) \mu \qquad (\alpha \neq \beta) \qquad \text{(I,4b)}$$

$$\mu^{-1} = \begin{vmatrix} I'_{xx} & -I'_{xy} & -I'_{xz} \\ -I'_{xy} & I'_{yy} & -I'_{yz} \\ -I'_{xz} & -I'_{yz} & I'_{zz} \end{vmatrix} \qquad \text{(I,5)}$$

and

$$I'_{\alpha\alpha} = I^{e}_{\alpha\alpha} + \sum_{s\sigma} a^{\alpha\alpha}_{s\sigma} Q_{s\sigma} + \sum_{s\sigma s'\sigma'} A'^{\alpha\alpha}_{s\sigma s'\sigma'} Q_{s\sigma} Q_{s'\sigma'} \qquad \text{(I,6a)}$$

$$I'_{\alpha\beta} = I^{e}_{\alpha\beta} - \sum_{s\sigma} a^{\alpha\beta}_{s\sigma} Q_{s\sigma} - \sum_{s\sigma s'\sigma'} A'^{\alpha\beta}_{s\sigma s'\sigma'} Q_{s\sigma} Q_{s'\sigma'} \qquad (\alpha \neq \beta) \qquad \text{(I,6b)}$$

where $I^{e}_{\alpha\alpha}$ and $I^{e}_{\alpha\beta}$ $(\alpha \neq \beta)$ are the equilibrium values of the moments and products of inertia respectively. Since we use the principal axes of inertia as our molecular framework, the values of $I^{e}_{\alpha\beta} (\alpha \neq \beta)$ vanish in Equation (I,6b).

Explicit evaluations of $a^{\alpha\beta}_{s\sigma}$ and $A'^{\alpha\beta}_{s\sigma s'\sigma'}$, which are constants for a given molecule, are given in the Appendix to the present Chapter.

We may now make use of the definition :

$$p_{\alpha} = \sum_{s\sigma} \sum_{s'\sigma'} \zeta^{\alpha}_{s\sigma s'\sigma'} Q_{s\sigma} p^{*}_{s'\sigma'} \qquad \text{(I,7)}$$

where the $\zeta^{\alpha}_{s\sigma s'\sigma'}$ are Coriolis coupling coefficients (see Appendix), to transform the Equation (I,3) into :

$$H = \frac{1}{2} \sum_{\alpha\beta} \left\{ [P_{\alpha} - p_{\alpha}] \mu_{\alpha\beta} [P_{\beta} - p_{\beta}] + \mu^{\frac{1}{4}} (p_{\alpha} \mu_{\alpha\beta} \mu^{-\frac{1}{2}} (p_{\beta} \mu^{\frac{1}{4}})) \right\}$$

$$+ \frac{1}{2} \sum_{s\sigma} \left\{ p^{*2}_{s\sigma} + \mu^{\frac{1}{4}} (p^{*}_{s\sigma} \mu^{-\frac{1}{2}} (p^{*}_{s\sigma} \mu^{\frac{1}{4}})) \right\} + V \qquad \text{(I,8)}$$

where, following the notation introduced by E.B.Wilson and J.B.Howard [11], an operator in parentheses acts only upon quantities also embraced by those parentheses and not upon the wave functions; brackets and braces, however, are used in the customary fashion. Had we started from the Wilson-Howard Hamiltonian and tried to obtain the analog of Equation (I,8), we should have obtained additional terms linear in p_α and $p^*_{s\sigma}$. One of the principal advantages of employing the Darling-Dennison formulation is the elimination of these additional operators. We now cast Equation (I,8) into a form more convenient for future computation :

$$H=\frac{1}{2}\sum_{\alpha\beta}\left\{\mu_{\alpha\beta}P_\alpha P_\beta-\left[p_\alpha\mu_{\alpha\beta}+\mu_{\alpha\beta}p_\alpha\right]P_\beta+p_\alpha\mu_{\alpha\beta}p_\beta\right\}$$

$$+\frac{1}{2}\sum_{s\sigma}p^{*2}_{s\sigma}+U+V \qquad (I,9)$$

with :

$$U=\frac{1}{2}\mu^{\frac{1}{4}}\sum_{\alpha\beta}(p_\alpha\mu_{\alpha\beta}\mu^{-\frac{1}{2}}(p_\beta\mu^{\frac{1}{4}}))-\frac{1}{8}\mu\sum_{s\sigma}\left[(p^{*2}_{s\sigma}\mu^{-1})\right.$$

$$\left.-\frac{3}{4}\mu(p^*_{s\sigma}\mu^{-1})^2\right] \qquad (I,10)$$

It is not possible to find directly the eigenvalues of the operator (I,9) and we therefore seek an expansion in orders of magnitude [5]

$$H=H_0+\lambda H_1+\lambda^2H_2+\cdots \qquad (I,11)$$

which will allow us to apply perturbation theory. Let us first expand the kinetic part $H-V$ of the Hamiltonian (I,9) : to accomplish this, we may write[3]

$$\mu_{\alpha\beta}=(I^e_{\alpha\alpha}I^e_{\beta\beta})^{-1}\left[I^e_{\alpha\beta}+\sum_{s\sigma}\Omega^{(1)\alpha\beta}_{s\sigma}Q_{s\sigma}+\sum_{s\sigma s'\sigma'}\Omega^{(2)\alpha\beta}_{s\sigma s'\sigma'}Q_{s\sigma}Q_{s'\sigma'}+\cdots\right] \qquad (I,12a)$$

$$\mu=(I^e_{\delta\delta}I^e_{\varepsilon\varepsilon}I^e_{\eta\eta})^{-1}\left[\varphi^{(0)}+\sum_{s\sigma}\varphi^{(1)}_{s\sigma}Q_{s\sigma}+\sum_{s\sigma s'\sigma'}\varphi^{(2)}_{s\sigma s'\sigma'}Q_{s\sigma}Q_{s'\sigma'}+\cdots\right] \qquad (I,12b)$$

and

$$\mu^{-1}=(I^{e}_{\delta\delta}I^{e}_{\varepsilon\varepsilon}I^{e}_{\eta\eta})\left[\delta^{(0)}+\sum_{s\sigma}\delta^{(1)}_{s\sigma}Q_{s\sigma}+\sum_{s\sigma s'\sigma'}\delta^{(2)}_{s\sigma s'\sigma'}Q_{s\sigma}Q_{s'\sigma'}+\cdots\right] \qquad (I,12c)$$

where δ , ε and η may be either x, y or z and where $\delta \neq \varepsilon \neq \eta$. Explicit values of the coefficients $\Omega^{(n)\alpha\beta}_{\cdots}$, $\varphi^{(n)}_{\cdots}$, $\delta^{(n)}_{\cdots}$ in terms of the equilibrium values of the moment of inertia $I^{e}_{\alpha\alpha}$ and of the rotation-vibration interaction coefficients $a^{\alpha\beta}_{s\sigma}$ and $A'^{\alpha\beta}_{s\sigma s'\sigma'}$, can be found in reference [1]. When Equation (I,7) is taken into account, these results make it possible to express the kinetic part of operators H_0 , H_1 , H_2 H_3 and H_4 in Equation (I,11) with respect to

- the rotational operators P_α (components of the total angular momentum),
- the vibrational operators $Q_{s\sigma}$, $p^{*}_{s\sigma}$ (normal coordinates and their conjugate momenta),
- and the molecular parameters $I^{e}_{\alpha\alpha}$, $a^{\alpha\beta}_{s\sigma}$, $A'^{\alpha\beta}_{s\sigma s'\sigma'}$, $\zeta^{\alpha}_{s\sigma s'\sigma'}$.

Actually, this computation is made simpler when the three following remarks are taken into consideration :

(1) Using the relations between rotation-vibration interaction coefficients established in reference [13], it is possible to express $A'^{\alpha\beta}_{s\sigma s'\sigma'}$ in terms of the other parameters : $a^{\alpha\beta}_{s\sigma}$, $\zeta^{\alpha}_{s\sigma s'\sigma'}$ and $I^{e}_{\alpha\alpha}$ which are therefore the only molecular parameters appearing in the kinetic part of H .

(2) Recently, Watson [14], making use of properties of the tensor $\mu_{\alpha\beta}$, has shown that, in Hamiltonian (I,9) the term U (defined by Equation (I,10)) can actually be written under a very simple form, namely:

$$U=-\frac{\hbar^2}{8}\sum_{\alpha}\mu_{\alpha\alpha} \qquad (I,13)$$

so that it is not necessary to compute the operators μ , μ^{-1} , $\mu^{-\frac{1}{2}}$, $\mu^{-\frac{1}{4}}$.

(3) Watson [14] gives, for the expansion of $\mu_{\alpha\beta}$ with respect to the normal coordinates $Q_{s\sigma}$, a general formula from which the coefficients $\Omega^{(n)}_{\cdots}$ appearing in (I,12a) can be conveniently derived. These

coefficients have been written by L.S.Rothman and S.A.Clough[4]; explicit expressions for $\Omega^{(1)\alpha\beta}_{s\sigma}$, $\Omega^{(2)\alpha\beta}_{s\sigma s'\sigma'}$, $\Omega^{(3)\alpha\beta}_{s\sigma s'\sigma' s''\sigma''}$ and $\Omega^{(4)\alpha\beta}_{s\sigma s'\sigma' s''\sigma'' s'''\sigma'''}$ are given by them as follows[5] :

$$\Omega^{(1)\alpha\beta}_{s} = 2 I^{e}_{\alpha\alpha} I^{e}_{\beta\beta} \left(\frac{\lambda_s}{\hbar^2}\right)^{\frac{1}{4}} \, {}^{1}R^{\alpha\beta}_{s} \tag{I,14a}$$

$$\Omega^{(2)\alpha\beta}_{ss'} = 2 I^{e}_{\alpha\alpha} I^{e}_{\beta\beta} \left(\frac{\lambda_s \lambda_{s'}}{\hbar^4}\right)^{\frac{1}{4}} \, {}^{2}R^{\alpha\beta}_{ss'} \tag{I,14b}$$

$$\Omega^{(3)\alpha\beta}_{ss's''} = 2 I^{e}_{\alpha\alpha} I^{e}_{\beta\beta} \left(\frac{\lambda_s \lambda_{s'} \lambda_{s''}}{\hbar^6}\right)^{\frac{1}{4}} \, {}^{3}R^{\alpha\beta}_{ss's''} \tag{I,14c}$$

$$\Omega^{(4)\alpha\beta}_{ss's''s'''} = 2 I^{e}_{\alpha\alpha} I^{e}_{\beta\beta} \left(\frac{\lambda_s \lambda_{s'} \lambda_{s''} \lambda_{s'''}}{\hbar^8}\right)^{\frac{1}{4}} \, {}^{4}R^{\alpha\beta}_{ss's''s'''} \tag{I,14d}$$

with

$${}^{1}R^{\alpha\beta}_{s} = -\frac{\mathcal{a}^{\alpha\beta}_{s}}{2 I^{e}_{\alpha\alpha}} \tag{I,15a}$$

$${}^{2}R^{\alpha\beta}_{ss'} = -\frac{3}{8} \sum_{\gamma} \left({}^{1}R^{\alpha\gamma}_{s} \, \mathcal{a}^{\gamma\beta}_{s'} + {}^{1}R^{\alpha\gamma}_{s'} \, \mathcal{a}^{\gamma\beta}_{s} \right) \tag{I,15b}$$

$${}^{3}R^{\alpha\beta}_{ss's''} = -\frac{2}{9} \sum_{\gamma} \left({}^{2}R^{\alpha\gamma}_{ss'} \, \mathcal{a}^{\gamma\beta}_{s''} + {}^{2}R^{\alpha\gamma}_{ss''} \, \mathcal{a}^{\gamma\beta}_{s'} + {}^{2}R^{\alpha\gamma}_{s's''} \, \mathcal{a}^{\gamma\beta}_{s} \right) \tag{I,15c}$$

$$\begin{aligned} {}^{4}R^{\alpha\beta}_{ss's''s'''} = -\frac{5}{32} \sum_{\gamma} \Big(& {}^{3}R^{\alpha\gamma}_{ss's''} \, \mathcal{a}^{\gamma\beta}_{s'''} + {}^{3}R^{\alpha\gamma}_{ss's'''} \, \mathcal{a}^{\gamma\beta}_{s''} \\ & + {}^{3}R^{\alpha\gamma}_{ss''s'''} \, \mathcal{a}^{\gamma\beta}_{s'} + {}^{3}R^{\alpha\gamma}_{s's''s'''} \, \mathcal{a}^{\gamma\beta}_{s} \Big) \end{aligned} \tag{I,15d}$$

and

$$\mathcal{a}^{\alpha\beta}_{s} = \frac{a^{\alpha\beta}_{s}}{I^{e}_{\beta\beta}} \left(\frac{\hbar^2}{\lambda_s}\right)^{\frac{1}{4}} \tag{I,16}$$

Finally, Equations (I,12a) and (I,13) are substituted in Equation (I,9) to yield H expanded in orders of magnitude according to the following terms :

$$H_0 = \frac{1}{2}\sum_{s\sigma}\sum_{\alpha}\left\{\frac{P_\alpha^2}{I_{\alpha\alpha}^e} + \hbar\lambda_s^{\frac{1}{2}}\left(\frac{p_{s\sigma}^2}{\hbar^2} + q_{s\sigma}^2\right)\right\} \tag{I,17a}$$

$$H_1 = \frac{1}{2}\sum_{s\sigma}\sum_{\alpha\beta}\left\{\frac{\Omega_{s\sigma}^{(1)\alpha\beta}}{I_{\alpha\alpha}^e I_{\beta\beta}^e}\left(\frac{\hbar^2}{\lambda_s}\right)^{\frac{1}{4}} q_{s\sigma}P_\alpha P_\beta - \frac{2p_\alpha P_\alpha}{I_{\alpha\alpha}^e}\right\} + V_1 \tag{I,17b}$$

$$H_2 = \frac{1}{2}\sum_{s\sigma s'\sigma'}\sum_{\alpha\beta}\left\{\frac{\Omega_{s\sigma s'\sigma'}^{(2)\alpha\beta}}{I_{\alpha\alpha}^e I_{\beta\beta}^e}\left(\frac{\hbar^4}{\lambda_s\lambda_{s'}}\right)^{\frac{1}{4}} q_{s\sigma}q_{s'\sigma'}P_\alpha P_\beta\right.$$

$$\left. - \frac{\Omega_{s\sigma}^{(1)\alpha\beta}}{I_{\alpha\alpha}^e I_{\beta\beta}^e}\left(\frac{\hbar^2}{\lambda_s}\right)^{\frac{1}{4}}(p_\alpha q_{s\sigma} + q_{s\sigma}p_\alpha)P_\beta + \frac{p_\alpha^2}{I_{\alpha\alpha}^e}\right\} + V_2 \tag{I,17c}$$

$$H_3 = \frac{1}{2}\sum_{s\sigma s'\sigma' s''\sigma''}\sum_{\alpha\beta}\left\{\frac{\Omega_{s\sigma s'\sigma' s''\sigma''}^{(3)\alpha\beta}}{I_{\alpha\alpha}^e I_{\beta\beta}^e}\left(\frac{\hbar^6}{\lambda_s\lambda_{s'}\lambda_{s''}}\right)^{\frac{1}{4}} q_{s\sigma}q_{s'\sigma'}q_{s''\sigma''}P_\alpha P_\beta\right.$$

$$- \frac{\Omega_{s\sigma s'\sigma'}^{(2)\alpha\beta}}{I_{\alpha\alpha}^e I_{\beta\beta}^e}\left(\frac{\hbar^4}{\lambda_s\lambda_{s'}}\right)^{\frac{1}{4}}(p_\alpha q_{s\sigma}q_{s'\sigma'} + q_{s\sigma}q_{s'\sigma'}p_\alpha)P_\beta$$

$$\left. + \frac{\Omega_{s\sigma}^{(1)\alpha\beta}}{I_{\alpha\alpha}^e I_{\beta\beta}^e}\left(\frac{\hbar^2}{\lambda_s}\right)^{\frac{1}{4}} p_\alpha q_{s\sigma}p_\beta + \Lambda_{s\sigma}^{(3)}\left(\frac{\hbar^2}{\lambda_s}\right)^{\frac{1}{4}} q_{s\sigma}\right\} + V_3 \tag{I,17d}$$

$$H_4 = \frac{1}{2}\sum_{\substack{s\sigma s'\sigma' s''\sigma''\\ s'''\sigma'''}}\sum_{\alpha\beta}\left\{\frac{\Omega_{s\sigma s'\sigma' s''\sigma'' s'''\sigma'''}^{(4)\alpha\beta}}{I_{\alpha\alpha}^e I_{\beta\beta}^e}\left(\frac{\hbar^8}{\lambda_s\lambda_{s'}\lambda_{s''}\lambda_{s'''}}\right)^{\frac{1}{4}} q_{s\sigma}q_{s'\sigma'}q_{s''\sigma''}q_{s'''\sigma'''}P_\alpha P_\beta\right.$$

$$- \frac{\Omega_{s\sigma s'\sigma' s''\sigma''}^{(3)\alpha\beta}}{I_{\alpha\alpha}^e I_{\beta\beta}^e}\left(\frac{\hbar^6}{\lambda_s\lambda_{s'}\lambda_{s''}}\right)^{\frac{1}{4}}(p_\alpha q_{s\sigma}q_{s'\sigma'}q_{s''\sigma''} + q_{s\sigma}q_{s'\sigma'}q_{s''\sigma''}p_\alpha)P_\beta$$

$$+\frac{\Omega^{(2)\alpha\beta}_{s\sigma s'\sigma'}}{I^e_{\alpha\alpha}I^e_{\beta\beta}}\left(\frac{\hbar^4}{\lambda_s\lambda_{s'}}\right)^{\frac{1}{4}}p_\alpha q_{s\sigma}q_{s'\sigma'}p_\beta+\Lambda^{(4)}_{s\sigma s'\sigma'}\left(\frac{\hbar^4}{\lambda_s\lambda_{s'}}\right)^{\frac{1}{4}}q_{s\sigma}q_{s'\sigma'}\Bigg\}+V_4 \quad (I,17e)$$

In these equations, λ_s is a force constant of the harmonic potential

$$V_0=\frac{1}{2}\sum_{s\sigma}\lambda_s Q^2_{s\sigma} \quad (I,18)$$

$q_{s\sigma}$ is a dimensionless normal coordinate and $p_{s\sigma}$ its conjugate momentum defined by :

$$q_{s\sigma}=\left(\frac{\lambda_s}{\hbar_2}\right)^{\frac{1}{4}}Q_{s\sigma} \quad , \quad p_{s\sigma}=-i\hbar\frac{\partial}{\partial q_{s\sigma}} \quad (I,19)$$

V_1 and V_2 are respectively the cubic and the quartic terms of the anharmonic potential and are given by

$$V_1=hc\sum_{\substack{s\sigma s'\sigma' s''\sigma''\\ s\sigma\leqslant s'\sigma'\leqslant s''\sigma''}} k_{s\sigma s'\sigma' s''\sigma''}q_{s\sigma}q_{s'\sigma'}q_{s''\sigma''} \quad (I,20)$$

$$V_2=hc\sum_{\substack{s\sigma s'\sigma' s''\sigma'' s'''\sigma'''\\ s\sigma\leqslant s'\sigma'\leqslant s''\sigma''\leqslant s'''\sigma'''}} k_{s\sigma s'\sigma' s''\sigma'' s'''\sigma'''}q_{s\sigma}q_{s'\sigma'}q_{s''\sigma''}q_{s'''\sigma'''} \quad (I,21)$$

Similar definitions involving the force constants $k_{s\sigma s'\sigma' s''\sigma'' s'''\sigma''' s''''\sigma''''}$ and $k_{s\sigma s'\sigma' s''\sigma'' s'''\sigma''' s''''\sigma'''' s'''''\sigma'''''}$ hold for V_3 and V_4 .

The quantities $\Lambda^{(3)}_{s\sigma}$ and $\Lambda^{(4)}_{s\sigma s'\sigma'}$, which appear in H_3 and H_4 originate from the term U in Equation (I,9) ; according to Equations (I,12a) and (I,13), they are defined by :

$$\Lambda^{(3)}_{s\sigma}=-\frac{\hbar^2}{4}\sum_\alpha\frac{\Omega^{(1)\alpha\alpha}_{s\sigma}}{\left(I^e_{\alpha\alpha}\right)^2} \quad (I,22)$$

$$\Lambda^{(4)}_{s\sigma s'\sigma'}=-\frac{\hbar^2}{4}\sum_\alpha\frac{\Omega^{(2)\alpha\alpha}_{s\sigma s'\sigma'}}{\left(I^e_{\alpha\alpha}\right)^2} \quad (I,23)$$

APPENDIX TO CHAPTER I

The expressions for the $a^{\alpha\beta}_{s\sigma}$ and $A'^{\alpha\beta}_{s\sigma s'\sigma'}$ as given in reference [10] are:

$$a^{\alpha\alpha}_{s\sigma} = 2\sum_i M_i^{\frac{1}{2}}(\beta_i'^{0}\, l^{\beta}_{is\sigma} + \gamma_i'^{0}\, l^{\gamma}_{is\sigma}) \qquad \text{(AI,1a)}$$

$$a^{\alpha\beta}_{s\sigma} = -\sum_i M_i^{\frac{1}{2}}(\alpha_i'^{0}\, l^{\beta}_{is\sigma} + \beta_i'^{0}\, l^{\alpha}_{is\sigma}) \qquad (\alpha\neq\beta) \qquad \text{(AI,1b)}$$

$$A^{\alpha\alpha}_{s\sigma s''\sigma''} = \sum_i (l^{\beta}_{is\sigma}\, l^{\beta}_{is''\sigma''} + l^{\gamma}_{is\sigma}\, l^{\gamma}_{is''\sigma''}) \qquad \text{(AI,2a)}$$

$$A^{\alpha\beta}_{s\sigma s''\sigma''} = -\sum_i l^{\alpha}_{is\sigma}\, l^{\beta}_{is''\sigma''} \qquad (\alpha\neq\beta) \qquad \text{(AI,2b)}$$

$$A'^{\alpha\beta}_{s\sigma s''\sigma''} = A^{\alpha\beta}_{s\sigma s''\sigma''} - \sum_{s'\sigma'} \zeta^{\alpha}_{s\sigma s'\sigma'}\, \zeta^{\beta}_{s''\sigma'' s'\sigma'} \qquad (\alpha\neq\beta \text{ and } \alpha=\beta) \qquad \text{(AI, 3)}$$

where M_i is the mass of the ith nucleus, $\alpha_i'^{0}$, $\beta_i'^{0}$, and $\gamma_i'^{0}$ are its Cartesian coordinates(relative to the center of mass of the molecule)in its equilibrium position, the $l^{\alpha}_{is\sigma}$ are coefficients of the matrix which transforms weighted Cartesian to normal coordinates, and the $\zeta^{\alpha}_{s\sigma s'\sigma'}$ are given by the expression

$$\zeta^{\alpha}_{s\sigma s'\sigma'} = \sum_i (l^{\beta}_{is\sigma}\, l^{\gamma}_{is'\sigma'} - l^{\beta}_{is'\sigma'}\, l^{\gamma}_{is\sigma}) \qquad (\alpha\beta\gamma = xyz, yzx, zxy) \qquad \text{(AI, 4)}$$

The relations established between the interaction coefficients $a^{\alpha\beta}_{s\sigma}$, $A^{\alpha\beta}_{s\sigma s'\sigma'}$, $\zeta^{\alpha}_{s\sigma s'\sigma'}$ in the reference [13], enable us to write $A'^{\alpha\alpha}_{s\sigma s''\sigma''}$ and $A'^{\alpha\beta}_{s\sigma s''\sigma''}$ as

$$A'^{\alpha\alpha}_{s\sigma s''\sigma''} = \sum_\delta \frac{a^{\alpha\delta}_{s\sigma}\, a^{\alpha\delta}_{s''\sigma''}}{4 I^{e}_{\delta\delta}} \qquad \text{(AI,5a)}$$

$$A'^{\alpha\beta}_{s\sigma s''\sigma''} + A'^{\alpha\beta}_{s''\sigma'' s\sigma} = \sum_\delta \frac{a^{\alpha\delta}_{s\sigma}\, a^{\beta\delta}_{s''\sigma''} + a^{\alpha\delta}_{s''\sigma''}\, a^{\beta\delta}_{s\sigma}}{4 I^{e}_{\delta\delta}} \qquad \text{(AI,5b)}$$

NOTES TO CHAPTER I

1 . According to the customary notation, λ is a parameter of smallness equal to unity, which denotes the order of magnitude of the different terms in H .

2 . The energies obtained by the Darling-Dennison and the Wilson-Howard methods are identical if one suitably normalizes the corresponding wave functions (compare reference [10]).

3 . The following equation is valid for both the conditions $\alpha=\beta$ and $\alpha\neq\beta$. Let us recall that $I^{e}_{\alpha\beta}$ vanishes for $\alpha\neq\beta$.

4 . The authors are indebted to L.S. Rothman and S.A. Clough for private communication of their results.

5 . In Equations (I,14),(I,15) and (I,16), subscripts $s, s', s'', \cdots$ are written for $s\sigma, s'\sigma', s''\sigma'', \cdots$.

CHAPTER II

FIRST CONTACT TRANSFORMATION (TO THIRD ORDER)

INTRODUCTION

The Hamiltonian H , as given by Equations (I,17), may be transformed into one which is much more convenient for a perturbation calculation[1] of the rotation-vibration energies. This will be performed by subjecting the Hamiltonian to successive contact transformations according to the method discussed by W.H.Shaffer, H.H.Nielsen and L.H.Thomas [15]. In the present Chapter, as a step in the calculation of the rotation-vibration energies of a molecule to the third order of approximation, we shall carry out a first contact transformation on the Hamiltonian

$$e^{i\lambda S} H e^{-i\lambda S} = H' = H_0' + \lambda H_1' + \lambda^2 H_2' + \lambda^3 H_3' + \lambda^4 H_4' \qquad \text{(II, 1)}$$

the S function being so chosen that the operator $H_0 + \lambda H_1'$ will only have diagonal matrix elements with respect to the vibrational quantum numbers v_s in the representation which diagonalizes H_0 . The terms of H' will then be grouped in such a manner that the terms of the same "true" magnitude stand together[2]

$$H' = h_0' + \lambda h_1' + \lambda^2 h_2' + \lambda^3 h_3' + \lambda^4 h_4' \qquad \text{(II, 2)}$$

Before giving explicit expressions for the operators h_0' , h_1' , h_2' and h_3' (the expression for h_4' is deferred to Chapter IV), we shall, in the two next paragraphs, make several generalizations in relation to the contact transformation and develop the notation employed hereafter.

GENERALITIES CONCERNING THE CONTACT TRANSFORMATION

The general expressions for the operators H_n' are obtained by equating the coefficients of like powers of λ in the equation (II,1) expanded as

$$(1+i\lambda S - \cdots)(H_0 + \lambda H_1 + \cdots)(1 - i\lambda S + \cdots) = H_0' + \lambda H_1' + \cdots \qquad \text{(II, 3)}$$

The term H_n' may be expressed as

$$H_n' = H_n + \left[iS, H_{n-1}'' \right] \qquad \text{(II, 4)}$$

where

$$H_{n-1}'' = H_{n-1} + \frac{1}{2}\left[iS, H_{n-2}''' \right] \qquad \text{(II, 5)}$$

and so on to

$$H_n^{(m)} = H_n + \frac{1}{m}\left[iS, H_{n-1}^{(m+1)} \right] \qquad \text{(II, 6)}$$

where (m) indicates m primes. It should be noted that $H_0^{(m)} = H_0$ for all values of m .

We are thus led to consider, in addition to H_1' , the operators H_1'' , H_1''' , and H_1'''', respectively, necessary to calculate H_2' ,H_3' and H_4'.

Now the operators iS and $H_{n-1}^{(m+1)}$ whose commutator appears in Equation (II,6), are each made up of combinations of the vibrational operators $p_{s\sigma}$ and $q_{s\sigma}$ and of rotational operators P_α . This may be expressed symbolically as $iS = \sum_k i(^k\mathcal{S})$ with $(^k\mathcal{S}) = (^k\mathcal{S}_V)(^k\mathcal{S}_R)$ and $H_{n-1}^{(m+1)} = \sum_k (^k\mathcal{H})$ with $(^k\mathcal{H}) = (^k\mathcal{H}_V)(^k\mathcal{H}_R)$ wherein the subscripts V and R refer, respectively, to vibrational and rotational operators. After dropping the k notation, one discovers that

$$\left[i\mathcal{S}, \mathcal{H} \right] = \left[i\mathcal{S}, \mathcal{H} \right]_V + \left[i\mathcal{S}, \mathcal{H} \right]_R \qquad \text{(II, 7)}$$

where

$$\left[i\mathcal{S}, \mathcal{H} \right]_V = \left[i\mathcal{S}_V, \mathcal{H}_V \right]\left(\frac{\mathcal{S}_R \mathcal{H}_R + \mathcal{H}_R \mathcal{S}_R}{2} \right) \qquad \text{(II, 8)}$$

$$\left[i\mathcal{S}, \mathcal{H} \right]_R = \left(\frac{\mathcal{S}_V \mathcal{H}_V + \mathcal{H}_V \mathcal{S}_V}{2} \right) \left[i\mathcal{S}_R, \mathcal{H}_R \right] \qquad \text{(II, 9)}$$

Now since $\mathcal{H}$ is an operator of order $n-1$, one finds that $\left[i\mathcal{S}, \mathcal{H} \right]_V$

is of order n whereas $[i\mathcal{S}, \mathcal{H}]_R$ is of order $n+1$. This comes about[3] since the $[P_\alpha, P_\beta]$ occurring in $[i\mathcal{S}_R, \mathcal{H}_R]$ are equivalent to $-i\hbar P_\gamma$ and are therefore of an order of magnitude smaller than $P_\alpha P_\beta$, whereas the $[p, q]$ occurring in $[i\mathcal{S}_V, \mathcal{H}_V]$ are equivalent to $-i\hbar$ and are therefore of the same order of magnitude as pq. Thus, after regrouping terms which are truly of the same order of magnitude, one finds for the transformed Hamiltonian,

$$h'_n = H_n + [iS, h''_{n-1}]_V + [iS, h''_{n-2}]_R \tag{II,10}$$

where one has, in analogy with Equations (II,5) and (II,6),

$$h''_n = H_n + \frac{1}{2}[iS, h'''_{n-1}]_V + \frac{1}{2}[iS, h'''_{n-2}]_R \tag{II,11}$$

and, in general,

$$h^{(m)}_n = H_n + \frac{1}{m}[iS, h^{(m+1)}_{n-1}]_V + \frac{1}{m}[iS, h^{(m+1)}_{n-2}]_R \tag{II,12}$$

NOTATION

In order to make our notation less cumbersome, we characterize the various normal modes and their conjugate momenta, with the subscripts $a, b, c, \ldots$ in place of $s\sigma$, $s'\sigma'$, $s''\sigma''$, ... and in this connection introduce the Kronecker delta $\delta_{ab} = \delta_{s_a s_b}\delta_{\sigma_a \sigma_b}$.

In the expression for H_n (Eq.(I,1)) we designate by ${}^{\alpha\beta\ldots}_{(n)}X^{de\ldots}_{ab\ldots}$ the coefficient of $\frac{1}{2}(q_a q_b \cdots p_d p_e \cdots + p_d p_e \cdots q_a q_b \cdots) P_\alpha P_\beta \cdots$ so that we write, for example,

$$H_0 = \sum_\alpha \left({}^{\alpha\alpha}_{(0)}X\right) P_\alpha^2 + \sum_a \left({}_{(0)}X^{aa}\right) p_a^2 + \sum_a \left({}_{(0)}X_{aa}\right) q_a^2 \tag{II,13}$$

The values of the required ${}^{\alpha\beta\ldots}_{(n)}X^{de\ldots}_{ab\ldots}$ are given in Table V of the Appendix to this Chapter. An analogous manner of writing is used for h'_n, h''_n, h'''_n and h''''_n in which ${}_{(n)}X$ is replaced by ${}_{(n)}Y$, ${}_{(n)}T$, ${}_{(n)}U$ and ${}_{(n)}W$, respectively.

The S function, which has been calculated by R.C.Herman and W.H. Shaffer [16] may be expressed as[4]

$$S = \sum_{\alpha\beta}\sum_{a}\left({}^{\alpha\beta}S^{a}\right)p_a P_\alpha P_\beta + \sum_{\alpha}\sum_{\substack{ab\\(s_a<s_b)}}\left[\left({}^{\alpha}S_{ab}\right)q_a q_b + \left({}^{\alpha}S^{ab}\right)p_a p_b\right]P_\alpha$$

$$+ \sum_{\substack{abc\\(a\leqslant b\leqslant c)}}\left(S^{abc}\right)p_a p_b p_c + \sum_{\substack{abc\\(a\leqslant b)}}\left(S^{c}_{ab}\right)\tfrac{1}{2}\left(q_a q_b p_c + p_c q_a q_b\right) \qquad \text{(II,14)}$$

the coefficients ${}^{\alpha\beta}S^{a}$, ${}^{\alpha}S^{ab}$, etc... being given in Table VI of the Appendix.

For the evaluation of $\left[i\mathcal{S}_V , \mathcal{H}_V\right]$ we use the method elsewhere described [18] which effects a considerable simplification of this computation. The corresponding calculation of $\left[i\mathcal{S}_R , \mathcal{H}_R\right]$ is somewhat more complicated and warrants discussion at this point. These rotational commutators are of the types

$$\left[\sum_{\alpha}\left({}^{\alpha}f\right)P_\alpha , \sum_{\alpha\beta\ldots}\left({}^{\alpha\beta\ldots}g\right)P_\alpha P_\beta\ldots\right]$$

and

$$\left[\sum_{\alpha\beta}\left({}^{\alpha\beta}f\right)P_\alpha P_\beta , \sum_{\alpha\beta\ldots}\left({}^{\alpha\beta\ldots}g\right)P_\alpha P_\beta\ldots\right]$$

If one takes cognizance of the identities

$$\left[\eta,\alpha\beta\gamma\right] = \left[\eta,\alpha\right]\beta\gamma + \alpha\left[\eta,\beta\right]\gamma + \alpha\beta\left[\eta,\gamma\right]$$

and

$$\left[\eta\theta,\alpha\beta\gamma\right] = \eta\left[\theta,\alpha\beta\gamma\right] + \left[\eta,\alpha\beta\gamma\right]\theta$$

he finds that

$$\left[\sum_{\alpha}\left({}^{\alpha}f\right)P_\alpha , \sum_{\alpha\beta\gamma\ldots\eta}\left({}^{\alpha\beta\gamma\ldots\eta}g\right)P_\alpha P_\beta P_\gamma\ldots P_\eta\right]$$

$$= i\hbar\sum_{\alpha\beta\gamma\ldots\eta}\left\langle\left({}^{\alpha''}f\right)\left({}^{\alpha'\beta\gamma\ldots\eta}g\right)\right\rangle P_\alpha P_\beta P_\gamma\ldots P_\eta \qquad \text{(II,15)}$$

and

$$\left[\sum_{\alpha\beta} \left({}^{\alpha\beta}f\right) P_\alpha P_\beta , \sum_{\alpha\beta\gamma\cdots\eta} \left({}^{\alpha\beta\gamma\cdots\eta}g\right) P_\alpha P_\beta P_\gamma \cdots P_\eta \right]$$

$$= i\hbar \sum_{\alpha\beta\gamma\cdots\eta\theta} \left\langle \left({}^{\alpha\beta''}f\right)\left({}^{\beta'\gamma\cdots\theta}g\right)\right\rangle P_\alpha P_\beta P_\gamma \cdots P_\eta P_\theta \qquad \text{(II,16)}$$

where the < > notation is explained in Table I and where the primed Greek superscripts (e.g., α, α' and α'') are defined such that the order $\alpha\alpha'\alpha''$ presents a cyclic permutation of xyz, i.e., if $\alpha = x$, $\alpha' = y$ and $\alpha'' = z$; if $\alpha = y$, $\alpha' = z$ and $\alpha'' = x$; and if $\alpha = z$, $\alpha' = x$ and $\alpha'' = y$. We may thus write $[P_{\alpha''}, P_{\alpha'}] = i\hbar P_\alpha$ for all values of α which proves very convenient indeed for the derivations of Equations (II,15) and (II,16).

We shall here always write the vibrational operators in symmetric form as $\frac{1}{2}(q_a q_b \cdots p_d p_e \cdots + p_d p_e \cdots q_a q_b \cdots)$ rather than in the form $q_a q_b \cdots p_d p_e \cdots$, which from the immediate point of view has the advantage of simplifying the calculations, and from a more fundamental standpoint assures that these operators be Hermitian and hence can be identified with physical observables. One wishes ultimately to symmetrize the rotational operators also, but at the present stage of the calculation this appears to complicate things to the extent that we present these operators, even those which arise from the term $\frac{1}{2}(\mathcal{S}_R \mathcal{H}_R + \mathcal{H}_R \mathcal{S}_R)$ of Equation (II,8), in the nonsymmetric form indicated by Equations (II,15) and (II,16) for the term $[i\mathcal{S}_R, \mathcal{H}_R]$. The required desymmetrization is effected by identities such as

$$\sum_{\alpha\beta\gamma} \left({}^{\alpha,\beta\gamma}\theta\right) \frac{1}{2}\left(P_\alpha P_\beta P_\gamma + P_\beta P_\gamma P_\alpha\right) = \sum_{\alpha\beta\gamma} \frac{1}{2}\left[\left({}^{\alpha,\beta\gamma}\theta\right) + \left({}^{\gamma,\alpha\beta}\theta\right)\right] P_\alpha P_\beta P_\gamma \qquad \text{(II,17)}$$

where the comma between the indices serves to separate, in terms

TABLE I . Coefficients appearing in rotational commutations

$$\langle({}^{\alpha''}f)({}^{\alpha'}g)\rangle = ({}^{\alpha''}f)({}^{\alpha'}g) - ({}^{\alpha'}f)({}^{\alpha''}g)$$

$$\langle({}^{\alpha''}f)({}^{\alpha'\beta}g)\rangle = ({}^{\alpha''}f)({}^{\alpha'\beta}g) - ({}^{\alpha'}f)({}^{\alpha''\beta}g) + ({}^{\beta''}f)({}^{\alpha\beta'}g) - ({}^{\beta'}f)({}^{\alpha\beta''}g)$$

$$\langle({}^{\alpha''}f)({}^{\alpha'\beta\gamma}g)\rangle = ({}^{\alpha''}f)({}^{\alpha'\beta\gamma}g) - ({}^{\alpha'}f)({}^{\alpha''\beta\gamma}g) + ({}^{\beta''}f)({}^{\alpha\beta'\gamma}g) - ({}^{\beta'}f)({}^{\alpha\beta''\gamma}g)$$
$$+ ({}^{\gamma''}f)({}^{\alpha\beta\gamma'}g) - ({}^{\gamma'}f)({}^{\alpha\beta\gamma''}g)$$

$$\langle({}^{\alpha''}f)({}^{\alpha'\beta\gamma\delta}g)\rangle = ({}^{\alpha''}f)({}^{\alpha'\beta\gamma\delta}g) - ({}^{\alpha'}f)({}^{\alpha''\beta\gamma\delta}g) + ({}^{\beta''}f)({}^{\alpha\beta'\gamma\delta}g) - ({}^{\beta'}f)({}^{\alpha\beta''\gamma\delta}g)$$
$$+ ({}^{\gamma''}f)({}^{\alpha\beta\gamma'\delta}g) - ({}^{\gamma'}f)({}^{\alpha\beta\gamma''\delta}g) + ({}^{\delta''}f)({}^{\alpha\beta\gamma\delta'}g) - ({}^{\delta'}f)({}^{\alpha\beta\gamma\delta''}g)$$

$$\langle({}^{\alpha\beta''}f)({}^{\beta'}g)\rangle = ({}^{\alpha''\beta}f)({}^{\alpha'}g) - ({}^{\alpha'\beta}f)({}^{\alpha''}g) + ({}^{\alpha\beta''}f)({}^{\beta'}g) - ({}^{\alpha\beta'}f)({}^{\beta''}g)$$

$$\langle({}^{\alpha\beta''}f)({}^{\beta'\gamma}g)\rangle = ({}^{\alpha''\gamma}f)({}^{\alpha'\beta}g) - ({}^{\alpha'\gamma}f)({}^{\alpha''\beta}g) + ({}^{\beta''\gamma}f)({}^{\alpha\beta'}g) - ({}^{\beta'\gamma}f)({}^{\alpha\beta''}g)$$
$$+ ({}^{\alpha\gamma''}f)({}^{\beta\gamma'}g) - ({}^{\alpha\gamma'}f)({}^{\beta\gamma''}g) + ({}^{\alpha\beta''}f)({}^{\beta'\gamma}g) - ({}^{\alpha\beta'}f)({}^{\beta''\gamma}g)$$

$$\langle({}^{\alpha\beta''}f)({}^{\beta'\gamma\delta}g)\rangle = ({}^{\alpha''\delta}f)({}^{\alpha'\beta\gamma}g) - ({}^{\alpha'\delta}f)({}^{\alpha''\beta\gamma}g) + ({}^{\beta''\delta}f)({}^{\alpha\beta'\gamma}g) - ({}^{\beta'\delta}f)({}^{\alpha\beta''\gamma}g)$$
$$+ ({}^{\gamma''\delta}f)({}^{\alpha\beta\gamma'}g) - ({}^{\gamma'\delta}f)({}^{\alpha\beta\gamma''}g) + ({}^{\alpha\delta''}f)({}^{\beta\gamma\delta'}g) - ({}^{\alpha\delta'}f)({}^{\beta\gamma\delta''}g)$$
$$+ ({}^{\alpha\beta''}f)({}^{\beta'\gamma\delta}g) - ({}^{\alpha\beta'}f)({}^{\beta''\gamma\delta}g) + ({}^{\alpha\gamma''}f)({}^{\beta\gamma'\delta}g) - ({}^{\alpha\gamma'}f)({}^{\beta\gamma''\delta}g)$$

$$\langle({}^{\alpha\beta''}f)({}^{\beta'\gamma\delta\varepsilon}g)\rangle = ({}^{\alpha''\varepsilon}f)({}^{\alpha'\beta\gamma\delta}g) - ({}^{\alpha'\varepsilon}f)({}^{\alpha''\beta\gamma\delta}g) + ({}^{\beta''\varepsilon}f)({}^{\alpha\beta'\gamma\delta}g)$$
$$- ({}^{\beta'\varepsilon}f)({}^{\alpha\beta''\gamma\delta}g) + ({}^{\gamma''\varepsilon}f)({}^{\alpha\beta\gamma'\delta}g) - ({}^{\gamma'\varepsilon}f)({}^{\alpha\beta\gamma''\delta}g)$$
$$+ ({}^{\delta''\varepsilon}f)({}^{\alpha\beta\gamma\delta'}g) - ({}^{\delta'\varepsilon}f)({}^{\alpha\beta\gamma\delta''}g) + ({}^{\alpha\varepsilon''}f)({}^{\beta\gamma\delta\varepsilon'}g)$$
$$- ({}^{\alpha\varepsilon'}f)({}^{\beta\gamma\delta\varepsilon''}g) + ({}^{\alpha\beta''}f)({}^{\beta'\gamma\delta\varepsilon}g) - ({}^{\alpha\beta'}f)({}^{\beta''\gamma\delta\varepsilon}g)$$
$$+ ({}^{\alpha\gamma''}f)({}^{\beta\gamma'\delta\varepsilon}g) - ({}^{\alpha\gamma'}f)({}^{\beta\gamma''\delta\varepsilon}g) + ({}^{\alpha\delta''}f)({}^{\beta\gamma\delta'\varepsilon}g)$$
$$- ({}^{\alpha\delta'}f)({}^{\beta\gamma\delta''\varepsilon}g)$$

arising from $[i\mathcal{S},\mathcal{H}]_V$, the indices pertaining to $\mathcal{S}$ (and written first) from those pertaining to $\mathcal{H}$. We shall in the future set

$$\frac{1}{2}\left({}^{\alpha,\beta\gamma}\theta + {}^{\gamma,\alpha\beta}\theta\right) = {}^{\{\alpha,\beta\gamma\}}\theta \tag{II,18}$$

and likewise

$$\frac{1}{2}\left({}^{\alpha\beta,\gamma}\theta + {}^{\beta\gamma,\alpha}\theta\right) = {}^{\{\alpha\beta,\gamma\}}\theta \tag{II,19a}$$

$$\frac{1}{2}\left({}^{\alpha,\beta}\theta + {}^{\beta,\alpha}\theta\right) = {}^{\{\alpha,\beta\}}\theta \tag{II,19b}$$

$$\frac{1}{2}\left({}^{\alpha\beta,\gamma\delta}\theta + {}^{\gamma\delta,\alpha\beta}\theta\right) = {}^{\{\alpha\beta,\gamma\delta\}}\theta \tag{II,19c}$$

$$\frac{1}{2}\left({}^{\alpha,\beta\gamma\delta}\theta + {}^{\delta,\alpha\beta\gamma}\theta\right) = {}^{\{\alpha,\beta\gamma\delta\}}\theta \tag{II,19d}$$

and

$$\frac{1}{2}\left({}^{\alpha\beta\gamma,\delta}\theta + {}^{\beta\gamma\delta,\alpha}\theta\right) = {}^{\{\alpha\beta\gamma,\delta\}}\theta \tag{II,20}$$

We wish to return to a description of the notation to be employed in summations involving the p_a and q_a. Γ_{abc} will represent the coefficient of a term in $q_a q_b q_c$ symmetric with respect to the three indices a, b, c and appearing in a sum of the form $\sum\limits_{\substack{abc\\(a\leq b\leq c)}}$ whereas $\Gamma_{ab,c}$ and $\Gamma_{ab,cd}$ will enter into summations, respectively, of the forms $\sum\limits_{\substack{abc\\(a\leq b)}}$ and $\sum\limits_{\substack{abcd\\(a\leq b;\, c\leq d)}}$

In the course of our computations it will prove expedient [18] to transform summations of the form

$$\sum_{\substack{abcd\\(a\leq b;\, c\leq d)}} \Gamma_{ab,cd}\, q_a q_b q_c q_d$$

into those of the type

$$\sum_{\substack{abcd\\(a\leq b\leq c\leq d)}} {}^{*}\Gamma_{abcd}\, q_a q_b q_c q_d$$

The relations connecting the $^{*}\Gamma_{abcd}$ and $\Gamma_{ab,cd}$ are

$$\underset{(a<b<c<d)}{^{*}\Gamma_{abcd}} = \Gamma_{ab,cd} + \Gamma_{ac,bd} + \Gamma_{ad,bc} + \Gamma_{bc,ad} + \Gamma_{bd,ac} + \Gamma_{cd,ab} \quad \text{(II,21a)}$$

$$\underset{(a<b<c)}{^{*}\Gamma_{aabc}} = \Gamma_{aa,bc} + \Gamma_{ab,ac} + \Gamma_{ac,ab} + \Gamma_{bc,aa} \quad \text{(II,21b)}$$

$$\underset{(a<b<c)}{^{*}\Gamma_{abbc}} = \Gamma_{ab,bc} + \Gamma_{ac,bb} + \Gamma_{bb,ac} + \Gamma_{bc,ab} \quad \text{(II,21c)}$$

and, in general, the other required $^{*}\Gamma$'s may also be obtained from Equation (II,21a) by equating the necessary indexes and retaining only one term from each group of identical terms thus produced. To summarize, we may write

$$\underset{(a\leqslant b\leqslant c\leqslant d)}{^{*}\Gamma_{abcd}} = \sideset{^*}{}\sum_{\substack{klmn \\ (klmn\equiv abcd)}} \underset{(k\leqslant l;\, m\leqslant n)}{\Gamma_{kl,mn}} \quad \text{(II,22a)}$$

where $\sideset{^*}{}\sum_{klmn}$ indicates a summation, subject to the condition $k\leqslant l\,; m\leqslant n$ over all distinct permutations of the four indexes. One could also replace the $\sideset{^*}{}\sum$ notation with that used in references [18], e.g.,

$$\underset{(a\leqslant b\leqslant c\leqslant d)}{^{*}\Gamma_{abcd}} = \sum_{\substack{klmn \\ (klmn\leqslant abcd)}} \left(\frac{\Gamma_{kl,mn}}{g_2^2(kl,mn)}\right)_{k\leqslant l,\, m\leqslant n} \quad \text{(II,22b)}$$

with

$$g_2^2(kl,mn) = 1 + \delta_{km} + \delta_{kn} + \delta_{lm} + \delta_{ln} + \delta_{km}\delta_{ln} \quad \text{(II,23)}$$

ZERO-ORDER HAMILTONIAN AND FIRST-ORDER CORRECTION

The transformed Hamiltonian of zero order, h_0', presents no problem since

$$h_0' = H_0' = H_0$$

One has for the first-order correction

$$h_1' = H_1 + \left[iS, H_0\right]_V \quad \text{(II,24)}$$

$$h_1' = \sum_{\alpha\beta} \sum_{a} \left({}^{\alpha\beta}_{(1)}Y_a\right) q_a P_\alpha P_\beta + \sum_{\alpha} \sum_{ab} \left({}^{\alpha}_{(1)}Y^{b}_{a}\right) \tfrac{1}{2}\left(q_a p_b + p_b q_a\right) P_\alpha + \sum_{\substack{abc \\ (a \leq b \leq c)}} \left({}_{(1)}Y_{abc}\right) q_a q_b q_c \qquad \text{(II,25)}$$

where, because of the choice of the S function [17], it is evident that ${}^{\alpha\beta}_{(1)}Y_a = {}_{(1)}Y_{abc} = 0$. We have chosen to write Equation (II,25) in its present form, however, because one has exactly analogous equations for h_1'', h_1''', and h_1'''' where ${}_{(1)}Y$ is replaced by ${}_{(1)}T$, ${}_{(1)}U$, and ${}_{(1)}W$, respectively. These coefficients are presented together with the ${}_{(1)}Y$'s in Table II. One verifies that

$$h_1' = -\sum_{\alpha} \frac{p^*_\alpha P_\alpha}{I^e_{\alpha\alpha}} \qquad \text{(II,26)}$$

where p^*_α is obtained from p_α by setting $s' = s$ in Equation (I,7) and represents a component of the internal angular momentum of the degenerate modes of vibration.

TABLE II . Coefficients appearing in h_1', h_1'', h_1''' and h_1''''

${}^{\alpha\beta}_{(1)}Y_a = 0$	${}^{\alpha}_{(1)}Y^{b}_{a} = -\delta_{s_a s_b} \dfrac{\zeta^{\alpha}_{ab}}{I^e_{\alpha\alpha}}$	${}_{(1)}Y_{abc} = 0$
${}^{\alpha\beta}_{(1)}T_a = -\dfrac{\hbar^{\frac{1}{2}}}{4} \dfrac{a^{\alpha\beta}_a}{I^e_{\alpha\alpha} I^e_{\beta\beta} \lambda^{\frac{1}{4}}_a}$	${}^{\alpha}_{(1)}T^{b}_{a} = -\dfrac{(1+\delta_{s_a s_b})\zeta^{\alpha}_{ab}\lambda^{\frac{1}{4}}_b}{2I^e_{\alpha\alpha}\lambda^{\frac{1}{4}}_a}$	${}_{(1)}T_{abc} = \dfrac{hc}{2} k_{abc}$
${}^{\alpha\beta}_{(1)}U_a = -\dfrac{\hbar^{\frac{1}{2}}}{3} \dfrac{a^{\alpha\beta}_a}{I^e_{\alpha\alpha} I^e_{\beta\beta} \lambda^{\frac{1}{4}}_a}$	${}^{\alpha}_{(1)}U^{b}_{a} = -\dfrac{(2+\delta_{s_a s_b})\zeta^{\alpha}_{ab}\lambda^{\frac{1}{4}}_b}{3I^e_{\alpha\alpha}\lambda^{\frac{1}{4}}_a}$	${}_{(1)}U_{abc} = \dfrac{2hc}{3} k_{abc}$
${}^{\alpha\beta}_{(1)}W_a = -\dfrac{3\hbar^{\frac{1}{2}}}{8} \dfrac{a^{\alpha\beta}_a}{I^e_{\alpha\alpha} I^e_{\beta\beta} \lambda^{\frac{1}{4}}_a}$	${}^{\alpha}_{(1)}W^{b}_{a} = -\dfrac{(3+\delta_{s_a s_b})\zeta^{\alpha}_{ab}\lambda^{\frac{1}{4}}_b}{4I^e_{\alpha\alpha}\lambda^{\frac{1}{4}}_a}$	${}_{(1)}W_{abc} = \dfrac{3hc}{4} k_{abc}$

SECOND-ORDER CORRECTION

The second-order correction,

$$h_2' = H_2 + \left[iS, h_1''\right]_V + \left[iS, H_0\right]_R \tag{II,27}$$

proves to be

$$\begin{aligned} h_2' = & \sum_{\alpha\beta\gamma\delta} \left({}^{\alpha\beta\gamma\delta}_{(2)}Y\right) P_\alpha P_\beta P_\gamma P_\delta + \sum_{\alpha\beta\gamma} \sum_{a} \left({}^{\alpha\beta\gamma}_{(2)}Y^{a}\right) p_a P_\alpha P_\beta P_\gamma \\ & + \sum_{\alpha\beta} \sum_{\substack{ab \\ a\leq b}} \left({}^{\alpha\beta}_{(2)}Y^{ab}\, p_a p_b + {}^{\alpha\beta}_{(2)}Y_{ab}\, q_a q_b\right) P_\alpha P_\beta \\ & + \sum_{\alpha} \sum_{\substack{abc \\ a\leq b\leq c}} \left({}^{\alpha}_{(2)}Y^{abc}\right) p_a p_b p_c P_\alpha \\ & + \sum_{\alpha} \sum_{\substack{abc \\ a\leq b}} \left({}^{\alpha}_{(2)}Y^{c}_{ab}\right) \tfrac{1}{2}\left(q_a q_b p_c + p_c q_a q_b\right) P_\alpha \\ & + \sum_{\substack{abcd \\ a\leq b;\, c\leq d}} \left({}_{(2)}Y^{cd}_{ab}\right) \tfrac{1}{2}\left(q_a q_b p_c p_d + p_c p_d q_a q_b\right) \\ & + \sum_{\substack{abcd \\ a\leq b\leq c\leq d}} \left({}_{(2)}Y_{abcd}\right) q_a q_b q_c q_d \end{aligned} \tag{II,28}$$

The coefficients ${}_{(2)}Y$ and the coefficients ${}_{(2)}T$ and ${}_{(2)}W$ which replace them in h_2'' and h_2''', respectively, are found in Table III.

THIRD-ORDER CORRECTION

The third-order correction,

$$h_3' = H_3 + \left[iS, h_2''\right]_V + \left[iS, h_1''\right]_R \tag{II,29}$$

is given by

$$\begin{aligned} h_3' = & \sum_{\alpha\beta\gamma\delta} \sum_{a} \left({}^{\alpha\beta\gamma\delta}_{(3)}Y_a\right) q_a P_\alpha P_\beta P_\gamma P_\delta \\ & + \sum_{\alpha\beta\gamma} \sum_{ab} \left({}^{\alpha\beta\gamma}_{(3)}Y^{b}_{a}\right) \tfrac{1}{2}\left(q_a p_b + p_b q_a\right) P_\alpha P_\beta P_\gamma \end{aligned}$$

$$+\sum_{\alpha\beta}\sum_{\substack{abc\\b\leqslant c}}\left({}^{\alpha\beta}_{(3)}Y_{a}^{bc}\right)\frac{1}{2}\left(q_a p_b p_c + p_b p_c q_a\right)P_\alpha P_\beta$$

$$+\sum_{\alpha\beta}\sum_{\substack{abc\\a\leqslant b\leqslant c}}\left({}^{\alpha\beta}_{(3)}Y_{abc}\right)q_a q_b q_c P_\alpha P_\beta$$

$$+\sum_{\alpha}\sum_{\substack{abcd\\b\leqslant c\leqslant d}}\left({}^{\alpha}_{(3)}Y_{a}^{bcd}\right)\frac{1}{2}\left(q_a p_b p_c p_d + p_b p_c p_d q_a\right)P_\alpha$$

$$+\sum_{\alpha}\sum_{\substack{abcd\\a\leqslant b\leqslant c}}\left({}^{\alpha}_{(3)}Y_{abc}^{d}\right)\frac{1}{2}\left(q_a q_b q_c p_d + p_d q_a q_b q_c\right)P_\alpha$$

$$+\sum_{\substack{abcde\\b\leqslant c\leqslant d\leqslant e}}\left({}_{(3)}Y_{a}^{bcde}\right)\frac{1}{2}\left(q_a p_b p_c p_d p_e + p_b p_c p_d p_e q_a\right)$$

$$+\sum_{\substack{abcde\\a\leqslant b\leqslant c;\, d\leqslant e}}\left({}_{(3)}Y_{abc}^{de}\right)\frac{1}{2}\left(q_a q_b q_c p_d p_e + p_d p_e q_a q_b q_c\right)$$

$$+\sum_{\substack{abcde\\a\leqslant b\leqslant c\leqslant d\leqslant e}}\left({}_{(3)}Y_{abcde}\right)q_a q_b q_c q_d q_e$$

$$+\sum_{a}\left({}_{(3)}Y_{a}\right)q_a \tag{II,30}$$

Once again one has an analogous expression for h_3'' in which he has replaced the ${}_{(3)}Y$ of h_3' by ${}_{(3)}T$. The necessary coefficients are to be found in Table IV.

Equations (II,13), (II,25), (II,28) and (II,30) define the zero-order transformed Hamiltonian as well as the first-, second-, and third-order corrections of this operator. We have also obtained an expression for the operators h_2'' and h_3'' which will be used in the Chapter IV to calculate the fourth-order correction, h_4'.

TABLE III . Coefficients appearing in h_2' , h_2'' , and h_2'''

$${}^{\alpha\beta\gamma\delta}_{(2)}Y = {}^{\alpha\beta,\gamma\delta}_{(T)}\gamma$$

$${}^{\alpha\beta\gamma}_{(2)}Y^{a} = {}^{\{\alpha,\beta\gamma\}}_{(T)}\gamma^{a} + {}^{\{\alpha\beta,\gamma\}}_{(T)}\varepsilon^{a} + {}^{\alpha\beta\gamma}\alpha^{a}$$

$${}^{\alpha\beta}_{(2)}Y^{ab} = {}^{\alpha\beta}_{(T)}\gamma^{ab} + \frac{1}{1+\delta_{ab}}\left({}^{\{\alpha,\beta\}}_{(T)}\varepsilon^{a,b} + {}^{\{\alpha,\beta\}}_{(T)}\varepsilon^{b,a}\right) + \left(1-\delta_{s_a s_b}\right){}^{\alpha\beta}\alpha^{ab}$$

$${}^{\alpha\beta}_{(2)}Y_{ab} = {}^{\alpha\beta}_{(2)}X_{ab} + {}^{\alpha\beta}_{(T)}\gamma_{ab} + \frac{1}{1+\delta_{ab}}\left({}^{\{\alpha,\beta\}}_{(T)}\varepsilon_{a,b} + {}^{\{\alpha,\beta\}}_{(T)}\varepsilon_{b,a}\right) + {}^{\alpha\beta}_{(T)}\eta_{ab} + \left(1-\delta_{s_a s_b}\right){}^{\alpha\beta}\alpha_{ab}$$

$${}^{\alpha}_{(2)}Y^{abc} = \sideset{}{^*}\sum_{\substack{lmn\\(lmn)\equiv(abc)\\(l\leqslant m)}} {}^{\alpha}_{(T)}\gamma^{lm,n}$$

$${}^{\alpha}_{(2)}Y^{c}_{ab} = {}^{\alpha}_{(2)}X^{c}_{ab} + {}^{\alpha}_{(T)}\gamma^{c}_{ab} + \frac{1}{1+\delta_{ab}}\left({}^{\alpha}_{(T)}\varepsilon^{c}_{a,b} + {}^{\alpha}_{(T)}\varepsilon^{c}_{b,a}\right) + {}^{\alpha}_{(T)}\eta^{c}_{ab}$$

$${}_{(2)}Y^{cd}_{ab} = {}_{(2)}X^{cd}_{ab} + {}_{(T)}\gamma^{cd}_{ab}$$

$${}_{(2)}Y_{abcd} = {}_{(2)}X_{abcd} + \sideset{}{^*}\sum_{\substack{klmn\\(klmn)\equiv(abcd)\\(k\leqslant l\ ;\ m\leqslant n)}} {}_{(T)}\gamma_{kl,mn}$$

The ${}_{(2)}T$'s and ${}_{(2)}U$'s are defined by analogous expressions obtained by replacing the coefficients

$$\alpha,\ {}_{(T)}\gamma\ ,\ {}_{(T)}\varepsilon\ ,\ {}_{(T)}\eta$$

by $\frac{1}{2}\alpha,\ \frac{1}{2}\left({}_{(U)}\gamma\right),\ \frac{1}{2}\left({}_{(U)}\varepsilon\right),\ \frac{1}{2}\left({}_{(U)}\eta\right)$

and $\frac{1}{3}\alpha,\ \frac{1}{3}\left({}_{(W)}\gamma\right),\ \frac{1}{3}\left({}_{(W)}\varepsilon\right),\ \frac{1}{3}\left({}_{(W)}\eta\right)$, respectively.

TABLE IV . Coefficients appearing in h_3' and h_3''

$$ {}^{\alpha\beta\gamma\delta}_{(3)}Y_a = {}^{\{\alpha,\beta\gamma\delta\}}_{(T)}\theta_a + {}^{\{\alpha\beta,\gamma\delta\}}_{(T)}\mu_a $$

$$ {}^{\alpha\beta\gamma}_{(3)}Y_a^b = {}^{\alpha\beta\gamma}_{(T)}\theta_a^b + {}^{\{\alpha,\beta\gamma\}}_{(T)}\mu_a^b + {}^{\{\alpha,\beta\gamma\}}_{(T)}\nu_a^b + {}^{\{\alpha\beta,\gamma\}}_{(T)}\xi_a^b + {}^{\alpha\beta\gamma}_{(T)}\gamma_a^b $$

$$ {}^{\alpha\beta}_{(3)}Y_a^{bc} = {}^{\alpha\beta}_{(T)}\mu_a^{bc} + {}^{\{\alpha,\beta\}}_{(T)}\xi_a^{bc} + {}^{\alpha\beta}_{(T)}\pi_a^{bc} + {}^{\alpha\beta}_{(T)}\gamma_a^{bc}\left(1-\delta_{s_b s_c}\right) + \frac{1}{1+\delta_{bc}}\left({}^{\alpha\beta}_{(T)}\theta_a^{b,c} + {}^{\alpha\beta}_{(T)}\theta_a^{c,b} + {}^{\{\alpha,\beta\}}_{(T)}\nu_a^{b,c} + {}^{\{\alpha,\beta\}}_{(T)}\nu_a^{c,b} + {}^{\alpha\beta}_{(T)}\varepsilon_a^{b,c} + {}^{\alpha\beta}_{(T)}\varepsilon_a^{c,b}\right) $$

$$ {}^{\alpha\beta}_{(3)}Y_{abc} = {}^{\alpha\beta}_{(3)}X_{abc} + {}^{\alpha\beta}_{(T)}\nu_{abc} + \sideset{}{^*}\sum_{\substack{lmn\\(lmn)\equiv(abc)}} \Big(\underset{(l\leq m)}{{}^{\alpha\beta}_{(T)}\theta_{lm,n}} + \underset{(m\leq n)}{{}^{\{\alpha,\beta\}}_{(T)}\mu_{l,mn}} + \underset{(s_m<s_n)}{{}^{\alpha\beta}_{(T)}\gamma_{l,mn}} \Big) $$

$$ {}^{\alpha}_{(3)}Y_a^{bcd} = \sideset{}{^*}\sum_{\substack{lmn\\(lmn)\equiv(bcd)}} \Big(\underset{(l\leq m)}{{}^{\alpha}_{(T)}\theta_a^{lm,n}} + \underset{(l\leq m)}{{}^{\alpha}_{(T)}\mu_a^{lm,n}} + \underset{(m\leq n)}{{}^{\alpha}_{(T)}\nu_a^{l,mn}} + \underset{(s_l<s_m)}{{}^{\alpha}_{(T)}\gamma_a^{lm,n}} \Big) $$

$$ {}^{\alpha}_{(3)}Y_{abc}^{d} = {}^{\alpha}_{(3)}X_{abc}^{d} + {}^{\alpha}_{(T)}\nu_{abc}^{d} + \sideset{}{^*}\sum_{\substack{lmn\\(lmn)\equiv(abc)}} \Big(\underset{(l\leq m)}{{}^{\alpha}_{(T)}\theta_{lm,n}^{d}} + \underset{(m\leq n)}{{}^{\alpha}_{(T)}\mu_{l,mn}^{d}} + \underset{(s_m<s_n)}{{}^{\alpha}_{(T)}\gamma_{l,mn}^{d}} \Big) $$

$$ {}_{(3)}Y_{abc}^{de} = {}_{(3)}X_{abc}^{de} + {}_{(T)}\nu_{abc}^{de} + \sideset{}{^*}\sum_{\substack{lmn\\(lmn)\equiv(abc)}} \Big\{ \underset{(l\leq m)}{{}_{(T)}\theta_{lm,n}^{de}} + \frac{1}{1+\delta_{de}}\Big(\underset{(l\leq m)}{{}_{(T)}\mu_{lm,n}^{d,e}} + \underset{(l\leq m)}{{}_{(T)}\mu_{lm,n}^{e,d}} \Big)\Big\} $$

$$ {}_{(3)}Y_a^{bcde} = \sideset{}{^*}\sum_{\substack{klmn\\(klmn)\equiv(bcde)}} \underset{(k\leq l;\ m\leq n)}{{}_{(T)}\theta_a^{kl,mn}} $$

$$ {}_{(3)}Y_{abcde} = {}_{(3)}X_{abcde} + \sideset{}{^*}\sum_{\substack{jklmn\\(jklmn)\equiv(abcde)}} \underset{(j\leq k;\ l\leq m\leq n)}{{}_{(T)}\theta_{jk,lmn}} $$

$$ {}_{(3)}Y_a = {}_{(3)}X_a + {}_{(T)}\theta_a + {}_{(T)}\mu_a $$

The ${}_{(3)}T$'s are defined by analogous expressions obtained by replacing the coefficients ${}_{(T)}\gamma$, ${}_{(T)}\varepsilon$, ${}_{(T)}\theta$, ${}_{(T)}\mu$, ${}_{(T)}\nu$, ${}_{(T)}\xi$, ${}_{(T)}\pi$ by $\frac{1}{2}\left({}_{(U)}\gamma\right)$, $\frac{1}{2}\left({}_{(U)}\varepsilon\right)$, $\frac{1}{2}\left({}_{(U)}\theta\right)$, $\frac{1}{2}\left({}_{(U)}\mu\right)$, $\frac{1}{2}\left({}_{(U)}\nu\right)$, $\frac{1}{2}\left({}_{(U)}\xi\right)$, $\frac{1}{2}\left({}_{(U)}\pi\right)$

APPENDIX TO CHAPTER II

Herein will be found Tables V to X giving expressions for the various coefficients used in the calculation.

The following observations should be made with regard to these tables:

(a) The expressions given in Table V are immediate consequences of equations (I,17)

(b) The coefficients A_{abc} and $B_{ab,c}$ which appear in Table VI, have been previously given in reference [18]. For the sake of completeness we restate them as

$$A_{abc} = -\frac{2}{\hbar^2} k_{abc} (\lambda_a \lambda_b \lambda_c)^{\frac{1}{2}} D_{abc} \qquad \text{(AII,1a)}$$

$$A_{aab} = \frac{2}{\hbar^2} k_{aab} \frac{\lambda_a}{\lambda_b^{\frac{1}{2}} (4\lambda_a - \lambda_b)} \qquad \text{(AII,1b)}$$

$$A_{aaa} = \frac{2}{3\hbar^2} \frac{k_{aaa}}{\lambda_a^{\frac{1}{2}}} \qquad \text{(AII,1c)}$$

$$B_{ab,c} = k_{abc} \lambda_c^{\frac{1}{2}} (\lambda_c - \lambda_a - \lambda_b) D_{abc} \qquad \text{(AII,2a)}$$

$$B_{aa,b} = \frac{k_{aab}(2\lambda_a - \lambda_b)}{\lambda_b^{\frac{1}{2}}(4\lambda_a - \lambda_b)} \qquad \text{(AII,2b)}$$

$$B_{ab,a} = \frac{k_{aab} \lambda_a^{\frac{1}{2}}}{4\lambda_a - \lambda_b} \qquad \text{(AII,2c)}$$

$$B_{aa,a} = \frac{k_{aaa}}{3\lambda_a^{\frac{1}{2}}} \qquad \text{(AII,2d)}$$

with

$$D_{abc} = \Big[(\lambda_a^{\frac{1}{2}} + \lambda_b^{\frac{1}{2}} + \lambda_c^{\frac{1}{2}})(\lambda_a^{\frac{1}{2}} - \lambda_b^{\frac{1}{2}} - \lambda_c^{\frac{1}{2}}) \times (\lambda_a^{\frac{1}{2}} - \lambda_b^{\frac{1}{2}} + \lambda_c^{\frac{1}{2}})(\lambda_a^{\frac{1}{2}} + \lambda_b^{\frac{1}{2}} - \lambda_c^{\frac{1}{2}}) \Big]^{-1} \qquad \text{(AII,3)}$$

It should be noted that these coefficients depend upon s_a, s_b, s_c

TABLE V . Coefficients of untransformed Hamiltonians

$$ {}^{\alpha\alpha}_{(0)}X = \frac{1}{2I^e_{\alpha\alpha}} \qquad {}_{(0)}X^{aa} = \frac{\lambda_a^{\frac{1}{2}}}{2\hbar} \qquad {}_{(0)}X_{aa} = \frac{\hbar\lambda_a^{\frac{1}{2}}}{2} $$

$$ {}^{\alpha\beta}_{(1)}X_a = -\frac{a_a^{\alpha\beta}\hbar^{\frac{1}{2}}}{2I^e_{\alpha\alpha}I^e_{\beta\beta}\lambda_a^{\frac{1}{4}}} \qquad {}^{\alpha}_{(1)}X_a^{\ b} = -\frac{\zeta^{\alpha}_{ab}\lambda_b^{\frac{1}{4}}}{I^e_{\alpha\alpha}\lambda_a^{\frac{1}{4}}} \qquad {}_{(1)}X_{abc} = hc\,k_{abc} $$

$$ {}^{\alpha\beta}_{(2)}X_{ab} = \frac{\hbar}{2I^e_{\alpha\alpha}I^e_{\beta\beta}\lambda_a^{\frac{1}{4}}\lambda_b^{\frac{1}{4}}}\;\frac{\Omega^{(2)\alpha\beta}_{ab}+\Omega^{(2)\alpha\beta}_{ba}}{1+\delta_{ab}} $$

$$ {}^{\alpha}_{(2)}X_{ab}^{\ c} = \sum_{\beta}\frac{h^{\frac{1}{2}}\lambda_c^{\frac{1}{4}}}{I^e_{\alpha\alpha}I^e_{\beta\beta}\lambda_a^{\frac{1}{4}}\lambda_b^{\frac{1}{4}}}\;\frac{\left(a_a^{\alpha\beta}\zeta^{\beta}_{bc}+a_b^{\alpha\beta}\zeta^{\beta}_{ac}\right)}{1+\delta_{ab}} $$

$$ {}_{(2)}X_{ab}^{cd} = \sum_{\alpha}\frac{\lambda_c^{\frac{1}{4}}\lambda_d^{\frac{1}{4}}}{I^e_{\alpha\alpha}\lambda_a^{\frac{1}{4}}\lambda_b^{\frac{1}{4}}}\;\frac{\left(\zeta^{\alpha}_{ac}\zeta^{\alpha}_{bd}+\zeta^{\alpha}_{bc}\zeta^{\alpha}_{ad}\right)}{\left(1+\delta_{ab}\right)\left(1+\delta_{cd}\right)} $$

$$ {}_{(2)}X_{abcd} = hc\,k_{abcd} $$

$$ {}^{\alpha\beta}_{(3)}X_{abc} = \frac{\hbar^{\frac{3}{2}}}{2I^e_{\alpha\alpha}I^e_{\beta\beta}(\lambda_a\lambda_b\lambda_c)^{\frac{1}{4}}}\;\sideset{^*}{}\sum_{\substack{lmn\\(lmn)\equiv(abc)}}\Omega^{(3)\alpha\beta}_{l,m,n} $$

$$ {}^{\alpha}_{(3)}X_{abc}^{\ d} = -\frac{\hbar\lambda_d^{\frac{1}{4}}}{I^e_{\alpha\alpha}(\lambda_a\lambda_b\lambda_c)^{\frac{1}{4}}}\;\sideset{^*}{}\sum_{\substack{lmn\\(lmn)\equiv(abc)\\(m\leqslant n)}}\;\sum_{\beta}\frac{\zeta^{\beta}_{ld}\left(\Omega^{(2)\beta\alpha}_{mn}+\Omega^{(2)\beta\alpha}_{nm}\right)}{I^e_{\beta\beta}(1+\delta_{mn})} $$

$$ {}_{(3)}X_{abc}^{de} = \frac{\hbar^{\frac{1}{2}}(\lambda_d\lambda_e)^{\frac{1}{4}}}{2(\lambda_a\lambda_b\lambda_c)^{\frac{1}{4}}}\;\sideset{^*}{}\sum_{\substack{jklmn\\(jk)\equiv(de)\\(lmn)\equiv(abc)}}\;\sum_{\alpha\beta}\frac{\Omega^{(1)\alpha\beta}_{l}\zeta^{\alpha}_{mj}\zeta^{\beta}_{nk}}{I^e_{\alpha\alpha}I^e_{\beta\beta}} $$

$$ {}_{(3)}X_a = -\frac{\hbar^{\frac{5}{2}}}{4\lambda_a^{\frac{1}{4}}}\sum_{bc}\sum_{\alpha\beta}\frac{\left(a_a^{\alpha\beta}\zeta^{\alpha}_{cb}\zeta^{\beta}_{bc}+2a_c^{\alpha\beta}\zeta^{\alpha}_{ab}\zeta^{\beta}_{bc}\right)}{I^e_{\alpha\alpha}I^e_{\beta\beta}}+\frac{\hbar^{\frac{1}{2}}}{2\lambda_a^{\frac{1}{4}}}\Lambda^{(3)}_a $$

$$ {}_{(3)}X_{abcde} = hc\,k_{abcde} $$

TABLE VI . Coefficients of the S function

$$^{\alpha\beta}S^{a} = \frac{a_{a}^{\alpha\beta}}{2 I_{\alpha\alpha}^{e} I_{\beta\beta}^{e} \hbar^{\frac{3}{2}} \lambda_{a}^{\frac{3}{4}}}$$

$$^{\alpha}S_{ab} = \frac{\zeta_{ab}^{\alpha}(\lambda_{a}+\lambda_{b})}{(\lambda_{a}-\lambda_{b}) I_{\alpha\alpha}^{e} \lambda_{a}^{\frac{1}{4}} \lambda_{b}^{\frac{1}{4}}}$$

$$^{\alpha}S^{ab} = \frac{2\zeta_{ab}^{\alpha} \lambda_{a}^{\frac{1}{4}} \lambda_{b}^{\frac{1}{4}}}{(\lambda_{a}-\lambda_{b}) I_{\alpha\alpha}^{e} \hbar^{2}}$$

$$S^{abc} = -\frac{2\pi c}{\hbar} A_{abc}$$

$$S_{ab}^{c} = -\frac{2\pi c}{\hbar} B_{ab,c}\left(1+\delta_{ac}+\delta_{bc}\right)$$

TABLE VII . Coefficients appearing in $\left[iS, H_{0}\right]_{R}$

$$^{\alpha\beta\gamma}\alpha^{a} = -\hbar \left\langle \left(^{\alpha\beta''}S^{a}\right)\left(^{\beta'\gamma}_{(0)}X\right)\right\rangle$$

$$^{\alpha\beta}\alpha_{ab} = -\hbar \left\langle \left(^{\alpha''}S_{ab}\right)\left(^{\alpha'\beta}_{(0)}X\right)\right\rangle$$

$$^{\alpha\beta}\alpha^{ab} = -\hbar \left\langle \left(^{\alpha''}S^{ab}\right)\left(^{\alpha'\beta}_{(0)}X\right)\right\rangle$$

TABLE VIII . Coefficients appearing in $[iS, h_1'']_V$

$$ {}^{\alpha\beta,\gamma\delta}_{(T)}\gamma = -\frac{1}{8}\sum_m \frac{a_m^{\alpha\beta}\, a_m^{\gamma\delta}}{I^e_{\alpha\alpha} I^e_{\beta\beta} I^e_{\gamma\gamma} I^e_{\delta\delta}\lambda_m} $$

$$ {}^{\alpha,\beta\gamma}_{(T)}\gamma^{a} = -\frac{1}{2\hbar^{\frac{1}{2}}}\sum_{\substack{m \\ (s_m\neq s_a)}} \frac{\zeta^{\alpha}_{am}\, a_m^{\beta\gamma}\,\lambda_a^{\frac{1}{4}}}{I^e_{\alpha\alpha} I^e_{\beta\beta} I^e_{\gamma\gamma}(\lambda_a-\lambda_m)} $$

$$ {}^{\alpha\beta}_{(T)}\gamma^{ab} = \frac{\pi c\hbar^{\frac{1}{2}}}{2}\sum_m \frac{A_{abm}\, a_m^{\alpha\beta}}{I^e_{\alpha\alpha} I^e_{\beta\beta}\lambda_m^{\frac{1}{4}}}\left(1+\delta_{am}+\delta_{bm}\right) $$

$$ {}^{\alpha\beta}_{(T)}\gamma_{ab} = \frac{\pi c\hbar^{\frac{1}{2}}}{2}\sum_m \frac{B_{ab,m}\, a_m^{\alpha\beta}}{I^e_{\alpha\alpha} I^e_{\beta\beta}\lambda_m^{\frac{1}{4}}}\left(1+\delta_{am}+\delta_{bm}\right) $$

$$ {}^{\alpha\beta,\gamma}_{(T)}\varepsilon^{a} = -\frac{1}{4\hbar^{\frac{1}{2}}}\sum_m \frac{a_m^{\alpha\beta}\,\zeta^{\gamma}_{ma}\,\lambda_a^{\frac{1}{4}}}{I^e_{\alpha\alpha} I^e_{\beta\beta} I^e_{\gamma\gamma}\lambda_m}\left(1+\delta_{s_a s_m}\right) $$

$$ {}^{\alpha,\beta}_{(T)}\varepsilon_{a,b} = \frac{\hbar}{2}\sum_{\substack{m \\ (s_m\neq s_a)}} \frac{\zeta^{\alpha}_{am}\,\zeta^{\beta}_{bm}(\lambda_a+\lambda_m)}{I^e_{\alpha\alpha} I^e_{\beta\beta}(\lambda_a\lambda_b)^{\frac{1}{4}}(\lambda_a-\lambda_m)}\left(1+\delta_{s_b s_m}\right) $$

$$ {}^{\alpha,\beta}_{(T)}\varepsilon^{a,b} = -\frac{1}{\hbar}\sum_{\substack{m \\ (s_m\neq s_a)}} \frac{\zeta^{\alpha}_{am}\,\zeta^{\beta}_{mb}(\lambda_a\lambda_b)^{\frac{1}{4}}}{I^e_{\alpha\alpha} I^e_{\beta\beta}(\lambda_a-\lambda_m)}\left(1+\delta_{s_m s_b}\right) $$

$$ {}^{\alpha}_{(T)}\gamma^{ab,c} = \pi c\sum_m \frac{A_{abm}\,\zeta^{\alpha}_{mc}\,\lambda_c^{\frac{1}{4}}}{I^e_{\alpha\alpha}\lambda_m^{\frac{1}{4}}}\left(1+\delta_{am}+\delta_{bm}\right)\left(1+\delta_{s_c s_m}\right) $$

$$ {}^{\alpha}_{(T)}\gamma^{c}_{ab} = \pi c\sum_m \frac{B_{ab,m}\,\zeta^{\alpha}_{mc}\,\lambda_c^{\frac{1}{4}}}{I^e_{\alpha\alpha}\lambda_m^{\frac{1}{4}}}\left(1+\delta_{am}+\delta_{bm}\right)\left(1+\delta_{s_c s_m}\right) $$

TABLE VIII . Continued

$$ {}^{\alpha}_{(T)}\varepsilon^{c}_{a,b} = -\pi c \sum_{m} \frac{B_{am,c}\,\zeta^{\alpha}_{bm}\,\lambda_m^{\frac{1}{4}}}{I^{e}_{\alpha\alpha}\,\lambda_b^{\frac{1}{4}}} (1+\delta_{ac}+\delta_{cm})(1+\delta_{am})(1+\delta_{s_m s_b}) $$

$$ {}^{\alpha\beta}_{(T)}\eta_{ab} = \frac{\pi c \hbar^{\frac{1}{2}}}{2} \sum_{m} \frac{a_m^{\alpha\beta}\,k_{abm}}{I^{e}_{\alpha\alpha} I^{e}_{\beta\beta}\,\lambda_m^{\frac{3}{4}}} (1+\delta_{am}+\delta_{bm}) $$

$$ {}^{\alpha}_{(T)}\eta^{c}_{ab} = 2\pi c \sum_{\substack{m \\ (s_m \neq s_c)}} \frac{\zeta^{\alpha}_{cm}\,\lambda_c^{\frac{1}{4}}\,\lambda_m^{\frac{1}{4}}\,k_{abm}}{I^{e}_{\alpha\alpha}(\lambda_c - \lambda_m)} (1+\delta_{am}+\delta_{bm}) $$

$$ {}_{(T)}\gamma^{cd}_{ab} = -\pi c^2 h \sum_{m} A_{cdm}\,k_{abm}(1+\delta_{cm}+\delta_{dm})(1+\delta_{am}+\delta_{bm}) $$

$$ {}_{(T)}\gamma_{ab,cd} = -\pi c^2 h \sum_{m} B_{ab,m}\,k_{cdm}(1+\delta_{am}+\delta_{bm})(1+\delta_{cm}+\delta_{dm}) $$

TABLE IX . Coefficients appearing in $[iS, h_1'']_R$

$$ {}^{\alpha\beta\gamma}_{(T)}\gamma^{b}_{a} = -\hbar \langle ({}^{\alpha\beta''}S^{b})({}^{\beta'\gamma}_{(1)}T_{a}) \rangle $$

$$ {}^{\alpha\beta}_{(T)}\gamma_{a,bc} = -\hbar \langle ({}^{\alpha''}S_{bc})({}^{\alpha'\beta}_{(1)}T_{a}) \rangle $$

$$ {}^{\alpha\beta}_{(T)}\gamma^{bc}_{a} = -\hbar \langle ({}^{\alpha''}S^{bc})({}^{\alpha'\beta}_{(1)}T_{a}) \rangle $$

$$ {}^{\alpha\beta}_{(T)}\varepsilon^{b,c}_{a} = -\hbar \langle ({}^{\alpha\beta''}S^{c})({}^{\beta'}_{(1)}T^{b}_{a}) \rangle $$

$$ {}^{\alpha}_{(T)}\gamma^{d}_{a,bc} = -\hbar \langle ({}^{\alpha''}S_{bc})({}^{\alpha'}_{(1)}T^{d}_{a}) \rangle $$

$$ {}^{\alpha}_{(T)}\gamma^{bc,d}_{a} = -\hbar \langle ({}^{\alpha''}S^{bc})({}^{\alpha'}_{(1)}T^{d}_{a}) \rangle $$

TABLE X . Coefficients appearing in $\left[iS, h''_2\right]_V$

$$ {}^{\alpha,\beta\gamma\delta}_{(T)}\theta_a = -\hbar \sum_{\substack{m\\(s_a\neq s_m)}} \left({}^{\alpha}S_{am}\right)\left({}^{\beta\gamma\delta}_{(2)}T^{m}\right) $$

$$ {}^{\alpha\beta\gamma}_{(T)}\theta_a^{b} = -\hbar \sum_{m} \left(S^{b}_{am}\right)\left(1+\delta_{am}\right)\left({}^{\alpha\beta\gamma}_{(2)}T^{m}\right) $$

$$ {}^{\alpha,\beta\gamma}_{(T)}\mu_a^{b} = -\hbar \sum_{\substack{m\\(s_a\neq s_m)}} \left({}^{\alpha}S_{am}\right)\left({}^{\beta\gamma}_{(2)}T^{bm}\right)\left(1+\delta_{bm}\right) $$

$$ {}^{\alpha\beta}_{(T)}\theta_a^{b,c} = -\hbar \sum_{m} \left(S^{b}_{am}\right)\left(1+\delta_{am}\right)\left({}^{\alpha\beta}_{(2)}T^{cm}\right)\left(1+\delta_{cm}\right) $$

$$ {}^{\alpha\beta,\gamma\delta}_{(T)}\mu_a = \hbar \sum_{m} \left({}^{\alpha\beta}S^{m}\right)\left({}^{\gamma\delta}_{(2)}T_{am}\right)\left(1+\delta_{am}\right) $$

$$ {}^{\alpha,\beta\gamma}_{(T)}\nu_a^{b} = \hbar \sum_{\substack{m\\(s_m\neq s_b)}} \left({}^{\alpha}S^{bm}\right)\left({}^{\beta\gamma}_{(2)}T_{am}\right)\left(1+\delta_{am}\right) $$

$$ {}^{\alpha\beta}_{(T)}\mu_a^{bc} = \hbar \sum_{m} \left(S^{bcm}\right)\left(1+\delta_{bm}+\delta_{cm}\right)\left({}^{\alpha\beta}_{(2)}T_{am}\right)\left(1+\delta_{am}\right) $$

$$ {}^{\alpha\beta}_{(T)}\theta_{ab,c} = \hbar \sum_{m} \left(S^{m}_{ab}\right)\left({}^{\alpha\beta}_{(2)}T_{cm}\right)\left(1+\delta_{cm}\right) $$

$$ {}^{\alpha\beta,\gamma}_{(T)}\xi_a^{b} = \hbar \sum_{m} \left({}^{\alpha\beta}S^{m}\right)\left({}^{\gamma}_{(2)}T^{b}_{am}\right)\left(1+\delta_{am}\right) $$

$$ {}^{\alpha,\beta}_{(T)}\mu_{a,bc} = -\hbar \sum_{\substack{m\\(s_a\neq s_m)}} \left({}^{\alpha}S_{am}\right)\left({}^{\beta}_{(2)}T^{m}_{bc}\right) $$

$$ {}^{\alpha,\beta}_{(T)}\nu_a^{b,c} = \hbar \sum_{\substack{m\\(s_b\neq s_m)}} \left({}^{\alpha}S^{bm}\right)\left({}^{\beta}_{(2)}T^{c}_{am}\right)\left(1+\delta_{am}\right) $$

$$ {}^{\alpha}_{(T)}\theta_a^{bc,d} = \hbar \sum_{m} \left(S^{bcm}\right)\left(1+\delta_{bm}+\delta_{cm}\right)\left({}^{\alpha}_{(2)}T^{d}_{am}\right)\left(1+\delta_{am}\right) $$

$$ {}^{\alpha}_{(T)}\theta^{d}_{ab,c} = \hbar \sum_{m} \left\{\left(S^{m}_{ab}\right)\left({}^{\alpha}_{(2)}T^{d}_{cm}\right) - \left(S^{d}_{cm}\right)\left({}^{\alpha}_{(2)}T^{m}_{ab}\right)\right\}\left(1+\delta_{cm}\right) $$

$$ {}^{\alpha,\beta}_{(T)}\xi_a^{bc} = -\hbar \sum_{\substack{m\\(s_m\neq s_a)}} \left({}^{\alpha}S_{am}\right)\left({}^{\beta}_{(2)}T^{bcm}\right)\left(1+\delta_{bm}+\delta_{cm}\right) $$

TABLE X . Continued

$${}^{\alpha}_{(T)}\mu_a^{bc,d} = -\hbar \sum_{m} \left(S^{d}_{am}\right)\left(1+\delta_{am}\right)\left({}^{\alpha}_{(2)}T^{bcm}\right)\left(1+\delta_{bm}+\delta_{cm}\right)$$

$${}^{\alpha\beta}_{(T)}\pi_a^{bc} = \hbar \sum_{m} \left({}^{\alpha\beta}S^{m}\right)\left({}_{(2)}T^{bc}_{am}\right)\left(1+\delta_{am}\right)$$

$${}^{\alpha}_{(T)}\mu^{d}_{a,bc} = -\hbar \sum_{\substack{m \\ (s_m \neq s_a)}} \left({}^{\alpha}S_{am}\right)\left({}_{(2)}T^{dm}_{bc}\right)\left(1+\delta_{dm}\right)$$

$${}^{\alpha}_{(T)}\nu_a^{b,cd} = \hbar \sum_{\substack{m \\ (s_m \neq s_b)}} \left({}^{\alpha}S^{bm}\right)\left({}_{(2)}T^{cd}_{am}\right)\left(1+\delta_{am}\right)$$

$${}_{(T)}\theta_a^{bc,de} = \hbar \sum_{m} \left(S^{bcm}\right)\left(1+\delta_{bm}+\delta_{cm}\right)\left({}_{(2)}T^{de}_{am}\right)\left(1+\delta_{am}\right)$$

$${}_{(T)}\theta^{de}_{ab,c} = \hbar \sum_{m} \left(S^{m}_{ab}\right)\left({}_{(2)}T^{de}_{cm}\right)\left(1+\delta_{cm}\right)$$

$${}_{(T)}\mu^{d,e}_{ab,c} = -\hbar \sum_{m} \left(S^{d}_{cm}\right)\left(1+\delta_{cm}\right)\left({}_{(2)}T^{em}_{ab}\right)\left(1+\delta_{em}\right)$$

$${}_{(T)}\theta_a = -\frac{\hbar^3}{4} \sum_{bcm} \left(1+\delta_{bm}\right)\left(1+\delta_{ac}\right)\Big[\left(S^{c}_{bm}\right)\left({}_{(2)}T^{bm}_{ac}\right)\left(1+\delta_{bm}\right)+\left(S^{b}_{bm}\right)\left({}_{(2)}T^{cm}_{ac}\right)\left(1+\delta_{cm}\right)\Big]$$

$${}^{\alpha\beta}_{(T)}\nu_{abc} = \hbar \sum_{m} \left({}^{\alpha\beta}S^{m}\right)\left({}_{(2)}T_{abcm}\right)\left(1+\delta_{am}+\delta_{bm}+\delta_{cm}\right)$$

$${}^{\alpha}_{(T)}\nu^{d}_{abc} = \hbar \sum_{\substack{m \\ (s_m \neq s_d)}} \left({}^{\alpha}S^{dm}\right)\left({}_{(2)}T_{abcm}\right)\left(1+\delta_{am}+\delta_{bm}+\delta_{cm}\right)$$

$${}_{(T)}\nu^{de}_{abc} = \hbar \sum_{m} \left(S^{dem}\right)\left(1+\delta_{dm}+\delta_{em}\right)\left({}_{(2)}T_{abcm}\right)\left(1+\delta_{am}+\delta_{bm}+\delta_{cm}\right)$$

$${}_{(T)}\mu_a = \frac{\hbar^3}{2}\Bigg\{\sum_{\substack{bmn \\ (m \leq n)}} \left(S^{bmn}\right)\left(1+\delta_{bm}+\delta_{bn}\right)\left({}_{(2)}T_{abmn}\right)\left(1+\delta_{mn}\right)\Big(1+\delta_{am}+\delta_{an}$$
$$+\delta_{bm}+\delta_{bn}+\delta_{am}\delta_{bn}+\delta_{an}\delta_{bm}-\delta_{abmn}\Big)\left(1+\delta_{ab}\right) - 2\sum_{\substack{mnp \\ (m \leq n \leq p)}} \left(S^{mnp}\right)$$
$$\left({}_{(2)}T_{amnp}\right)\left(1+\delta_{am}+\delta_{an}+\delta_{ap}\right)\left(1+\delta_{mn}+\delta_{np}+3\delta_{mnp}\right)\Bigg\}$$

$${}_{(T)}\theta_{ab,cde} = \hbar \sum_{m} \left(S^{m}_{ab}\right)\left({}_{(2)}T_{cdem}\right)\left(1+\delta_{cm}+\delta_{dm}+\delta_{em}\right)$$

and possibly upon $\sigma_a, \sigma_b, \sigma_c$, and that the c in $B_{ab,c}$ pertains to a momentum whereas the subscripts a and b pertain to normal coordinates.

(c) Starting from the coefficients ${}_{(T)}\gamma, {}_{(T)}\varepsilon, {}_{(T)}\eta$ given in Table VIII, one will obtain ${}_{(U)}\gamma, {}_{(U)}\varepsilon, {}_{(U)}\eta$, by multiplying by 4/3 and replacing $\delta_{s_i s_j}$ by $\delta_{s_i s_j}/2$; he will obtain ${}_{(W)}\gamma, {}_{(W)}\varepsilon$ and ${}_{(W)}\eta$ upon multiplying by 3/2 and replacing $\delta_{s_i s_j}$ by $\delta_{s_i s_j}/3$. This is an immediate result of the values of ${}_{(1)}T, {}_{(1)}U$ and ${}_{(1)}W$ given in Table II.

(d) The formulas of Tables IX and X which define ${}_{(T)}\gamma, {}_{(T)}\varepsilon, {}_{(T)}\theta, {}_{(T)}\mu, {}_{(T)}\nu, {}_{(T)}\xi, {}_{(T)}\pi$ permit one to calculate ${}_{(U)}\gamma, {}_{(U)}\varepsilon$ etc.., by replacing ${}_{(1)}T$ and ${}_{(2)}T$ with ${}_{(1)}U$ and ${}_{(2)}U$.

NOTES TO CHAPTER II

1 . The results stated in Chapters II to IV can be used only when no resonance due to an accidental degeneracy occurs in the molecule.

2 . The operator $h'_0 + \lambda h'_1$, as $H'_0 + \lambda H'_1$, only has diagonal matrix elements in the quantum numbers v_s . Moreover, the matrix elements of h'_2 , which are nondiagonal in v_s , contribute to the energy in an order higher than the third, when no accidental resonance occurs.

3 . This point will be discussed in more detail in Chapter VII.

4 . It is to be recalled that the S function given by (II,14) is so chosen that its diagonal matrix elements in the vibrational quantum numbers v_s vanish [17].

CHAPTER III

SECOND CONTACT TRANSFORMATION (TO THIRD ORDER)

INTRODUCTION

As a step in the calculation of the rotation-vibration energies of a molecule to fourth-order of approximation, it is desirable to carry out on the once transformed Hamiltonian H' as given by Equation (II,2) a second contact transformation

$$e^{i\lambda^2 \mathcal{S}} H' e^{-i\lambda^2 \mathcal{S}} = H^{\dagger} = H_0^{\dagger} + \lambda H_1^{\dagger} + \lambda^2 H_2^{\dagger} + \cdots \qquad \text{(III,1)}$$

in such a manner that $H_0^{\dagger} + \lambda H_1^{\dagger} + \lambda^2 H_2^{\dagger}$ will be diagonal with respect to the quantum numbers v_s in the representation which diagonalizes H_0' i.e. H_0. It is significant, however, to note that while the operator $H_0^{\dagger} + \lambda H_1^{\dagger} + \lambda^2 H_2^{\dagger}$ will be diagonal in the quantum numbers v_s, it need not,in the case of the two dimensionally isotropic oscillator, be diagonal in the quantum number l_s, or in the case of the three dimensionally isotropic oscillator,be diagonal in the quantum numbers l_s and m_s. It may, of course, also have matrix elements which are non-diagonal in the rotational quantum number K.

We shall concern ourselves in the present Chapter with this second contact transformation, using the notation defined in Chapter II. Taking into account the true order of magnitude of the terms, we shall finally express the twice transformed Hamiltonian $H^{\dagger}$ as[1]

$$H^{\dagger} = h_0^{\dagger} + \lambda h_1^{\dagger} + \lambda^2 h_2^{\dagger} + \cdots \qquad \text{(III,2)}$$

Explicit expressions for the operators $h_0^{\dagger}$, $h_1^{\dagger}$, $h_2^{\dagger}$ and $h_3^{\dagger}$ will be given, the expression for $h_4^{\dagger}$ being deferred to the next Chapter.

SECOND CONTACT TRANSFORMATION

It is readily verified that in replacing $e^{i\lambda^2 \mathcal{S}}$ by its expansion $e^{i\lambda^2 \mathcal{S}} = 1 + i\lambda^2 \mathcal{S} - \frac{\lambda^4}{2} \mathcal{S}^2 + \cdots$ (and its inverse by $e^{-i\lambda^2 \mathcal{S}} = 1 - i\lambda^2 \mathcal{S} - \frac{\lambda^4}{2} \mathcal{S}^2 + \cdots$) the operators $H_n^\dagger$, appearing in the doubly transformed Hamiltonian [Equation (III,1)], will be the following

$$H_0^\dagger = h_0' \qquad \text{(III,3a)}$$

$$H_1^\dagger = h_1' \qquad \text{(III,3b)}$$

$$H_2^\dagger = h_2' + i\left[\mathcal{S}, h_0'\right] \qquad \text{(III,3c)}$$

$$H_3^\dagger = h_3' + i\left[\mathcal{S}, h_1'\right] \qquad \text{(III,3d)}$$

$$H_4^\dagger = h_4' + i\left[\mathcal{S}, h_2'\right] - \tfrac{1}{2}\left[\mathcal{S}, \left[\mathcal{S}, h_0'\right]\right] \qquad \text{(III,3e)}$$

Again, as in Chapter II, the effect of the transformation is actually that certain terms in $H^\dagger$, which formally belong to the operator $H^\dagger_{m+1}$, become of the order of magnitude of H_m . It then is expedient to regroup the components of the twice transformed Hamiltonian so as to obtain in the notation of Chapter II

$$h_0^\dagger = h_0' \qquad \text{(III,4a)}$$

$$h_1^\dagger = h_1' \qquad \text{(III,4b)}$$

$$h_2^\dagger = h_2' + \left[i\mathcal{S}, h_0'\right]_V \qquad \text{(III,4c)}$$

$$h_3^\dagger = h_3' + \left[i\mathcal{S}, h_1'\right]_V + \left[i\mathcal{S}, h_0'\right]_R \qquad \text{(III,4d)}$$

$$h_4^\dagger = h_4' + \left[i\mathcal{S}, h_2^{\dagger\dagger}\right]_V + \left[i\mathcal{S}, h_1'\right]_R \qquad \text{(III,4e)}$$

where

$$h_2^{\dagger\dagger} = h_2' + \tfrac{1}{2}\left[i\mathcal{S}, h_0'\right]_V = \tfrac{1}{2}\left(h_2' + h_2^\dagger\right) \qquad \text{(III,4f)}$$

TRANSFORMATION FUNCTION, $\mathcal{S}$

It is desired to perform a contact transformation on H', so that to second order of approximation $H^{\dagger}$ will have elements diagonal in the vibration quantum numbers v_s only. The second-order Hamiltonian, h_2', has matrix elements some of which are diagonal in v_s and some of which are non-diagonal in v_s. The function $\mathcal{S}$ must therefore be chosen, according to (III,4c), in such a manner that the commutator $-[i\mathcal{S}, h_0']_v$ will have no matrix elements which are diagonal in v_s, but will have the same off diagonal matrix elements with respect to v_s as does h_2'. According to (III,4c), the commutator $-[i\mathcal{S}, h_0']$ will, therefore, offset the non-vanishing non-diagonal matrix elements of h_2' so that $h_2^{\dagger}$ will be left with elements only which are diagonal in the quantum numbers v_s.

The various terms occurring in h_2' are listed in Table XI together with the corresponding basic $\mathcal{S}$ functions required to remove the non-diagonal elements of these terms. The total function $\mathcal{S}$ required to effect the entire transformation will, therefore, be equal to the sum of the individual $\mathcal{S}$ functions given in the second column of Table XI. It is found to be the following

$$\begin{aligned}
\mathcal{S} = & \sum_{\alpha\beta\gamma} \sum_{a} {}^{\alpha\beta\gamma}\mathcal{S}_a \, q_a P_\alpha P_\beta P_\gamma \\
& + \sum_{\alpha\beta} \sum_{ab} {}^{\alpha\beta}\mathcal{S}_a^b \, \tfrac{1}{2}(q_a p_b + p_b q_a) P_\alpha P_\beta \\
& + \sum_{\alpha} \Big[\sum_{\substack{abc \\ (a\leqslant b)}} {}^{\alpha}\mathcal{S}_c^{ab} \, \tfrac{1}{2}(p_a p_b q_c + q_c p_a p_b) \\
& \qquad + \sum_{\substack{abc \\ (a\leqslant b\leqslant c)}} {}^{\alpha}\mathcal{S}_{abc} \, q_a q_b q_c \Big] P_\alpha \\
& + \sum_{\substack{abcd \\ (a\leqslant b\leqslant c)}} \Big[\mathcal{S}_d^{abc} \, \tfrac{1}{2}(p_a p_b p_c q_d + q_d p_a p_b p_c) \\
& \qquad + \mathcal{S}_{abc}^{d} \, \tfrac{1}{2}(q_a q_b q_c p_d + p_d q_a q_b q_c) \Big]
\end{aligned} \quad \text{(III,5)}$$

TABLE XI . The $\mathcal{S}$ function

h_2'	$\mathcal{S}$
$\sum_{\alpha\beta\gamma}\sum_{a} {}^{\alpha\beta\gamma}_{(2)}Y^{a}\, p_a P_\alpha P_\beta P_\gamma$	$\sum_{\alpha\beta\gamma}\sum_{a} {}^{\alpha\beta\gamma}\mathcal{S}_{a}\, q_a P_\alpha P_\beta P_\gamma$
$\sum_{\alpha\beta}\sum_{\substack{ab\\a\leq b}} {}^{\alpha\beta}_{(2)}Y^{ab}\, p_a p_b P_\alpha P_\beta$	$\sum_{\alpha\beta}\sum_{ab} {}^{\odot\alpha\beta}\mathcal{S}^{b}_{a}\, \frac{1}{2}(q_a p_b + p_b q_a) P_\alpha P_\beta$
$\sum_{\alpha\beta}\sum_{\substack{ab\\a\leq b}} {}^{\alpha\beta}_{(2)}Y_{ab}\, q_a q_b P_\alpha P_\beta$	$\sum_{\alpha\beta}\sum_{ab} {}^{\Box\alpha\beta}\mathcal{S}^{b}_{a}\, \frac{1}{2}(q_a p_b + p_b q_a) P_\alpha P_\beta$
$\sum_{\alpha}\sum_{\substack{abc\\a\leq b\leq c}} {}^{\alpha}_{(2)}Y^{abc}\, p_a p_b p_c P_\alpha$	$\sum_{\alpha}\sum_{\substack{abc\\a\leq b}} {}^{\Box\alpha}\mathcal{S}^{ab}_{c}\, \frac{1}{2}(p_a p_b q_c + q_c p_a p_b) P_\alpha$ $+\sum_{\alpha}\sum_{\substack{abc\\a\leq b\leq c}} {}^{\Box\alpha}\mathcal{S}_{abc}\, q_a q_b q_c P_\alpha$
$\sum_{\alpha}\sum_{\substack{abc\\a\leq b}} {}^{\alpha}_{(2)}Y^{c}_{ab}\, \frac{1}{2}(q_a q_b p_c$ $+ p_c q_a q_b) P_\alpha$	$\sum_{\alpha}\sum_{\substack{abc\\a\leq b}} {}^{\odot\alpha}\mathcal{S}^{ab}_{c}\, \frac{1}{2}(p_a p_b q_c + q_c p_a p_b) P_\alpha$ $+\sum_{\alpha}\sum_{\substack{abc\\a\leq b\leq c}} {}^{\odot\alpha}\mathcal{S}_{abc}\, q_a q_b q_c P_\alpha$
$\sum_{\substack{abcd\\a\leq b;\, c\leq d}} {}_{(2)}Y^{cd}_{ab}\, \frac{1}{2}(q_a q_b p_c p_d$ $+ p_c p_d q_a q_b)$	$\sum_{\substack{abcd\\a\leq b\leq c}} {}^{\odot}\mathcal{S}^{abc}_{d}\, \frac{1}{2}(p_a p_b p_c q_d + q_d p_a p_b p_c)$ $+\sum_{\substack{abcd\\a\leq b\leq c}} {}^{\odot}\mathcal{S}^{d}_{abc}\, \frac{1}{2}(q_a q_b q_c p_d + p_d q_a q_b q_c)$
$\sum_{\substack{abcd\\a\leq b\leq c\leq d}} {}_{(2)}Y_{abcd}\, q_a q_b q_c q_d$	$\sum_{\substack{abc\\a\leq b\leq c}} {}^{\Box}\mathcal{S}^{abc}_{d}\, \frac{1}{2}(p_a p_b p_c q_d + q_d p_a p_b p_c)$ $+\sum_{\substack{abc\\a\leq b\leq c}} {}^{\Box}\mathcal{S}^{d}_{abc}\, \frac{1}{2}(q_a q_b q_c p_d + p_d q_a q_b q_c)$

The coefficients ${}^{\square\alpha\beta}\mathcal{S}^{b}_{a}$, ${}^{\square\alpha}\mathcal{S}^{ab}_{c}$, ${}^{\square\alpha}\mathcal{S}_{abc}$, ${}^{\square}\mathcal{S}^{abc}_{d}$ and ${}^{\square}\mathcal{S}^{d}_{abc}$ and the coefficients ${}^{\odot\alpha\beta}\mathcal{S}^{b}_{a}$, ${}^{\odot\alpha}\mathcal{S}^{ab}_{c}$, ${}^{\odot\alpha}\mathcal{S}_{abc}$, ${}^{\odot}\mathcal{S}^{abc}_{d}$ and ${}^{\odot}\mathcal{S}^{d}_{abc}$ which occur in Table XI do not occur explicitly in the function $\mathcal{S}$. They do appear in the function $\mathcal{S}$ nevertheless, in that ${}^{\alpha\beta}\mathcal{S}^{b}_{a}$, ${}^{\alpha}\mathcal{S}^{ab}_{c}$, ${}^{\alpha}\mathcal{S}_{abc}$ $\mathcal{S}^{abc}_{d}$ and $\mathcal{S}^{d}_{abc}$ are sums, respectively of ${}^{\square\alpha\beta}\mathcal{S}^{b}_{a} + {}^{\odot\alpha\beta}\mathcal{S}^{b}_{a}$, ${}^{\square\alpha}\mathcal{S}^{ab}_{c} + {}^{\odot\alpha}\mathcal{S}^{ab}_{c}$, etc. It may be verified that ${}^{\square\alpha\beta}\mathcal{S}^{b}_{a}$, ${}^{\square\alpha}\mathcal{S}^{ab}_{c}$, etc., in Table XI are equal, respectively, to the last term of the quantities ${}^{\alpha\beta}\mathcal{S}^{b}_{a}$, ${}^{\alpha}\mathcal{S}^{ab}_{c}$, etc., defined in Table XII. Similarly it may be shown that the terms ${}^{\odot\alpha\beta}\mathcal{S}^{b}_{a}$, ${}^{\odot\alpha}\mathcal{S}^{ab}_{c}$, etc., in Table XI are, respectively, equal to the remaining terms of ${}^{\alpha\beta}\mathcal{S}^{b}_{a}$, ${}^{\alpha}\mathcal{S}^{ab}_{c}$, etc.

The coefficients ${}^{\alpha\beta\ldots}\mathcal{S}^{cd\ldots}_{ab\ldots}$ depend upon three kinds of quantities in addition to Planck's constant $\hbar$. These are :

(a) the coefficients ${}^{\alpha\beta\ldots}_{(2)}Y^{cd\ldots}_{ab\ldots}$ from the second-order Hamiltonian, h'_2

(b) functions involving the Kronecker symbols, δ_{ij}

(c) the coefficients $\mathcal{A}_{ab\ldots}$ and $\mathcal{B}_{ab\ldots}$. The first and second types of terms, i.e., types (a) and (b) depend upon both the indices s and σ commonly used to designate the normal coordinates. Thus for example $\delta_{ij} = 1$ only when $s_i = s_j$ and $\sigma_i = \sigma_j$. The third type, i.e., the coefficients $\mathcal{A}_{ab\ldots}$ and $\mathcal{B}_{ab\ldots}$, on the other hand, depend only upon the indexes s. The notation, $\mathcal{A}_{ab\ldots}$, $\mathcal{B}_{ab\ldots}$, used is adopted as a simplification of the somewhat more cumbersome, $\mathcal{A}_{s_a s_b\ldots}$ and $\mathcal{B}_{s_a s_b\ldots}$.

As an example of how to use Table XII it is instructive to evaluate the different values of ${}^{\alpha\beta}\mathcal{S}^{b}_{a}$. Consider (a), that $a \equiv (s_a, \sigma_a)$, $b \equiv (s_b, \sigma_b)$ where $s_a \neq s_b$; then,

$$ {}^{\alpha\beta}\mathcal{S}^{b}_{a} = {}^{\alpha\beta}_{(2)}Y^{ab}\,\mathcal{A}_{b,a} - \frac{1}{\hbar^2}\,{}^{\alpha\beta}_{(2)}Y_{ab}\,\mathcal{A}_{a,b} $$

TABLE XII . Coefficients of the $\mathcal{S}$ function

$$^{\alpha\beta\gamma}\mathcal{S}_a = {}^{\alpha\beta\gamma}_{(2)}Y^a\, \mathcal{A}_a$$

$$^{\alpha\beta}\mathcal{S}^b_a = (1+\delta_{ab})\left[{}^{\alpha\beta}_{(2)}Y^{ab}\, \mathcal{A}_{b,a} - \frac{1}{\hbar^2}\, {}^{\alpha\beta}_{(2)}Y_{ab}\, \mathcal{A}_{a,b}\right]$$

$$^{\alpha}\mathcal{S}^{ab}_c = -\frac{1}{\hbar^2}\, {}^{\alpha}_{(2)}Y^c_{ab}\, \mathcal{A}_{abc} + \frac{1}{\hbar^2}\left[\frac{1+\delta_{bc}}{1+\delta_{ab}}\right]{}^{\alpha}_{(2)}Y^a_{bc}\, \mathcal{B}_{ac,b}$$

$$+\frac{1}{\hbar^2}\left[\frac{1+\delta_{ac}}{1+\delta_{ab}}\right]{}^{\alpha}_{(2)}Y^b_{ac}\, \mathcal{B}_{bc,a} - (1+\delta_{ca}+\delta_{cb})\, {}^{\alpha}_{(2)}Y^{abc}\, \mathcal{B}_{ab,c}$$

$$^{\alpha}\mathcal{S}_{abc} = -\left[\frac{1}{1+\delta_{ab}+\delta_{ac}}\right]{}^{\alpha}_{(2)}Y^a_{bc}\, \mathcal{B}_{bc,a}$$

$$-\left[\frac{1}{1+\delta_{ba}+\delta_{bc}}\right]{}^{\alpha}_{(2)}Y^b_{ac}\, \mathcal{B}_{ac,b}$$

$$-\left[\frac{1}{1+\delta_{ca}+\delta_{cb}}\right]{}^{\alpha}_{(2)}Y^c_{ab}\, \mathcal{B}_{ab,c} - \hbar^2\, {}^{\alpha}_{(2)}Y^{abc}\, \mathcal{A}_{abc}$$

$$\mathcal{S}^{abc}_d = -\frac{1}{\hbar^2}\left[\frac{1+\delta_{ad}}{1+\delta_{ba}+\delta_{ca}}\right]\left({}_{(2)}Y^{bc}_{ad}\, \mathcal{A}_{bcd,a} + {}_{(2)}Y^{ad}_{bc}\, \mathcal{B}_{bcd,a}\right)$$

$$-\frac{1}{\hbar^2}\left[\frac{1+\delta_{bd}}{1+\delta_{ab}+\delta_{cb}}\right]\left({}_{(2)}Y^{ac}_{bd}\, \mathcal{A}_{acd,b} + {}_{(2)}Y^{bd}_{ac}\, \mathcal{B}_{acd,b}\right)$$

$$-\frac{1}{\hbar^2}\left[\frac{1+\delta_{cd}}{1+\delta_{ac}+\delta_{bc}}\right]\left({}_{(2)}Y^{ab}_{cd}\, \mathcal{A}_{abd,c} + {}_{(2)}Y^{cd}_{ab}\, \mathcal{B}_{abd,c}\right)$$

$$+\frac{1}{\hbar^4}(1+\delta_{ad}+\delta_{bd}+\delta_{cd})\, {}_{(2)}Y_{abcd}\, \mathcal{B}_{abc,d}$$

TABLE XII . Continued

$$\mathcal{S}^{d}_{abc} = \left[\frac{1+\delta_{ad}}{1+\delta_{ba}+\delta_{ca}}\right]\left({}_{(2)}Y^{ad}_{bc}\,\mathcal{A}_{bcd,a}+{}_{(2)}Y^{bc}_{ad}\,\mathcal{B}_{bcd,a}\right)$$

$$+\left[\frac{1+\delta_{bd}}{1+\delta_{ab}+\delta_{cb}}\right]\left({}_{(2)}Y^{bd}_{ac}\,\mathcal{A}_{acd,b}+{}_{(2)}Y^{ac}_{bd}\,\mathcal{B}_{acd,b}\right)$$

$$+\left[\frac{1+\delta_{cd}}{1+\delta_{ac}+\delta_{bc}}\right]\left({}_{(2)}Y^{cd}_{ab}\,\mathcal{A}_{abd,c}+{}_{(2)}Y^{ab}_{cd}\,\mathcal{B}_{abd,c}\right)$$

$$-\frac{1}{\hbar^2}\left(1+\delta_{ad}+\delta_{bd}+\delta_{cd}\right){}_{(2)}Y_{abcd}\,\mathcal{A}_{abc,d}$$

Consider next (b), $a\equiv(s_a,\sigma_a)$, $b\equiv(s_a,\sigma_b)$ where $\sigma_a\neq\sigma_b$; then ${}^{\alpha\beta}\mathcal{S}^{b}_{a}=\left({}^{\alpha\beta}_{(2)}Y^{ab}-\frac{1}{\hbar^2}\,{}^{\alpha\beta}_{(2)}Y_{ab}\right)\mathcal{A}_{a,a}$. Consider finally (c), $a\equiv b=(s_a,\sigma_a)$; then

$${}^{\alpha\beta}\mathcal{S}^{b}_{a}=2\left({}^{\alpha\beta}_{(2)}Y^{aa}-\frac{1}{\hbar^2}\,{}^{\alpha\beta}_{(2)}Y_{aa}\right)\mathcal{A}_{a,a}$$

The definition of the coefficients $\mathcal{A}_{ab\ldots}$ and $\mathcal{B}_{ab\ldots}$ themselves, are given in Table XIII. These definitions hold for all values of the indices s. Certain terms in the relations defining $\mathcal{A}_{ab\ldots}$ and $\mathcal{B}_{ab\ldots}$ are enclosed in double parentheses, i.e., (( )). This is designated to indicate that such a term must be omitted where the indices s are such that the denominator vanishes. Thus, for example, if $s_a=s_b$ the second term in the coefficient $\mathcal{A}_{a,b}$ must be omitted, if $s_a=s_d$ and $s_b=s_c$ the last two terms in $\mathcal{A}_{abc,d}$ must be omitted, etc. One may readily verify that the quantities $\mathcal{A}_{abc}$ and $\mathcal{B}_{ab,c}$ are related to the A_{abc} and $B_{ab,c}$ used in the first contact transformation, discussed in Chapter II, by the formulas

$$A_{abc}=-\left(k_{abc}/\hbar^2\right)\mathcal{A}_{abc} \qquad B_{ab,c}=-k_{abc}\,\mathcal{B}_{ab,c}$$

TABLE XIII . Definition of the $\mathcal{A}_{ab\ldots}$ and $\mathcal{B}_{ab\ldots}$ coefficients

$$\mathcal{A}_{a} = \frac{1}{\lambda_a^{\frac{1}{2}}}$$

$$\mathcal{A}_{a,b} = \frac{1}{2}\left[\frac{1}{\lambda_a^{\frac{1}{2}}+\lambda_b^{\frac{1}{2}}} - \left(\left(\frac{1}{\lambda_a^{\frac{1}{2}}-\lambda_b^{\frac{1}{2}}}\right)\right)\right]$$

$$\mathcal{A}_{abc} = \frac{1}{4}\left[\frac{1}{\lambda_a^{\frac{1}{2}}+\lambda_b^{\frac{1}{2}}+\lambda_c^{\frac{1}{2}}} - \frac{1}{-\lambda_a^{\frac{1}{2}}+\lambda_b^{\frac{1}{2}}+\lambda_c^{\frac{1}{2}}} - \frac{1}{\lambda_a^{\frac{1}{2}}-\lambda_b^{\frac{1}{2}}+\lambda_c^{\frac{1}{2}}} - \frac{1}{\lambda_a^{\frac{1}{2}}+\lambda_b^{\frac{1}{2}}-\lambda_c^{\frac{1}{2}}}\right]$$

$$\mathcal{B}_{ab,c} = -\frac{1}{4}\left[\frac{1}{\lambda_a^{\frac{1}{2}}+\lambda_b^{\frac{1}{2}}+\lambda_c^{\frac{1}{2}}} + \frac{1}{-\lambda_a^{\frac{1}{2}}+\lambda_b^{\frac{1}{2}}+\lambda_c^{\frac{1}{2}}} + \frac{1}{\lambda_a^{\frac{1}{2}}-\lambda_b^{\frac{1}{2}}+\lambda_c^{\frac{1}{2}}} - \frac{1}{\lambda_a^{\frac{1}{2}}+\lambda_b^{\frac{1}{2}}-\lambda_c^{\frac{1}{2}}}\right]$$

$$\begin{aligned}\mathcal{A}_{abc,d} = \frac{1}{8}\Bigg[&\frac{1}{\lambda_a^{\frac{1}{2}}+\lambda_b^{\frac{1}{2}}+\lambda_c^{\frac{1}{2}}+\lambda_d^{\frac{1}{2}}} + \frac{1}{-\lambda_a^{\frac{1}{2}}+\lambda_b^{\frac{1}{2}}+\lambda_c^{\frac{1}{2}}+\lambda_d^{\frac{1}{2}}} + \frac{1}{\lambda_a^{\frac{1}{2}}-\lambda_b^{\frac{1}{2}}+\lambda_c^{\frac{1}{2}}+\lambda_d^{\frac{1}{2}}}\\ &+ \frac{1}{\lambda_a^{\frac{1}{2}}+\lambda_b^{\frac{1}{2}}-\lambda_c^{\frac{1}{2}}+\lambda_d^{\frac{1}{2}}} - \frac{1}{\lambda_a^{\frac{1}{2}}+\lambda_b^{\frac{1}{2}}+\lambda_c^{\frac{1}{2}}-\lambda_d^{\frac{1}{2}}} + \left(\left(\frac{1}{\lambda_a^{\frac{1}{2}}-\lambda_b^{\frac{1}{2}}-\lambda_c^{\frac{1}{2}}+\lambda_d^{\frac{1}{2}}}\right)\right)\\ &+ \left(\left(\frac{1}{-\lambda_a^{\frac{1}{2}}+\lambda_b^{\frac{1}{2}}-\lambda_c^{\frac{1}{2}}+\lambda_d^{\frac{1}{2}}}\right)\right) + \left(\left(\frac{1}{-\lambda_a^{\frac{1}{2}}-\lambda_b^{\frac{1}{2}}+\lambda_c^{\frac{1}{2}}+\lambda_d^{\frac{1}{2}}}\right)\right)\Bigg]\end{aligned}$$

$$\begin{aligned}\mathcal{B}_{abc,d} = \frac{1}{8}\Bigg[&\frac{1}{\lambda_a^{\frac{1}{2}}+\lambda_b^{\frac{1}{2}}+\lambda_c^{\frac{1}{2}}+\lambda_d^{\frac{1}{2}}} - \frac{1}{-\lambda_a^{\frac{1}{2}}+\lambda_b^{\frac{1}{2}}+\lambda_c^{\frac{1}{2}}+\lambda_d^{\frac{1}{2}}} - \frac{1}{\lambda_a^{\frac{1}{2}}-\lambda_b^{\frac{1}{2}}+\lambda_c^{\frac{1}{2}}+\lambda_d^{\frac{1}{2}}}\\ &- \frac{1}{\lambda_a^{\frac{1}{2}}+\lambda_b^{\frac{1}{2}}-\lambda_c^{\frac{1}{2}}+\lambda_d^{\frac{1}{2}}} + \frac{1}{\lambda_a^{\frac{1}{2}}+\lambda_b^{\frac{1}{2}}+\lambda_c^{\frac{1}{2}}-\lambda_d^{\frac{1}{2}}} + \left(\left(\frac{1}{\lambda_a^{\frac{1}{2}}-\lambda_b^{\frac{1}{2}}-\lambda_c^{\frac{1}{2}}+\lambda_d^{\frac{1}{2}}}\right)\right)\\ &+ \left(\left(\frac{1}{-\lambda_a^{\frac{1}{2}}+\lambda_b^{\frac{1}{2}}-\lambda_c^{\frac{1}{2}}+\lambda_d^{\frac{1}{2}}}\right)\right) + \left(\left(\frac{1}{-\lambda_a^{\frac{1}{2}}-\lambda_b^{\frac{1}{2}}+\lambda_c^{\frac{1}{2}}+\lambda_d^{\frac{1}{2}}}\right)\right)\Bigg]\end{aligned}$$

Table XIV gives the actual values of the coefficients $\mathcal{A}_{ab\ldots}$ and $\mathcal{B}_{ab\ldots}$ corresponding to all the different possible values of the indexes s . The expressions, as stated, are consistent with the general formulation given in Table XIII.

SECOND-ORDER TWICE TRANSFORMED HAMILTONIAN

The second-order twice transformed Hamiltonian, $h_2^\dagger$, is given by (III,4c) and is found to be equal to

$$
\begin{aligned}
h_2^\dagger = & \sum_{\alpha\beta\gamma\delta} {}^{\alpha\beta\gamma\delta}_{(2)}Y \, P_\alpha P_\beta P_\gamma P_\delta \\
& + \sum_{\alpha\beta} \sum_{\substack{ab \\ s_a = s_b \\ \sigma_a \leqslant \sigma_b}} \left(\hbar^2 \, {}^{\alpha\beta}_{(2)}Y^{ab} + {}^{\alpha\beta}_{(2)}Y_{ab} \right) \frac{1}{2} \left(q_a q_b + \frac{p_a p_b}{\hbar^2} \right) P_\alpha P_\beta \\
& + \sum_{\substack{abcd \\ s_a = s_b < s_c = s_d \\ \sigma_a \leqslant \sigma_b ;\, \sigma_c \leqslant \sigma_d}} \left[{}_{(2)}Y_{abcd} + \hbar^2 \left({}_{(2)}Y^{ab}_{cd} + {}_{(2)}Y^{cd}_{ab} \right) \right] \frac{1}{4} \left(q_a q_b + \frac{p_a p_b}{\hbar^2} \right) \left(q_c q_d + \frac{p_c p_d}{\hbar^2} \right) \\
& + \sum_{\substack{abcd \\ s_a = s_c < s_b = s_d}} \frac{1}{4} \, {}_{(2)}Y^{cd}_{ab} \left(q_a p_c - q_c p_a \right) \left(q_b p_d - q_d p_b \right) \\
& + \sum_a \frac{1}{8} \Bigg\{ \left(\hbar^2 \, {}_{(2)}Y^{aa}_{aa} + 3 \, {}_{(2)}Y_{aaaa} \right) \left(q_a^2 + \frac{p_a^2}{\hbar^2} \right)^2 \\
& \quad + \sum_{\substack{b \\ \left(\substack{s_a = s_b \\ \sigma_a < \sigma_b} \right)}} \left(\hbar^2 \, {}_{(2)}Y^{bb}_{aa} + \hbar^2 \, {}_{(2)}Y^{aa}_{bb} + \hbar^2 \, {}_{(2)}Y^{ab}_{ab} + 3 \, {}_{(2)}Y_{aabb} \right) \left(q_a^2 + \frac{p_a^2}{\hbar^2} \right) \left(q_b^2 + \frac{p_b^2}{\hbar^2} \right) \Bigg\} \\
& + \sum_{\substack{ab \\ \left(\substack{s_a = s_b \\ \sigma_a < \sigma_b} \right)}} \frac{1}{4} \left(\hbar^2 \, {}_{(2)}Y^{bb}_{aa} + \hbar^2 \, {}_{(2)}Y^{aa}_{bb} - \hbar^2 \, {}_{(2)}Y^{ab}_{ab} - {}_{(2)}Y_{aabb} \right) \left(q_a \frac{p_b}{\hbar} - q_b \frac{p_a}{\hbar} \right)^2 + \mathcal{H}_2^\dagger \qquad \text{(III.6)}
\end{aligned}
$$

where $\mathcal{H}_2^\dagger$ contains terms multiplied by coefficients such as ${}_{(2)}Y^{ab}_{aa}$, ${}_{(2)}Y^{aa}_{ab}$, ${}_{(2)}Y_{aaab}$, ${}_{(2)}Y^{bc}_{aa}$, ${}_{(2)}Y^{aa}_{bc}$, ${}_{(2)}Y^{ab}_{ac}$, ${}_{(2)}Y_{aabc}$ with $s_a = s_b = s_c$.

TABLE XIV . Values for the $\mathcal{A}_{ab\ldots}$ and $\mathcal{B}_{ab\ldots}$ coefficients

$$\mathcal{A}_{s} = \lambda_s^{-\frac{1}{2}}$$

$$\begin{cases} \mathcal{A}_{s,s'} = -\lambda_{s'}^{\frac{1}{2}}\left(\lambda_s - \lambda_{s'}\right)^{-1} & (s \neq s') \\ \mathcal{A}_{s,s} = \frac{1}{4}\lambda_s^{-\frac{1}{2}} \end{cases}$$

$$\begin{cases} \mathcal{A}_{ss's''} = 2\left(\lambda_s \lambda_{s'} \lambda_{s''}\right)^{\frac{1}{2}} D_{ss's''} & (s \neq s' \neq s'') \\ \mathcal{A}_{sss'} = -2\lambda_s \lambda_{s'}^{-\frac{1}{2}}\left(4\lambda_s - \lambda_{s'}\right)^{-1} & (s = s') \\ \mathcal{A}_{sss} = -\frac{2}{3}\lambda_s^{-\frac{1}{2}} \end{cases}$$

$$\begin{cases} \mathcal{B}_{ss';s''} = -\lambda_{s''}^{\frac{1}{2}}\left(\lambda_{s''} - \lambda_s - \lambda_{s'}\right) D_{ss's''} & (s \neq s' \neq s'') \\ \mathcal{B}_{ss,s'} = -\left(2\lambda_s - \lambda_{s'}\right)\lambda_{s'}^{-\frac{1}{2}}\left(4\lambda_s - \lambda_{s'}\right)^{-1} & (s \neq s') \\ \mathcal{B}_{ss',s} = -\lambda_s^{\frac{1}{2}}\left(4\lambda_s - \lambda_{s'}\right)^{-1} & (s \neq s') \\ \mathcal{B}_{ss,s} = -\frac{1}{3}\lambda_s^{-\frac{1}{2}} \end{cases}$$

$$\begin{cases} \mathcal{A}_{ss's'';s'''} = \lambda_{s'''}^{\frac{1}{2}}\left\{-16\lambda_s\lambda_{s'}\lambda_{s''} + \left(\lambda_{s'''} - \lambda_s - \lambda_{s'} - \lambda_{s''}\right)\left[D_{ss's''}^{-1} + \lambda_{s'''}\left(\lambda_{s'''} - 2\lambda_s - 2\lambda_{s'} - 2\lambda_{s''}\right)\right]\right\} D_{ss's''s'''} & (s \neq s' \neq s'' \neq s''') \\ \mathcal{A}_{sss';s''} = -\lambda_{s''}^{\frac{1}{2}}\left\{\mathcal{D}_{s,s's''}^{-1} - 2\lambda_s\left(4\lambda_s - 3\lambda_{s'} - \lambda_{s''}\right)\right\} \times \mathcal{D}_{s,s's''}\left(\lambda_{s'} - \lambda_{s''}\right)^{-1} & (s \neq s' \neq s'') \\ \mathcal{A}_{ss's'',s} = \lambda_s^{\frac{1}{2}}\left(4\lambda_s - \lambda_{s'} - \lambda_{s''}\right)\mathcal{D}_{s,s's''} & (s \neq s' \neq s'') \\ \mathcal{A}_{sss';s'} = \frac{1}{8}\left(\lambda_s - 2\lambda_{s'}\right)\lambda_{s'}^{-\frac{1}{2}}\left(\lambda_s - \lambda_{s'}\right)^{-1} & (s \neq s') \\ \mathcal{A}_{sss,s'} = -\lambda_{s'}^{\frac{1}{2}}\left(7\lambda_s - \lambda_{s'}\right)\left(9\lambda_s - \lambda_{s'}\right)^{-1}\left(\lambda_s - \lambda_{s'}\right)^{-1} & (s \neq s') \\ \mathcal{A}_{sss';s} = \lambda_s^{\frac{1}{2}}\left(3\lambda_s - \lambda_{s'}\right)\left(9\lambda_s - \lambda_{s'}\right)^{-1}\left(\lambda_s - \lambda_{s'}\right)^{-1} & (s \neq s') \\ \mathcal{A}_{sss,s} = \frac{5}{32}\lambda_s^{-\frac{1}{2}} \end{cases}$$

TABLE XIV . Continued

$$\mathcal{B}_{ss's'',s'''} = 2\left(\lambda_s \lambda_{s'} \lambda_{s''}\right)^{\frac{1}{2}}\left[D_{ss's''}{}^{-1} - \lambda_{s'''}\left(3\lambda_{s'''} - 2\lambda_s - 2\lambda_{s'} - 2\lambda_{s''}\right)\right] D_{ss's''s'''} \qquad (s \neq s' \neq s'' \neq s''')$$

$$\mathcal{B}_{sss',s''} = -2\lambda_s \lambda_{s'}^{\frac{1}{2}}\left(4\lambda_s - \lambda_{s'} - 3\lambda_{s''}\right) \times \mathcal{D}_{s,s's''}\left(\lambda_{s'} - \lambda_{s''}\right)^{-1} \qquad (s \neq s' \neq s'')$$

$$\mathcal{B}_{ss's'',s} = 2\left(\lambda_s \lambda_{s'} \lambda_{s''}\right)^{\frac{1}{2}} \mathcal{D}_{s,s's''} \qquad (s \neq s' \neq s'')$$

$$\mathcal{B}_{sss',s'} = -\frac{1}{8}\lambda_s \lambda_{s'}^{-\frac{1}{2}}\left(\lambda_s - \lambda_{s'}\right)^{-1} \qquad (s \neq s')$$

$$\mathcal{B}_{sss,s'} = -6\lambda_s^{\frac{1}{2}}\left(9\lambda_s - \lambda_{s'}\right)^{-1}\left(\lambda_s - \lambda_{s'}\right)^{-1} \qquad (s \neq s')$$

$$\mathcal{B}_{sss',s} = 2\lambda_s \lambda_{s'}^{\frac{1}{2}}\left(9\lambda_s - \lambda_{s'}\right)^{-1}\left(\lambda_s - \lambda_{s'}\right)^{-1} \qquad (s \neq s')$$

$$\mathcal{B}_{sss,s} = -\frac{3}{32}\lambda_s^{-\frac{1}{2}}$$

$$D_{ss's''} = \left[\left(\lambda_s^{\frac{1}{2}} + \lambda_{s'}^{\frac{1}{2}} + \lambda_{s''}^{\frac{1}{2}}\right)\left(\lambda_s^{\frac{1}{2}} - \lambda_{s'}^{\frac{1}{2}} - \lambda_{s''}^{\frac{1}{2}}\right) \times\left(\lambda_s^{\frac{1}{2}} - \lambda_{s'}^{\frac{1}{2}} + \lambda_{s''}^{\frac{1}{2}}\right)\left(\lambda_s^{\frac{1}{2}} + \lambda_{s'}^{\frac{1}{2}} - \lambda_{s''}^{\frac{1}{2}}\right)\right]^{-1}$$

$$\mathcal{D}_{s,s's''} = \left[\left(2\lambda_s^{\frac{1}{2}} + \lambda_{s'}^{\frac{1}{2}} + \lambda_{s''}^{\frac{1}{2}}\right)\left(2\lambda_s^{\frac{1}{2}} - \lambda_{s'}^{\frac{1}{2}} - \lambda_{s''}^{\frac{1}{2}}\right) \times\left(2\lambda_s^{\frac{1}{2}} - \lambda_{s'}^{\frac{1}{2}} + \lambda_{s''}^{\frac{1}{2}}\right)\left(2\lambda_s^{\frac{1}{2}} + \lambda_{s'}^{\frac{1}{2}} - \lambda_{s''}^{\frac{1}{2}}\right)\right]^{-1}$$

$$\begin{aligned} D_{ss's''s'''} = \Big[&\left(\lambda_s^{\frac{1}{2}} + \lambda_{s'}^{\frac{1}{2}} + \lambda_{s''}^{\frac{1}{2}} + \lambda_{s'''}^{\frac{1}{2}}\right)\left(\lambda_s^{\frac{1}{2}} - \lambda_{s'}^{\frac{1}{2}} - \lambda_{s''}^{\frac{1}{2}} - \lambda_{s'''}^{\frac{1}{2}}\right) \\ &\times\left(\lambda_s^{\frac{1}{2}} - \lambda_{s'}^{\frac{1}{2}} + \lambda_{s''}^{\frac{1}{2}} + \lambda_{s'''}^{\frac{1}{2}}\right)\left(\lambda_s^{\frac{1}{2}} + \lambda_{s'}^{\frac{1}{2}} - \lambda_{s''}^{\frac{1}{2}} + \lambda_{s'''}^{\frac{1}{2}}\right) \\ &\times\left(\lambda_s^{\frac{1}{2}} + \lambda_{s'}^{\frac{1}{2}} + \lambda_{s''}^{\frac{1}{2}} - \lambda_{s'''}^{\frac{1}{2}}\right)\left(\lambda_s^{\frac{1}{2}} - \lambda_{s'}^{\frac{1}{2}} - \lambda_{s''}^{\frac{1}{2}} + \lambda_{s'''}^{\frac{1}{2}}\right) \\ &\times\left(\lambda_s^{\frac{1}{2}} + \lambda_{s'}^{\frac{1}{2}} - \lambda_{s''}^{\frac{1}{2}} - \lambda_{s'''}^{\frac{1}{2}}\right)\left(\lambda_s^{\frac{1}{2}} - \lambda_{s'}^{\frac{1}{2}} + \lambda_{s''}^{\frac{1}{2}} - \lambda_{s'''}^{\frac{1}{2}}\right)\Big]^{-1} \end{aligned}$$

It is, perhaps, of value, in view of eventual further applications, to note that the fourth term on the right-hand side of Equation

TABLE XV . Coefficients appearing in $h_3^{\dagger}$

$$ {}^{\alpha\beta\gamma\delta}_{(3)}Z_a = {}^{\alpha\beta\gamma\delta}_{(3)}Y_a + {}^{\alpha\beta\gamma\delta}\eta_a + {}^{\{\alpha\beta\gamma,\delta\}}\alpha_a $$

$$ {}^{\alpha\beta\gamma}_{(3)}Z_a^b = {}^{\alpha\beta\gamma}_{(3)}Y_a^b + {}^{\alpha\beta\gamma}\eta_a^b + {}^{\{\alpha\beta,\gamma\}}\alpha_a^b $$

$$ {}^{\alpha\beta}_{(3)}Z_a^{bc} = {}^{\alpha\beta}_{(3)}Y_a^{bc} + {}^{\alpha\beta}\eta_a^{bc} + {}^{\{\alpha,\beta\}}\alpha_a^{bc} + \frac{\left({}^{\{\alpha,\beta\}}\beta_a^{b,c} + {}^{\{\alpha,\beta\}}\beta_a^{c,b}\right)}{1+\delta_{bc}} $$

$$ {}^{\alpha\beta}_{(3)}Z_{abc} = {}^{\alpha\beta}_{(3)}Y_{abc} + {}^{\alpha\beta}\eta_{abc} + \sum^{*}_{\substack{lmn \\ (lmn)\equiv(abc)}} {}^{\{\alpha,\beta\}}\alpha_{lm,n} \quad (l \leqslant m) $$

$$ {}^{\alpha}_{(3)}Z_a^{bcd} = {}^{\alpha}_{(3)}Y_a^{bcd} + {}^{\alpha}\alpha_a^{bcd} + \sum^{*}_{\substack{lmn \\ (lmn)\equiv(abc)}} {}^{\alpha}\beta_a^{lm,n} \quad ((l \leqslant m) $$

$$ {}^{\alpha}_{(3)}Z_{abc}^{d} = {}^{\alpha}_{(3)}Y_{abc}^{d} + {}^{\alpha}\alpha_{abc}^{d} + \sum^{*}_{\substack{lmn \\ (lmn)\equiv abc)}} {}^{\alpha}\beta_{lm,n}^{d} \quad (l \leqslant m) $$

$$ {}_{(3)}Z_a^{bcde} = {}_{(3)}Y_a^{bcde} $$

$$ {}_{(3)}Z_{abcde} = {}_{(3)}Y_{abcde} $$

$$ {}_{(3)}Z_{abc}^{de} = {}_{(3)}Y_{abc}^{de} $$

$$ {}_{(3)}Z_a = {}_{(3)}Y_a $$

TABLE XVI . Coefficients appearing in $[i\mathcal{S}, h_0']_R$

$$ {}^{\alpha\beta\gamma\delta}\eta_a = \hbar \left\langle\left({}^{\alpha\beta''}_{(0)}X\right)\left({}^{\beta'\gamma\delta}\mathcal{S}_a\right)\right\rangle \qquad (\alpha=\beta'') $$

$$ {}^{\alpha\beta\gamma}\eta_a^b = -\hbar \left\langle\left({}^{\alpha\beta''}\mathcal{S}_a^b\right)\left({}^{\beta'\gamma}_{(0)}X\right)\right\rangle \qquad (\beta'=\gamma) $$

$$ {}^{\alpha\beta}\eta_c^{ab} = -\hbar \left\langle\left({}^{\alpha''}\mathcal{S}_c^{ab}\right)\left({}^{\alpha'\beta}_{(0)}X\right)\right\rangle \qquad (\alpha'=\beta) $$

$$ {}^{\alpha\beta}\eta_{abc} = -\hbar \left\langle\left({}^{\alpha''}\mathcal{S}_{abc}\right)\left({}^{\alpha'\beta}_{(0)}X\right)\right\rangle \qquad (\alpha'=\beta) $$

TABLE XVII . Coefficients appearing in $\left[i\mathcal{S}, h'_1\right]_V$

$$ {}^{\alpha\beta\gamma,\delta}\alpha_a = -\hbar \sum_{\substack{m \\ (s_m = s_a)}} {}^{\alpha\beta\gamma}\mathcal{S}_m \; {}^{\delta}_{(1)}Y^m_a $$

$$ {}^{\alpha\beta,\gamma}\alpha^b_a = \hbar\left(\sum_{\substack{m \\ (s_m = s_b)}} {}^{\alpha\beta}\mathcal{S}^m_a \; {}^{\gamma}_{(1)}Y^b_m - \sum_{\substack{m \\ s_m = s_a}} {}^{\alpha\beta}\mathcal{S}^b_m \; {}^{\gamma}_{(1)}Y^m_a \right) $$

$$ {}^{\alpha,\beta}\alpha^{bc}_a = -\hbar \sum_{\substack{m \\ (s_m = s_a)}} {}^{\alpha}\mathcal{S}^{bc}_m \; {}^{\beta}_{(1)}Y^m_a $$

$$ {}^{\alpha,\beta}\beta^{b,c}_a = \hbar \sum_{\substack{m \\ (s_m = s_c)}} {}^{\alpha}\mathcal{S}^{bm}_a (1+\delta_{bm}) \; {}^{\beta}_{(1)}Y^c_m $$

$$ {}^{\alpha,\beta}\alpha_{ab,c} = -\hbar \sum_{\substack{m \\ (s_m = s_c)}} {}^{\alpha}\mathcal{S}_{abm} (1+\delta_{am}+\delta_{bm}) \; {}^{\beta}_{(1)}Y^m_c $$

$$ {}^{\alpha}\alpha^{bcd}_a = -\hbar \sum_{\substack{m \\ (s_m = s_a)}} \mathcal{S}^{bcd}_m \; {}^{\alpha}_{(1)}Y^m_a $$

$$ {}^{\alpha}\beta^{bc,d}_a = \hbar \sum_{\substack{m \\ (s_m = s_d)}} \mathcal{S}^{bcm}_a (1+\delta_{bm}+\delta_{cm}) \; {}^{\alpha}_{(1)}Y^d_m $$

$$ {}^{\alpha}\alpha^{d}_{abc} = \hbar \sum_{\substack{m \\ (s_m = s_d)}} \mathcal{S}^m_{abc} \; {}^{\alpha}_{(1)}Y^d_m $$

$$ {}^{\alpha}\beta^{d}_{ab,c} = -\hbar \sum_{\substack{m \\ (s_m = s_c)}} \mathcal{S}^d_{abm} (1+\delta_{am}+\delta_{bm}) \; {}^{\alpha}_{(1)}Y^m_c $$

(III,6) may also be written as

$$ \sum_{\substack{abcd \\ s_a = s_c < s_b = s_d \\ \sigma_a < \sigma_c ;\, \sigma_b < \sigma_d}} (\hbar/2)^2 \left\{ {}_{(2)}Y^{cd}_{ab} + {}_{(2)}Y^{ab}_{cd} - {}_{(2)}Y^{bc}_{ad} - {}_{(2)}Y^{ad}_{bc} \right\} \times \left(q_a \frac{p_c}{\hbar} - q_c \frac{p_a}{\hbar} \right)\left(q_b \frac{p_d}{\hbar} - q_d \frac{p_b}{\hbar} \right) \qquad \text{(III,7)} $$

THIRD-ORDER TWICE TRANSFORMED HAMILTONIAN

The third-order twice transformed Hamiltonian, $h^{\dagger}_3$, is given by

Equation (III,4d) and is found to be equal to

$$
\begin{aligned}
h_3^{\dagger} = & \sum_{\alpha\beta\gamma\delta} \sum_{a} {}^{\alpha\beta\gamma\delta}_{(3)}Z_a \, q_a P_\alpha P_\beta P_\gamma P_\delta \\
& + \sum_{\alpha\beta\gamma} \sum_{ab} {}^{\alpha\beta\gamma}_{(3)}Z_a^{b} \, \tfrac{1}{2}(q_a p_b + p_b q_a) P_\alpha P_\beta P_\gamma \\
& + \sum_{\alpha\beta} \sum_{\substack{abc \\ b \leqslant c}} {}^{\alpha\beta}_{(3)}Z_a^{bc} \, \tfrac{1}{2}(q_a p_b p_c + p_b p_c q_a) P_\alpha P_\beta \\
& + \sum_{\alpha\beta} \sum_{\substack{abc \\ a \leqslant b \leqslant c}} {}^{\alpha\beta}_{(3)}Z_{abc} \, q_a q_b q_c P_\alpha P_\beta \\
& + \sum_{\alpha} \sum_{\substack{abcd \\ b \leqslant c \leqslant d}} {}^{\alpha}_{(3)}Z_a^{bcd} \, \tfrac{1}{2}(q_a p_b p_c p_d + p_b p_c p_d q_a) P_\alpha \\
& + \sum_{\alpha} \sum_{\substack{abcd \\ a \leqslant b \leqslant c}} {}^{\alpha}_{(3)}Z_{abc}^{d} \, \tfrac{1}{2}(q_a q_b q_c p_d + p_d q_a q_b q_c) P_\alpha \\
& + \sum_{\substack{abcde \\ b \leqslant c \leqslant d \leqslant e}} {}_{(3)}Z_a^{bcde} \, \tfrac{1}{2}(q_a p_b p_c p_d p_e + p_b p_c p_d p_e q_a) \\
& + \sum_{\substack{abcde \\ a \leqslant b \leqslant c \\ d \leqslant e}} {}_{(3)}Z_{abc}^{de} \, \tfrac{1}{2}(q_a q_b q_c p_d p_e + p_d p_e q_a q_b q_c) \\
& + \sum_{\substack{abcde \\ a \leqslant b \leqslant c \leqslant d \leqslant e}} {}_{(3)}Z_{abcde} \, q_a q_b q_c q_d q_e + \sum_{a} {}_{(3)}Z_a q_a . \qquad \text{(III.8)}
\end{aligned}
$$

The coefficients ${}_{(3)}Z$ are given in Table XV as functions of the coefficients ${}_{(3)}Y$, occurring in h_3' in Chapter II, and of certain new quantities, α, β and η listed in Tables XVI and XVII. The coefficients α and β listed in Table XVII are coefficients appearing in $[i\mathcal{S}, h_1']_V$ and the η's listed in Table XVI are coefficients appearing in $[i\mathcal{S}, h_0']_R$

NOTE TO CHAPTER III

1 . The operator $h_0^{\dagger} + \lambda h_1^{\dagger} + \lambda^2 h_2^{\dagger}$, as $H_0^{\dagger} + \lambda H_1^{\dagger} + \lambda^2 H_2^{\dagger}$, is diagonal with respect to the quantum numbers v_s. The matrix elements of $h_3^{\dagger}$ which are nondiagonal in v_s only contribute to the energy to the sixth order of approximation.

CHAPTER IV

FIRST AND SECOND CONTACT TRANSFORMATIONS

(FOURTH ORDER TERMS)

INTRODUCTION

In the preceding Chapter, the twice transformed Hamiltonian has been expanded as

$$H^{\dagger} = h_0^{\dagger} + \lambda h_1^{\dagger} + \lambda^2 h_2^{\dagger} + \lambda^3 h_3^{\dagger} + \lambda^4 h_4^{\dagger} \qquad \text{(IV,1)}$$

and explicit expressions given for the first terms until $h_3^{\dagger}$. In the present Chapter, we shall complete the calculation of the Hamiltonian to fourth order of approximation, by computing the operator $h_4^{\dagger}$, which will finally be expressed in the following form

$$\begin{aligned}
h_4^{\dagger} = & \sum_{\alpha\beta\gamma\delta\varepsilon\eta} {}_{(4)}Z^{\alpha\beta\gamma\delta\varepsilon\eta} P_\alpha P_\beta P_\gamma P_\delta P_\varepsilon P_\eta + \sum_{\alpha\beta\gamma\delta} \sum_{\substack{ab \\ a\leqslant b}} \Big[{}_{(4)}^{\alpha\beta\gamma\delta}Z_{ab}\, q_a q_b \\
& \qquad + {}_{(4)}^{\alpha\beta\gamma\delta}Z^{ab}\, p_a p_b \Big] P_\alpha P_\beta P_\gamma P_\delta \\
& + \sum_{\alpha\beta} \Big\{ \sum_{\substack{abcd \\ a\leqslant b\leqslant c\leqslant d}} \Big[{}_{(4)}^{\alpha\beta}Z_{abcd}\, q_a q_b q_c q_d + {}_{(4)}^{\alpha\beta}Z^{abcd}\, p_a p_b p_c p_d \Big] \\
& + \sum_{\substack{abcd \\ a\leqslant b;\, c\leqslant d}} {}_{(4)}^{\alpha\beta}Z_{ab}^{cd}\, \tfrac{1}{2}(q_a q_b p_c p_d + p_c p_d q_a q_b) + {}_{(4)}^{\alpha\beta}Z \Big\} P_\alpha P_\beta \\
& + \sum_{\substack{abcdef \\ a\leqslant b\leqslant c\leqslant d\leqslant e\leqslant f}} \Big[{}_{(4)}Z_{abcdef}\, q_a q_b q_c q_d q_e q_f + {}_{(4)}Z^{abcdef}\, p_a p_b p_c p_d p_e p_f \Big] \\
& + \sum_{\substack{abcdef \\ a\leqslant b\leqslant c\leqslant d \\ e\leqslant f}} \Big[{}_{(4)}Z_{abcd}^{ef}\, \tfrac{1}{2}(q_a q_b q_c q_d p_e p_f + p_e p_f q_a q_b q_c q_d) \\
& \qquad + {}_{(4)}Z_{ef}^{abcd}\, \tfrac{1}{2}(q_e q_f p_a p_b p_c p_d + p_a p_b p_c p_d q_e q_f) \Big] \\
& + \sum_{\substack{ab \\ a\leqslant b}} \Big[{}_{(4)}Z_{ab}\, q_a q_b + {}_{(4)}Z^{ab}\, p_a p_b \Big] + h_4^{\dagger *}
\end{aligned} \qquad \text{(IV,2)}$$

with

$$h_4^{\dagger *} = \sum_{\alpha\beta\gamma\delta\varepsilon} \sum_{a} {}^{\alpha\beta\gamma\delta\varepsilon}_{(4)}Z^{a}\, p_a P_\alpha P_\beta P_\gamma P_\delta P_\varepsilon + \sum_{\alpha\beta\gamma} \Big[\sum_{\substack{abc \\ a \leq b \leq c}} {}^{\alpha\beta\gamma}_{(4)}Z^{abc}\, p_a p_b p_c$$

$$+ \sum_{\substack{abc \\ a \leq b}} {}^{\alpha\beta\gamma}_{(4)}Z^{c}_{ab}\, \tfrac{1}{2}(q_a q_b p_c + p_c q_a q_b) \Big] P_\alpha P_\beta P_\gamma$$

$$+ \sum_{\alpha} \Big[\sum_{\substack{abcde \\ a \leq b \leq c \leq d}} {}^{\alpha}_{(4)}Z^{e}_{abcd}\, \tfrac{1}{2}(q_a q_b q_c q_d p_e + p_e q_a q_b q_c q_d)$$

$$+ \sum_{\substack{abcde \\ a \leq b;\, c \leq d \leq e}} {}^{\alpha}_{(4)}Z^{cde}_{ab}\, \tfrac{1}{2}(q_a q_b p_c p_d p_e + p_c p_d p_e q_a q_b)$$

$$+ \sum_{\substack{abcde \\ a \leq b \leq c \leq d \leq e}} {}^{\alpha}_{(4)}Z^{abcde}\, p_a p_b p_c p_d p_e + \sum_{a} {}^{\alpha}_{(4)}Z^{a}\, p_a \Big] P_\alpha \qquad \text{(IV,3)}$$

The first step will be to compute h_4' which appears in the expansion (II,2) of the once-transformed Hamiltonian H'.

FOURTH-ORDER ONCE-TRANSFORMED HAMILTONIAN

Equation (II,10) enables us to write the fourth-order once-transformed Hamiltonian as

$$h_4' = H_4 + [\, iS\, ,\, h_3''\,]_V + [\, iS\, ,\, h_2''\,]_R \qquad \text{(IV,4)}$$

The operator h_4' can be written in a form which is entirely similar to the one given by Equations (IV,2) and (IV,3) for $h_4^\dagger$ except that the coefficient of

$$\tfrac{1}{2}(q_a q_b \cdots p_d p_e \cdots + p_d p_e \cdots q_a q_b \cdots) P_\alpha P_\beta \cdots$$

which is designated by ${}^{\alpha\beta}_{(4)}Z^{de\ldots}_{ab\ldots}$ in $h_4^\dagger$ will be designated by ${}^{\alpha\beta}_{(4)}Y^{de\ldots}_{ab\ldots}$ in h_4' in accordance with the general formulation used in the previous chapters. Expressions for the ${}_{(4)}Y$ coefficients are given in Table XVIII as functions of the following quantities :

(a) the coefficients ${}_{(4)}X$ originating from the fourth-order untransformed Hamiltonian, H_4, and listed in Table XIX.

TABLE XVIII . Coefficients of the once-transformed Hamiltonian h_4'

$$^{\alpha\beta\gamma\delta\varepsilon\eta}_{(4)}Y = \{\alpha\beta,\gamma\delta\varepsilon\eta\}\overline{\omega}$$

$$^{\alpha\beta\gamma\delta}_{(4)}Y_{ab} = {}^{\alpha\beta\gamma\delta}\overline{\omega}_{ab} + (1+\delta_{ab})^{-1}\left[{}^{\{\alpha,\beta\gamma\delta\}}\rho_{a,b} + {}^{\{\alpha,\beta\gamma\delta\}}\rho_{b,a}\right] + {}^{\{\alpha\beta,\gamma\delta\}}\sigma_{ab} + \underset{(s_a \neq s_b)}{{}^{\alpha\beta\gamma\delta}\theta_{ab}}$$

$$^{\alpha\beta\gamma\delta}_{(4)}Y^{ab} = {}^{\alpha\beta\gamma\delta}\overline{\omega}^{ab} + (1+\delta_{ab})^{-1}\left[{}^{\{\alpha,\beta\gamma\delta\}}\rho^{a,b} + {}^{\{\alpha,\beta\gamma\delta\}}\rho^{b,a}\right] + {}^{\{\alpha\beta,\gamma\delta\}}\sigma^{ab} + \underset{(s_a \neq s_b)}{{}^{\alpha\beta\gamma\delta}\theta^{ab}} + (1+\delta_{ab})^{-1}\left({}^{\alpha\beta\gamma\delta}\mu^{a,b} + {}^{\alpha\beta\gamma\delta}\mu^{b,a}\right)$$

$$^{\alpha\beta}_{(4)}Y_{abcd} = {}^{\alpha\beta}_{(4)}X_{abcd} + \sideset{^*}{}\sum_{\substack{jklm \\ (jklm)\equiv(abcd)}} \left[\underset{(j\leq k;\ l\leq m)}{{}^{\alpha\beta}\overline{\omega}_{jk,lm}} + \underset{(k\leq l\leq m)}{{}^{\{\alpha,\beta\}}\rho_{j,klm}} + \underset{(s_j<s_k;\ l\leq m)}{{}^{\alpha\beta}\theta_{jk,lm}}\right] + {}^{\alpha\beta}\sigma_{abcd}$$

$$^{\alpha\beta}_{(4)}Y^{abcd} = \sideset{^*}{}\sum_{\substack{jklm \\ (jklm)\equiv(abcd)}} \left[\underset{(j\leq k;\ l\leq m)}{{}^{\alpha\beta}\overline{\omega}^{jk,lm}} + \underset{(k\leq l\leq m)}{{}^{\{\alpha,\beta\}}\rho^{j,klm}} + \underset{(k\leq l\leq m)}{{}^{\alpha\beta}\mu^{j,klm}} + \underset{(s_j<s_k;\ l\leq m)}{{}^{\alpha\beta}\theta^{jk,lm}}\right] + {}^{\alpha\beta}\sigma^{abcd}$$

$$^{\alpha\beta}_{(4)}Y^{cd}_{ab} = {}^{\alpha\beta}\overline{\omega}^{cd}_{ab} + {}^{\alpha\beta}\sigma^{cd}_{ab} + {}^{\alpha\beta}\psi^{cd}_{ab} + (1+\delta_{ab})^{-1}\left[{}^{\{\alpha,\beta\}}\tau^{cd}_{a,b} + {}^{\{\alpha,\beta\}}\tau^{cd}_{b,a}\right] + (1+\delta_{cd})^{-1}\left[{}^{\{\alpha,\beta\}}\phi^{c,d}_{ab} + {}^{\{\alpha,\beta\}}\phi^{d,c}_{ab}\right]$$

$$+ \sideset{^*}{}\sum_{\substack{jklm \\ (jk)\equiv(ab) \\ (lm)\equiv(cd)}} {}^{\alpha\beta}\rho^{l,m}_{j,k} + (1+\delta_{cd})^{-1}\left({}^{\alpha\beta}\nu^{c,d}_{ab} + {}^{\alpha\beta}\nu^{d,c}_{ab}\right) + \underset{(s_a\neq s_b)\ (s_c\neq s_d)}{{}^{\alpha\beta}\theta^{cd}_{ab} + {}^{\alpha\beta}\mu^{cd}_{ab}}$$

TABLE XVIII . Continued

$$ {}^{\alpha\beta}_{(4)}Y = {}^{\alpha\beta}\overline{\omega} + {}^{\alpha\beta}\rho + {}^{\{\alpha,\beta\}}\sigma + {}^{\{\alpha,\beta\}}\tau + {}^{\alpha\beta}\phi + {}^{\alpha\beta}\theta $$

$$ {}_{(4)}Y_{abcdef} = {}_{(4)}X_{abcdef} + \sideset{}{^*}\sum_{\substack{ijklmn \\ (ijklmn)\equiv(abcdef)}} \overline{\omega}_{ij,klmn} \quad \begin{pmatrix} i\leq j \\ k\leq l\leq m\leq n \end{pmatrix} $$

$$ Y^{abcdef} = \sideset{}{^*}\sum_{\substack{ijklmn \\ (ijklmn)\equiv(abcdef)}} \overline{\omega}^{ij,klmn} \quad \begin{pmatrix} i\leq j \\ k\leq l\leq m\leq n \end{pmatrix} $$

$$ {}_{(4)}Y^{ef}_{abcd} = {}_{(4)}X^{ef}_{abcd} + \sigma^{ef}_{abcd} + \sideset{}{^*}\sum_{\substack{jklm \\ (jklm)\equiv(abcd)}} \overline{\omega}^{ef}_{jk,lm} \quad (j\leq k;\, l\leq m) $$

$$ + \sideset{}{^*}\sum_{\substack{ijklmn \\ (ijkl)\equiv(abcd) \\ (mn)\equiv(ef)}} \rho^{m,n}_{i,jkl} \quad (j\leq k\leq l) $$

$$ {}_{(4)}Y^{cdef}_{ab} = \overline{\omega}^{cdef}_{ab} + \sideset{}{^*}\sum_{\substack{jklm \\ (jklm)\equiv(cdef)}} \sigma^{jk,lm}_{ab} \begin{pmatrix} j<k \\ l\leq m \end{pmatrix} + \sideset{}{^*}\sum_{\substack{ijklm \\ (ijkl)\equiv(cdef) \\ (mn)\equiv(ab)}} \rho^{i,jkl}_{m,n} \quad (j\leq k\leq l) $$

$$ {}_{(4)}Y_{ab} = {}_{(4)}X_{ab} + \sigma_{ab} + (1+\delta_{ab})^{-1}(\overline{\omega}_{a,b} + \overline{\omega}_{b,a}) + \rho_{ab} $$

$$ {}_{(4)}Y^{ab} = \sigma^{ab} + (1+\delta_{ab})^{-1}(\overline{\omega}^{a,b} + \overline{\omega}^{b,a} + \rho^{a,b} + \rho^{b,a}) $$

(b) the coefficients $\overline{\omega}, \rho, \sigma, \tau$, ϕ and ψ appearing in $[iS, h_3'']_V$ and listed in Table XX

(c) the coefficients θ , μ and ν appearing in $[iS, h_2'']_R$ and listed in Table XXI.

TABLE XIX . Coefficients of the untransformed Hamiltonian H_4

$$ {}^{\alpha\beta}_{(4)}X_{abcd} = \frac{\hbar^2}{2 I^e_{\alpha\alpha} I^e_{\beta\beta} (\lambda_a \lambda_b \lambda_c \lambda_d)^{\frac{1}{4}}} \sideset{^*}{}\sum_{\substack{klmn \\ (klmn)\equiv(abcd)}} \Omega^{(4)\,\alpha\beta}_{k,l,m,n} $$

$$ {}_{(4)}X^{ef}_{abcd} = \frac{\hbar}{2} \left(\frac{\lambda_e \lambda_f}{\lambda_a \lambda_b \lambda_c \lambda_d} \right)^{\frac{1}{4}} \sideset{^*}{}\sum_{\substack{ijklmn \\ (ijkl)\equiv(abcd) \\ (mn)\equiv(ef)}} \sum_{\alpha\beta} \frac{\Omega^{(2)\,\alpha\beta}_{i,j} \zeta^{\alpha}_{km} \zeta^{\beta}_{ln}}{I^e_{\alpha\alpha} I^e_{\beta\beta}} $$

$$ {}_{(4)}X_{ab} = \frac{\hbar^3}{4(\lambda_a\lambda_b)^{\frac{1}{4}}(1+\delta_{ab})} \sum_{cd} \sum_{\alpha\beta} \frac{1}{I^e_{\alpha\alpha} I^e_{\beta\beta}} \Big\{ \zeta^{\beta}_{dc} \Big[\zeta^{\alpha}_{cd} (\Omega^{(2)\alpha\beta}_{ab} + \Omega^{(2)\alpha\beta}_{ba}) $$

$$ + 2\zeta^{\alpha}_{ad} (\Omega^{(2)\alpha\beta}_{bc} + \Omega^{(2)\alpha\beta}_{cb}) + 2\zeta^{\alpha}_{bd} (\Omega^{(2)\alpha\beta}_{ac} + \Omega^{(2)\alpha\beta}_{ca}) \Big] $$

$$ + 2\zeta^{\beta}_{ac} \zeta^{\alpha}_{bd} (\Omega^{(2)\alpha\beta}_{cd} + \Omega^{(2)\alpha\beta}_{dc}) \Big\} + \frac{\hbar}{2(\lambda_a\lambda_b)^{\frac{1}{4}}} \Lambda^{(4)}_{ab} $$

$$ {}_{(4)}X_{abcdef} = hck_{abcdef} $$

FOURTH-ORDER TWICE TRANSFORMED HAMILTONIAN

The fourth-order twice-transformed Hamiltonian is given by the relation

$$ h_4^{\dagger} = h_4' + \frac{1}{2}[iS, h_2']_V + [iS, h_1']_R + \frac{1}{2}[iS, h_2^{\dagger}]_V \qquad \text{(IV,5)} $$

according to Equations (III,4e) and (III,4f) in Chapter III. As we know, $h_4^{\dagger}$ can be written in the form of Equation (IV,2). The coefficients ${}_{(4)}Z$ appearing in this equation are given in Table XXII as functions of the quantities ${}_{(4)}Y$, of the coefficients γ , ε , η , ξ, π , ω , β , and of the coefficient α . The coefficients ${}_{(4)}Y$ have been discussed above and are given in Table XVIII. The

TABLE XX. Coefficients appearing in $\left[iS\,,\,h_3''\right]_V$

$$^{\alpha\beta,\gamma\delta\varepsilon\eta}\overline{\omega} = \hbar \sum_m {}^{\alpha\beta}S^{m}\; {}^{\gamma\delta\varepsilon\eta}_{(3)}T_m$$

$$^{\alpha\beta\gamma\delta}\overline{\omega}^{ab} = \hbar \sum_m S^{abm}\left(1+\delta_{am}+\delta_{bm}\right) {}^{\alpha\beta\gamma\delta}_{(3)}T_m$$

$$^{\alpha\beta\gamma\delta}\overline{\omega}_{ab} = \hbar \sum_m S^{m}_{ab}\; {}^{\alpha\beta\gamma\delta}_{(3)}T_m$$

$$^{\alpha,\beta\gamma\delta}\rho_{a,b} = -\hbar \sum_{\substack{m\\(s_m\neq s_a)}} {}^{\alpha}S_{am}\; {}^{\beta\gamma\delta}_{(3)}T^{m}_{b}$$

$$^{\alpha,\beta\gamma\delta}\rho^{a,b} = \hbar \sum_{\substack{m\\(s_m\neq s_a)}} {}^{\alpha}S^{am}\; {}^{\beta\gamma\delta}_{(3)}T^{b}_{m}$$

$$^{\alpha\beta,\gamma\delta}\sigma^{ab} = \hbar \sum_m {}^{\alpha\beta}S^{m}\; {}^{\gamma\delta}_{(3)}T^{ab}_{m}$$

$$^{\alpha\beta}\overline{\omega}^{ab,cd} = \hbar \sum_m S^{abm}\left(1+\delta_{am}+\delta_{bm}\right) {}^{\alpha\beta}_{(3)}T^{cd}_{m}$$

$$^{\alpha\beta}\overline{\omega}^{cd}_{ab} = \hbar \sum_m S^{m}_{ab}\; {}^{\alpha\beta}_{(3)}T^{cd}_{m}$$

$$^{\alpha\beta}\rho^{c,d}_{a,b} = -\hbar \sum_m S^{c}_{am}\left(1+\delta_{am}\right) {}^{\alpha\beta}_{(3)}T^{dm}_{b}\left(1+\delta_{dm}\right)$$

$$^{\alpha\beta}\overline{\omega} = -\frac{\hbar^3}{4} \sum_{k\ell m}\left(1+\delta_{km}\right)\Big[S^{\ell}_{km}\; {}^{\alpha\beta}_{(3)}T^{km}_{\ell}\left(1+\delta_{km}\right) + S^{k}_{km}\; {}^{\alpha\beta}_{(3)}T^{\ell m}_{\ell}\left(1+\delta_{\ell m}\right)\Big]$$

$$^{\alpha\beta,\gamma\delta}\sigma_{ab} = \hbar \sum_m {}^{\alpha\beta}S^{m}\; {}^{\gamma\delta}_{(3)}T_{abm}\left(1+\delta_{am}+\delta_{bm}\right)$$

$$^{\alpha\beta}\sigma^{cd}_{ab} = \hbar \sum_m S^{mcd}\left(1+\delta_{cm}+\delta_{dm}\right) {}^{\alpha\beta}_{(3)}T_{mab}\left(1+\delta_{am}+\delta_{bm}\right)$$

$$^{\alpha\beta}\rho = \frac{\hbar^3}{2} \sum_{\substack{\ell m n\\ \ell\leqslant m\leqslant n}} S^{\ell m n}\; {}^{\alpha\beta}_{(3)}T_{\ell m n}\left(1+\delta_{\ell m}+\delta_{mn}+3\,\delta_{\ell m n}\right)$$

$$^{\alpha\beta}\overline{\omega}_{ab,cd} = \hbar \sum_m S^{m}_{ab}\; {}^{\alpha\beta}_{(3)}T_{mcd}\left(1+\delta_{cm}+\delta_{dm}\right)$$

$$^{\alpha,\beta}\tau^{cd}_{a,b} = -\hbar \sum_{\substack{m\\(s_m\neq s_a)}} {}^{\alpha}S_{am}\; {}^{\beta}_{(3)}T^{mcd}_{b}\left(1+\delta_{cm}+\delta_{dm}\right)$$

TABLE XX . Continued

$$^{\alpha,\beta}\sigma = -\frac{\hbar^3}{4} \sum_{\substack{lmn \\ (s_l \neq s_m)}} {}^{\alpha}S_{lm}\ {}^{\beta}_{(3)}T^{lmn}_{n} \left(1+\delta_{ln}+\delta_{mn}\right)$$

$$^{\alpha,\beta}\rho^{a,bcd} = \hbar \sum_{\substack{m \\ (s_m \neq s_a)}} {}^{\alpha}S^{am}\ {}^{\beta}_{(3)}T^{bcd}_{m}$$

$$^{\alpha,\beta}\rho_{a,bcd} = -\hbar \sum_{\substack{m \\ (s_m \neq s_a)}} {}^{\alpha}S_{am}\ {}^{\beta}_{(3)}T^{m}_{bcd}$$

$$^{\alpha,\beta}\phi^{c,d}_{ab} = \hbar \sum_{\substack{m \\ (s_m \neq s_c)}} {}^{\alpha}S^{cm}\ {}^{\beta}_{(3)}T^{d}_{abm} \left(1+\delta_{am}+\delta_{bm}\right)$$

$$^{\alpha,\beta}\tau = \frac{\hbar^3}{4} \sum_{\substack{lmn \\ (s_l \neq s_m)}} {}^{\alpha}S^{lm}\ {}^{\beta}_{(3)}T^{n}_{lmn} \left(1+\delta_{ln}+\delta_{mn}\right)$$

$$^{\alpha\beta}\sigma^{abcd} = \hbar \sum_{m} {}^{\alpha\beta}S^{m}\ {}_{(3)}T^{abcd}_{m}$$

$$\overline{\omega}^{ab,cdef} = \hbar \sum_{m} S^{abm} \left(1+\delta_{am}+\delta_{bm}\right) {}_{(3)}T^{cdef}_{m}$$

$$\overline{\omega}^{cdef}_{ab} = \hbar \sum_{m} S^{m}_{ab}\ {}_{(3)}T^{cdef}_{m}$$

$$\rho^{c,def}_{a,b} = -\hbar \sum_{m} S^{c}_{am} \left(1+\delta_{am}\right) {}_{(3)}T^{mdef}_{b} \left(1+\delta_{md}+\delta_{me}+\delta_{mf}\right)$$

$$^{\alpha\beta}\psi^{cd}_{ab} = \hbar \sum_{m} {}^{\alpha\beta}S^{m}\ {}_{(3)}T^{cd}_{abm} \left(1+\delta_{am}+\delta_{bm}\right)$$

$$\sigma^{cd,ef}_{ab} = \hbar \sum_{m} S^{cdm} \left(1+\delta_{cm}+\delta_{dm}\right) {}_{(3)}T^{ef}_{abm} \left(1+\delta_{am}+\delta_{bm}\right)$$

$$\overline{\omega}^{ef}_{ab,cd} = \hbar \sum_{m} S^{m}_{ab}\ {}_{(3)}T^{ef}_{cdm} \left(1+\delta_{cm}+\delta_{dm}\right)$$

$$\rho^{e,f}_{a,bcd} = -\hbar \sum_{m} S^{e}_{am} \left(1+\delta_{am}\right) {}_{(3)}T^{mf}_{bcd} \left(1+\delta_{mf}\right)$$

$$^{\alpha\beta}\sigma_{abcd} = \hbar \sum_{m} {}^{\alpha\beta}S^{m}\ {}_{(3)}T_{mabcd} \left(1+\delta_{am}+\delta_{bm}+\delta_{cm}+\delta_{dm}\right)$$

$$\sigma^{ef}_{abcd} = \hbar \sum_{m} S^{efm} \left(1+\delta_{em}+\delta_{fm}\right) {}_{(3)}T_{mabcd} \left(1+\delta_{am}+\delta_{bm}+\delta_{cm}+\delta_{dm}\right)$$

TABLE XX . Continued

$$\bar{\omega}_{ab,cdef} = \hbar \sum_m S_{ab}^{m} \; {}_{(3)}T_{mcdef}\left(1+\delta_{cm}+\delta_{dm}+\delta_{em}+\delta_{fm}\right)$$

$$^{\alpha\beta}\phi = \hbar \sum_m {}^{\alpha\beta}S^{m} \; {}_{(3)}T_m$$

$$\sigma^{ab} = \hbar \sum_m S^{abm}\left(1+\delta_{am}+\delta_{bm}\right) {}_{(3)}T_m$$

$$\sigma_{ab} = \hbar \sum_m S_{ab}^{m} \; {}_{(3)}T_m$$

$$\bar{\omega}^{a,b} = -\frac{\hbar^3}{8}\sum_{klm}\Big\{\Big[S_{lm}^{l} \; {}_{(3)}T_k^{abkm}(1+\delta_{lm}) + S_{km}^{l} \; {}_{(3)}T_l^{abkm}(1+\delta_{km})\Big]$$

$$\times(1+\delta_{am}+\delta_{bm}+\delta_{km})(1+\delta_{ak}+\delta_{bk})(1+\delta_{ab}) + 2S_{km}^{a} \; {}_{(3)}T_l^{bklm}(1+\delta_{km})$$

$$\times(1+\delta_{bm}+\delta_{km}+\delta_{lm})(1+\delta_{bk}+\delta_{bl})(1+\delta_{kl})\Big\}$$

$$\rho^{a,b} = \frac{\hbar^3}{4}\sum_{\substack{klm\\ k\leq l\leq m}} S^{klm} \; {}_{(3)}T_{klm}^{ab}\left(1+\delta_{kl}+\delta_{lm}+3\delta_{klm}\right)(1+\delta_{ab})$$

$$+\frac{\hbar^3}{4}\sum_{klm} S^{akm} \; {}_{(3)}T_{klm}^{bl}(1+\delta_{am}+\delta_{km})(1+\delta_{ak})(1+\delta_{km}+\delta_{lm})$$

$$\times(1+\delta_{kl})(1+\delta_{bl})$$

$$\bar{\omega}_{a,b} = -\frac{\hbar^3}{8}\sum_{klm}\Big[S_{km}^{l} \; {}_{(3)}T_{abl}^{km}(1+\delta_{km}) + S_{km}^{k} \; {}_{(3)}T_{abl}^{lm}(1+\delta_{lm})\Big]$$

$$\times(1+\delta_{km})(1+\delta_{al}+\delta_{bl})(1+\delta_{ab})$$

$$\rho_{ab} = \frac{\hbar^3}{2}\sum_{\substack{klm\\ k\leq l\leq m}} S^{klm} \; {}_{(3)}T_{abklm}\left(1+\delta_{kl}+\delta_{lm}+3\delta_{klm}\right)$$

$$\times\left(1+\delta_{ak}+\delta_{al}+\delta_{am}+\delta_{bk}+\delta_{bl}+\delta_{bm}+\delta_{ak}\delta_{bl}+\delta_{ak}\delta_{bm}+\delta_{al}\delta_{bm}\right)$$

TABLE XXI . Coefficients appearing in $[iS, h_2'']_R$

${}^{\alpha\beta\gamma\delta}\theta_{ab}$	$= -\hbar \langle {}^{\alpha''}S_{ab} \; {}^{\alpha'\beta\gamma\delta}_{(2)}T \rangle$
${}^{\alpha\beta\gamma\delta}\theta^{ab}$	$= -\hbar \langle {}^{\alpha''}S^{ab} \; {}^{\alpha'\beta\gamma\delta}_{(2)}T \rangle$
${}^{\alpha\beta\gamma\delta}\mu^{a,b}$	$= -\hbar \langle {}^{\alpha\beta''}S^{a} \; {}^{\beta'\gamma\delta}_{(2)}T^{b} \rangle$
${}^{\alpha\beta}\theta^{cd}_{ab}$	$= -\hbar \langle {}^{\alpha''}S_{ab} \; {}^{\alpha'\beta}_{(2)}T^{cd} \rangle$
${}^{\alpha\beta}\theta^{ab,cd}$	$= -\hbar \langle {}^{\alpha''}S^{ab} \; {}^{\alpha'\beta}_{(2)}T^{cd} \rangle$
${}^{\alpha\beta}\theta_{ab,cd}$	$= -\hbar \langle {}^{\alpha''}S_{ab} \; {}^{\alpha'\beta}_{(2)}T_{cd} \rangle$
${}^{\alpha\beta}\mu^{cd}_{ab}$	$= -\hbar \langle {}^{\alpha''}S^{cd} \; {}^{\alpha'\beta}_{(2)}T_{ab} \rangle$
${}^{\alpha\beta}\nu^{c,d}_{ab}$	$= -\hbar \langle {}^{\alpha\beta''}S^{c} \; {}^{\beta'}_{(2)}T^{d}_{ab} \rangle$
${}^{\alpha\beta}\mu^{a,bcd}$	$= -\hbar \langle {}^{\alpha\beta''}S^{a} \; {}^{\beta'}_{(2)}T^{bcd} \rangle$
${}^{\alpha\beta}\theta$	$= -\frac{\hbar^3}{4} \sum_{ab} \langle {}^{\alpha\beta''}S^{a} \; {}^{\beta'}_{(2)}T^{b}_{ab} \rangle (1 + \delta_{ab})$

coefficients γ, ε, η, ξ, π, ω and β appear in $[i\mathcal{S}, h_2']_V$ and are listed in Table XXIII. The coefficients α appear in $[i\mathcal{S}, h_1']_R$ and are listed in Table XXIV. It will be demonstrated in the next section that $\frac{1}{2}[i\mathcal{S}, h_2^\dagger]_V$ does not contribute to the coefficients ${}_{(4)}Z$ listed in Table XXII.

COMPUTATION OF THE ENERGY

The operator $h_0^\dagger + \lambda h_1^\dagger + \lambda^2 h_2^\dagger$ has been obtained in such a manner that it is diagonal in all the vibrational quantum numbers, v_s. The operators $h_3^\dagger$ and $h_4^\dagger$ have, of course, elements which are non-diagonal in the quantum numbers v_s. Quite generally, however, these matrix elements will contribute to the energy in an order higher than the fourth[1] and in computing the energy to the fourth order of approx-

TABLE XXII . Coefficients of the twice-transformed Hamiltonian $h_4^\dagger$

$$ {}^{\alpha\beta\gamma\delta\varepsilon\eta}_{(4)}Z = {}^{\alpha\beta\gamma\delta\varepsilon\eta}_{(4)}Y + \frac{1}{2}\,{}^{\{\alpha\beta\gamma,\delta\varepsilon\eta\}}\gamma $$

$$ {}^{\alpha\beta\gamma\delta}_{(4)}Z_{ab} = {}^{\alpha\beta\gamma\delta}_{(4)}Y_{ab} + \frac{1}{2}\,{}^{\{\alpha,\beta\gamma\delta\}}\gamma_{ab} + \frac{{}^{\{\alpha\beta,\gamma\delta\}}\varepsilon_{a,b} + {}^{\{\alpha\beta,\gamma\delta\}}\varepsilon_{b,a}}{2(1+\delta_{ab})} + \frac{1}{2}\,{}^{\{\alpha\beta\gamma,\delta\}}\eta_{ab} $$

$$ {}^{\alpha\beta\gamma\delta}_{(4)}Z^{ab} = {}^{\alpha\beta\gamma\delta}_{(4)}Y^{ab} + \frac{1}{2}\,{}^{\{\alpha,\beta\gamma\delta\}}\gamma^{ab} + \frac{{}^{\{\alpha\beta,\gamma\delta\}}\varepsilon^{a,b} + {}^{\{\alpha\beta,\gamma\delta\}}\varepsilon^{b,a}}{2(1+\delta_{ab})} + \frac{1}{2}\,{}^{\{\alpha\beta\gamma,\delta\}}\eta^{ab} $$

$$ {}^{\alpha\beta}_{(4)}Z_{abcd} = {}^{\alpha\beta}_{(4)}Y_{abcd} + \frac{1}{2}\,\sideset{}{^*}\sum_{\substack{jklm\\(jklm)\equiv(abcd)}} \Big[\underset{(j\leq k;\,l\leq m)}{{}^{\{\alpha,\beta\}}\gamma_{jk,lm}} + \underset{(k\leq l\leq m)}{{}^{\alpha\beta}\varepsilon_{j,klm}} + \underset{(j\leq k\leq l)}{{}^{\alpha\beta}\eta_{jkl,m}}\Big] $$

$$ {}^{\alpha\beta}_{(4)}Z^{abcd} = {}^{\alpha\beta}_{(4)}Y^{abcd} + \frac{1}{2}\,\sideset{}{^*}\sum_{\substack{jklm\\(jklm)\equiv(abcd)}} \Big[\underset{(j\leq k;\,l\leq m)}{{}^{\{\alpha,\beta\}}\gamma^{jk,lm}} + \underset{(k\leq l\leq m)}{{}^{\alpha\beta}\varepsilon^{j,klm}}\Big] $$

$$ {}^{\alpha\beta}_{(4)}Z_{ab}^{cd} = {}^{\alpha\beta}_{(4)}Y_{ab}^{cd} + \frac{1}{2}\,{}^{\{\alpha,\beta\}}\gamma_{ab}^{cd} + \frac{1}{2}\,{}^{\{\alpha,\beta\}}\eta_{ab}^{cd} + \frac{{}^{\alpha\beta}\xi_{a,b}^{cd} + {}^{\alpha\beta}\xi_{b,a}^{cd} + {}^{\alpha\beta}\omega_{a,b}^{cd} + {}^{\alpha\beta}\omega_{b,a}^{cd}}{2(1+\delta_{ab})} + \frac{{}^{\alpha\beta}\pi_{ab}^{c,d} + {}^{\alpha\beta}\pi_{ab}^{d,c} + {}^{\alpha\beta}\beta_{ab}^{c,d} + {}^{\alpha\beta}\beta_{ab}^{d,c}}{2(1+\delta_{cd})} + \sideset{}{^*}\sum_{\substack{jklm\\(jk)\equiv(ab)\\(lm)\equiv(cd)}} \Big[\frac{1}{2}\,{}^{\{\alpha,\beta\}}\varepsilon_{j,k}^{l,m} + {}^{\alpha\beta}\alpha_{j,k}^{l,m}\Big] $$

$$ {}^{\alpha\beta}_{(4)}Z = {}^{\alpha\beta}_{(4)}Y + \frac{1}{2}\,{}^{\{\alpha,\beta\}}\gamma + \frac{1}{2}\,{}^{\{\alpha,\beta\}}\varepsilon + \frac{1}{2}\,{}^{\alpha\beta}\eta + \frac{1}{2}\,{}^{\alpha\beta}\xi $$

TABLE XXII . (Continued)

$$ {}_{(4)}Z_{abcdef} = {}_{(4)}Y_{abcdef} + \frac{1}{2} \overset{*}{\sum_{\substack{ijklmn \\ (ijklmn)\equiv(abcdef)}}} \gamma_{ijk,lmn} \quad \begin{pmatrix} i\leq j\leq k \\ l\leq m\leq n \end{pmatrix} $$

$$ {}_{(4)}Z^{abcdef} = {}_{(4)}Y^{abcdef} $$

$$ {}_{(4)}Z^{ef}_{abcd} = {}_{(4)}Y^{ef}_{abcd} + \frac{1}{2} \overset{*}{\sum_{\substack{ijklmn \\ (ijkl)\equiv(abcd) \\ (mn)\equiv(ef)}}} \eta^{m,n}_{ij,kl} \begin{pmatrix} i\leq j \\ k\leq l \end{pmatrix} + \frac{1}{2} \overset{*}{\sum_{\substack{jklm \\ (jklm)\equiv(abcd)}}} \left[\varepsilon^{ef}_{j,klm} + \gamma^{ef}_{jkl,m} \right] \quad (k\leq l\leq m) \quad (j\leq k\leq l) $$

$$ {}_{(4)}Z^{abcd}_{ef} = {}_{(4)}Y^{abcd}_{ef} + \frac{1}{2} \overset{*}{\sum_{\substack{jklmn \\ (ij)=(ef) \\ (klmn)\equiv(abcd)}}} \gamma^{kl,mn}_{i,j} \begin{pmatrix} k\leq l \\ m\leq n \end{pmatrix} + \frac{1}{2} \overset{*}{\sum_{\substack{jklm \\ (jklm)\equiv(abcd)}}} \varepsilon^{j,klm}_{ef} \quad (k\leq l\leq m) $$

$$ {}_{(4)}Z_{ab} = {}_{(4)}Y_{ab} + \frac{\gamma_{a,b} + \gamma_{b,a} + \varepsilon_{a,b} + \varepsilon_{b,a}}{2(1+\delta_{ab})} $$

$$ {}_{(4)}Z^{ab} = {}_{(4)}Y^{ab} + \frac{\gamma^{a,b} + \gamma^{b,a}}{2(1+\delta_{ab})} $$

imation we may restrict our considerations to only the matrix elements of $h_0^\dagger + \lambda h_1^\dagger + \lambda^2 h_2^\dagger + \lambda^3 h_3^\dagger + \lambda^4 h_4^\dagger$ which are diagonal in the quantum numbers v_s . We can, moreover, neglect that part of $h_4^\dagger$ designated in Equations (IV,2) as $h_4^{\dagger *}$ in computing these matrix elements because $h_4^{\dagger *}$ contains only operators which are odd in the vibrational coordinates, q , and their conjugate momenta, p . They can, therefore, have no matrix elements which are diagonal with respect to the vibrational quantum numbers, v_s . We have, for this reason[2], not listed in Table XXII the coefficients ${}_{(4)}Z$, which would occur in $h_4^{\dagger *}$; neither

TABLE XXIII . Coefficients appearing in $\left[i\mathcal{S}, h_2' \right]_V$

$$ {}^{\alpha\beta\gamma,\delta\varepsilon\eta}\gamma = -\hbar \sum_m {}^{\delta\varepsilon\eta}\mathcal{S}_m \; {}^{\alpha\beta\gamma}_{(2)}Y^m $$

$$ {}^{\alpha,\beta\gamma\delta}\gamma^{ab} = -\hbar \sum_m {}^{\beta\gamma\delta}\mathcal{S}_m \; {}^{\alpha}_{(2)}Y^{abm}(1+\delta_{am}+\delta_{bm}) $$

$$ {}^{\alpha,\beta\gamma\delta}\gamma_{ab} = -\hbar \sum_m {}^{\beta\gamma\delta}\mathcal{S}_m \; {}^{\alpha}_{(2)}Y^{m}_{ab} $$

$$ {}^{\alpha\beta,\gamma\delta}\varepsilon_{a,b} = \hbar \sum_m {}^{\gamma\delta}\mathcal{S}^{m}_{b} \; {}^{\alpha\beta}_{(2)}Y_{am}(1+\delta_{am}) $$

$$ {}^{\alpha\beta,\gamma\delta}\varepsilon^{a,b} = -\hbar \sum_m {}^{\gamma\delta}\mathcal{S}^{b}_{m} \; {}^{\alpha\beta}_{(2)}Y^{am}(1+\delta_{am}) $$

$$ {}^{\alpha\beta\gamma,\delta}\eta^{ab} = -\hbar \sum_m {}^{\delta}\mathcal{S}^{ab}_{m} \; {}^{\alpha\beta\gamma}_{(2)}Y^{m} $$

$$ {}^{\alpha,\beta}\gamma^{ab,cd} = -\hbar \sum_m {}^{\beta}\mathcal{S}^{cd}_{m} \; {}^{\alpha}_{(2)}Y^{abm}(1+\delta_{am}+\delta_{bm}) $$

$$ {}^{\alpha,\beta}\gamma^{cd}_{ab} = -\hbar \sum_m {}^{\beta}\mathcal{S}^{cd}_{m} \; {}^{\alpha}_{(2)}Y^{m}_{ab} $$

$$ {}^{\alpha,\beta}\varepsilon^{c,d}_{a,b} = \hbar \sum_m {}^{\beta}\mathcal{S}^{dm}_{b}(1+\delta_{dm}) \; {}^{\alpha}_{(2)}Y^{c}_{am}(1+\delta_{am}) $$

$$ {}^{\alpha,\beta}\gamma = \frac{\hbar^3}{4} \sum_{klm} (1+\delta_{km}) \left[{}^{\beta}\mathcal{S}^{km}_{l}(1+\delta_{km}) \; {}^{\alpha}_{(2)}Y^{l}_{km} + {}^{\beta}\mathcal{S}^{lm}_{l}(1+\delta_{lm}) \; {}^{\alpha}_{(2)}Y^{k}_{km} \right] $$

$$ {}^{\alpha\beta\gamma,\delta}\eta_{ab} = -\hbar \sum_m {}^{\delta}\mathcal{S}_{abm}(1+\delta_{am}+\delta_{bm}) \; {}^{\alpha\beta\gamma}_{(2)}Y^{m} $$

$$ {}^{\alpha,\beta}\eta^{cd}_{ab} = -\hbar \sum_m {}^{\beta}\mathcal{S}_{abm}(1+\delta_{am}+\delta_{bm}) \; {}^{\alpha}_{(2)}Y^{cdm}(1+\delta_{cm}+\delta_{dm}) $$

$$ {}^{\alpha,\beta}\varepsilon = -\frac{\hbar^3}{2} \sum_{\substack{lmn \\ l \leqslant m \leqslant n}} {}^{\beta}\mathcal{S}_{lmn} \; {}^{\alpha}_{(2)}Y^{lmn}(1+\delta_{lm}+\delta_{mn}+3\,\delta_{lmn}) $$

$$ {}^{\alpha,\beta}\gamma_{ab,cd} = -\hbar \sum_m {}^{\beta}\mathcal{S}_{cdm}(1+\delta_{cm}+\delta_{dm}) \; {}^{\alpha}_{(2)}Y^{m}_{ab} $$

TABLE XXIII . Continued

$$ {}^{\alpha\beta}\xi^{cd}_{a,b} = \hbar \sum_m \mathcal{S}^{cdm}_{b}(1+\delta_{cm}+\delta_{dm})\, {}^{\alpha\beta}_{(2)}Y_{am}(1+\delta_{am}) $$

$$ {}^{\alpha\beta}\eta = \frac{\hbar^3}{4} \sum_{lmn} \mathcal{S}^{lmn}_{n}(1+\delta_{lm}+\delta_{mn})(1+\delta_{ln})\, {}^{\alpha\beta}_{(2)}Y_{lm}(1+\delta_{lm}) $$

$$ {}^{\alpha\beta}\varepsilon^{a,bcd} = -\hbar \sum_m \mathcal{S}^{bcd}_{m}\, {}^{\alpha\beta}_{(2)}Y^{am}(1+\delta_{am}) $$

$$ {}^{\alpha\beta}\varepsilon_{a,bcd} = \hbar \sum_m \mathcal{S}^{m}_{bcd}\, {}^{\alpha\beta}_{(2)}Y_{am}(1+\delta_{am}) $$

$$ {}^{\alpha\beta}\pi^{c,d}_{ab} = -\hbar \sum_m \mathcal{S}^{d}_{abm}(1+\delta_{am}+\delta_{bm})\, {}^{\alpha\beta}_{(2)}Y^{cm}(1+\delta_{cm}) $$

$$ {}^{\alpha\beta}\xi = -\frac{\hbar^3}{4} \sum_{lmn} \mathcal{S}^{n}_{lmn}(1+\delta_{lm}+\delta_{mn})(1+\delta_{ln})\, {}^{\alpha\beta}_{(2)}Y^{lm}(1+\delta_{lm}) $$

$$ {}^{\alpha\beta}\beta^{c,d}_{ab} = -\hbar \sum_m {}^{\alpha\beta}\mathcal{S}^{d}_{m}\, {}_{(2)}Y^{mc}_{ab}(1+\delta_{mc}) $$

$$ {}^{\alpha\beta}\omega^{cd}_{a,b} = \hbar \sum_m {}^{\alpha\beta}\mathcal{S}^{m}_{b}\, {}_{(2)}Y^{cd}_{ma}(1+\delta_{ma}) $$

$$ {}^{\alpha\beta}\eta_{abc,d} = \hbar \sum_m {}^{\alpha\beta}\mathcal{S}^{m}_{d}\, {}_{(2)}Y_{abcm}(1+\delta_{am}+\delta_{bm}+\delta_{cm}) $$

$$ \gamma^{cd,ef}_{a,b} = \hbar \sum_m \mathcal{S}^{efm}_{b}(1+\delta_{em}+\delta_{fm})\, {}_{(2)}Y^{cd}_{am}(1+\delta_{am}) $$

$$ \varepsilon^{c,def}_{ab} = -\hbar \sum_m \mathcal{S}^{def}_{m}\, {}_{(2)}Y^{cm}_{ab}(1+\delta_{cm}) $$

$$ \gamma^{ef}_{abc,d} = \hbar \sum_m \mathcal{S}^{efm}_{d}(1+\delta_{em}+\delta_{fm})\, {}_{(2)}Y_{abcm}(1+\delta_{am}+\delta_{bm}+\delta_{cm}) $$

$$ \varepsilon^{ef}_{a,bcd} = \hbar \sum_m \mathcal{S}^{m}_{bcd}\, {}_{(2)}Y^{ef}_{am}(1+\delta_{am}) $$

$$ \eta^{e,f}_{ab,cd} = -\hbar \sum_m \mathcal{S}^{f}_{cdm}(1+\delta_{cm}+\delta_{dm})\, {}_{(2)}Y^{em}_{ab}(1+\delta_{em}) $$

$$ \gamma_{abc,def} = \hbar \sum_m \mathcal{S}^{m}_{def}\, {}_{(2)}Y_{abcm}(1+\delta_{am}+\delta_{bm}+\delta_{cm}) $$

TABLE XXIII . Continued

$$\gamma^{a,b} = \frac{\hbar^3}{8}\sum_{klm}\Big\{\mathcal{S}_k^{klm}\ {}_{(2)}Y_{lm}^{ab}(1+\delta_{km}+\delta_{lm})(1+\delta_{kl})(1+\delta_{ab})(1+\delta_{lm})$$

$$+2\left[\mathcal{S}_k^{alm}\ {}_{(2)}Y_{lm}^{bk}(1+\delta_{lm})+\mathcal{S}_l^{alm}\ {}_{(2)}Y_{km}^{bk}(1+\delta_{km})\right]$$

$$\times(1+\delta_{am}+\delta_{lm})(1+\delta_{al})(1+\delta_{bk})\Big\}$$

$$\varepsilon_{a,b} = -\frac{\hbar^3}{8}\sum_{klm}\Big\{\mathcal{S}_{klm}^{k}\ {}_{(2)}Y_{ab}^{lm}(1+\delta_{km}+\delta_{lm})(1+\delta_{kl})(1+\delta_{ab})(1+\delta_{lm})$$

$$+2\left[\mathcal{S}_{alm}^{k}\ {}_{(2)}Y_{bk}^{lm}(1+\delta_{lm})+\mathcal{S}_{alm}^{l}\ {}_{(2)}Y_{bk}^{km}(1+\delta_{km})\right]$$

$$\times(1+\delta_{am}+\delta_{lm})(1+\delta_{al})(1+\delta_{bk})\Big\}$$

$$\gamma_{a,b} = \frac{\hbar^3}{8}\sum_{klm}\mathcal{S}_k^{klm}\ {}_{(2)}Y_{ablm}(1+\delta_{km}+\delta_{lm})(1+\delta_{kl})(1+\delta_{am}$$

$$+\delta_{bm}+\delta_{lm})(1+\delta_{al}+\delta_{bl})(1+\delta_{ab})$$

$$+\frac{\hbar^3}{2}\sum_{\substack{klm\\k\leq l\leq m}}\mathcal{S}_b^{klm}\ {}_{(2)}Y_{aklm}(1+\delta_{ak}+\delta_{al}+\delta_{am})$$

$$\times(1+\delta_{kl}+\delta_{lm}+3\delta_{klm})$$

TABLE XXIV . Coefficients appearing in $\left[i\mathcal{S},h_1'\right]_R$

$${}^{\alpha\beta}\alpha_{a,b}^{c,d} = -\hbar\left\langle {}^{\alpha\beta''}\mathcal{S}_a^c\ \ {}^{\beta'}_{(1)}Y_b^d\right\rangle\ (s_b=s_d)$$

$${}^{\alpha\beta}\alpha = -\frac{\hbar^3}{2}\sum_{\substack{ab\\(s_a=s_b)}}\left\langle {}^{\alpha\beta''}\mathcal{S}_a^b\ \ {}^{\beta'}_{(1)}Y_b^a\right\rangle = 0$$

have we listed in Table XVIII the coefficients ${}_{(4)}Y$, which would appear in $h_4'^{*}$. We can, furthermore, neglect the last term, $\frac{1}{2}[i\mathcal{S}, h_2^\dagger]_v$ in Equation (IV,5) which can contribute only to $h_4^{\dagger *}$. This may be seen from the fact that while $h_2^\dagger$ has only matrix elements diagonal with respect to v_s, the operator $\mathcal{S}$ has only off-diagonal matrix elements[3] in the quantum numbers v_s.

Bearing in mind, now, that all the operators occurring in the computation are diagonal with respect to both J and M, we arrive at the following three observations.

(a) We are interested only in matrix elements which can be written :

$$(J, M, v_1, v_2, \dots v_s, \dots | h_0^\dagger + h_1^\dagger + h_2^\dagger + h_3^\dagger + h_4^\dagger | J, M, v_1, v_2, \dots v_s, \dots) \qquad \text{(IV,6)}$$

(b) These matrix elements (IV,6) can be either diagonal or nondiagonal with respect to the remaining quantum numbers, i.e., the rotational quantum number K and the angular momentum quantum numbers, ℓ_s and m_s, due to vibration (ℓ_s has a meaning only when the vibration s is two- or threefold degenerate and m_s has a meaning only when the vibration s is threefold degenerate). It is well to recall in this connection that, in addition to matrix elements which are diagonal with respect to all quantum numbers, $h_0^\dagger$ has matrix elements which are nondiagonal in K when a molecule is asymmetric ; that $h_1^\dagger$ has matrix elements which are nondiagonal in K and m_s for spherical top molecules ; and that $h_2^\dagger$ has matrix elements which are nondiagonal in K for asymmetric molecules, matrix elements which are nondiagonal in K and ℓ_s for axially symmetric molecules, matrix elements which are nondiagonal in ℓ_s for linear molecules, and matrix elements which are nondiagonal in K, ℓ_s and m_s for spherical top molecules

(c) For given values of the quantum numbers, $J, M, v_1, v_2, \dots v_s, \dots$, the matrix elements (IV,6) can be arranged in matrices (submatrices of the general energy matrix) of which the eigenvalues will give values E_{VR} of the rotation-vibration energy, corrected to fourth order. These values will then be obtained by solving the secular equation

$$\det\Big\{(J,K,M,\ldots v_s, \ell_s, m_s \ldots | h^\dagger_0 + h^\dagger_1 + h^\dagger_2 + h^\dagger_3 + h^\dagger_4 | J,K',M,\ldots v_s, \ell'_s, m'_s \ldots) - \delta_{KK'} \ldots \delta_{\ell_s \ell'_s} \ldots \delta_{m_s m'_s} \ldots E_{VR}\Big\} = 0 \qquad \text{(IV,7)}$$

where δ is the Kronecker symbol.

NOTES TO CHAPTER IV

1 . This statement is no longer true in the event of a resonance due to an accidental degeneracy.

2 . For the same reason we have not listed ${}^{\alpha}_{(4)}X^{e}_{abcde}$ in Table XIX.

3 . It can be shown that the condition for $h^\dagger_0 + \lambda h^\dagger_1 + \lambda^2 h^\dagger_2$ to be diagonal with respect to v_s determines only the matrix elements of $\mathcal{S}$ which are nondiagonal in v_s while the matrix elements of $\mathcal{S}$ which are diagonal in v_s can be chosen arbitrarily. If the operator $\mathcal{S}$ were chosen in such a manner that its matrix elements which are diagonal in v_s are nonvanishing, these diagonal matrix elements would, nevertheless, not contribute to the energy. For more details on this point see Reference [17].

CHAPTER V

CONTACT TRANSFORMATIONS IN THE CASE OF A FERMI RESONANCE

INTRODUCTION

In carrying out the contact transformations referred to in the preceding Chapters, it is seen that the function S, and indeed also $\mathcal{S}$, contains terms with resonance denominators $n\omega_s + m\omega_{s'} - p\omega_{s''}$. When two frequencies[1], for example $n\omega_s + m\omega_{s'}$ and $p\omega_{s''}$, are accidentally nearly equal, the usual methods of perturbation theory fail. Most important of the resonances are the Fermi resonance, which stems from a term in the anharmonic portion of the potential function expansion for the molecule of the form $hc\, k_{s\sigma s'\sigma' s'\sigma'}\, q_{s\sigma}\, q_{s'\sigma'}^2$, and the Coriolis-type resonance, which arises from a term in the kinetic energy of the form

$$\zeta^{\alpha}_{s\sigma s'\sigma'}\left[\left(\frac{\omega_s}{\omega_{s'}}\right)^{\frac{1}{2}} q_{s'\sigma'}\frac{p_{s\sigma}}{\hbar} - \left(\frac{\omega_{s'}}{\omega_s}\right)^{\frac{1}{2}} q_{s\sigma}\frac{p_{s'\sigma'}}{\hbar}\right]\frac{P_\alpha}{I^{e}_{\alpha\alpha}}$$

Nielsen [19] has shown that, in treating the second-order problem of the energies of the molecule, it is still effective to transform the Hamiltonian H_1 by a contact transformation, but the contact transformation must this time be so chosen that it leaves the first-order transformed Hamiltonian containing, besides the degenerate Coriolis-interaction terms, terms which have the same matrix elements nondiagonal in v_s and $v_{s'}$ as those of H_1 which lead to resonance, all other matrix elements nondiagonal in v_s and $v_{s'}$ being equal to zero. The resonating frequencies are then regarded as degenerate, and the energies in these instances are then obtained by the methods of the degenerate-perturbation theory. The transformation functions will differ essentially from the functions S used originally in that they do not contain the specific resonance denominator in question.

We wish to study how the procedure must be generalized when energies of a molecule are to be considered to fourth order. Here, not only must the function S be altered, but the function $\mathcal{S}$ occurring in the second contact transformation must also be changed so as not to contain coefficients with the offending resonance denominators. In the present Chapter, we confine ourselves to the case of Fermi resonance between a non-degenerate frequency ω_1 and the first overtone of a doubly degenerate frequency $2\omega_2$. Then, the S and $\mathcal{S}$ functions will be replaced by new functions referred to as ${}_{(F)}S$ and ${}_{(F)}\mathcal{S}$ respectively.

Actually, here the calculation is carried through only to a third order of approximation. The formulation is, however, available to extend the computation through a fourth order if it is deemed necessary.

We consider the case of a nondegenerate vibration $\lambda_1^{\frac{1}{2}}$ in Fermi resonance with a twofold degenerate vibration $2\lambda_2^{\frac{1}{2}}$, where $\lambda_1^{\frac{1}{2}} \simeq 2\lambda_2^{\frac{1}{2}}$. The matrix elements $(v_1, v_2, \ell_2 | v_1 \pm 1, v_2 \mp 2, \ell_2)$ of the Fermi operator $hc \sum_{\sigma=1,2} k_{122} q_1 q_{2\sigma}^2$, which usually contribute to the vibration energy in the second order, will now contribute already in the first order.

${}_{(F)}S$ FUNCTION

In the first contact transformation which is performed on the Hamiltonian we use instead of the usual S function a new function, ${}_{(F)}S$. Instead of h_1', we then obtain a new operator ${}_{(F)}h_1'$. The function ${}_{(F)}S$ is chosen in such a way that ${}_{(F)}h_1'$ will have the same matrix elements $(v_1, v_2, \ell_2 | v_1 \pm 1, v_2 \mp 2, \ell_2)$ as

$hc \sum_{\sigma=1,2} k_{122} q_1 q_{2\sigma}^2$, the same matrix elements diagonal on v_s as H_1 , its other matrix elements being, however, zero. Proceeding in this fashion, Nielsen has shown that

$$ {}_{(F)}h_1' = H_1 + \left[i \; {}_{(F)}S , H_0 \right]_V \tag{V,1a} $$

$$ {}_{(F)}h_1' = -\sum_{\alpha} \frac{p_\alpha^* P_\alpha}{I_{\alpha\alpha}^e} + \frac{hck_{122}}{4} \sum_\sigma \left\{ \left(q_{2\sigma}\frac{p_{2\sigma}}{\hbar} + \frac{p_{2\sigma}}{\hbar} q_{2\sigma} \right)\frac{p_1}{\hbar} - \left(\frac{p_{2\sigma}^2}{\hbar^2} - q_{2\sigma}^2 \right) q_1 \right\} \tag{V,1b} $$

This equation replaces the Equation (II,26) in Chapter II. The function ${}_{(F)}S$ is given by an equation identical to (II,14), where the coefficients S^{abc} and S_{ab}^{c} must, however, be replaced by ${}_{(F)}S^{abc}$ and ${}_{(F)}S_{ab}^{c}$, respectively. These new coefficients are tabulated in Table XXV.

OPERATORS ${}_{(F)}h_1'$, ${}_{(F)}h_1''$, ${}_{(F)}h_1'''$

The new once transformed first-order Hamiltonian can also be written as

$$ \begin{aligned} {}_{(F)}h_1' = & \sum_{\alpha\beta} \sum_{a} {}_{(F)(1)}Y_a^{\alpha\beta} q_a P_\alpha P_\beta \\ & + \sum_{\alpha} \sum_{ab} {}_{(F)(1)}Y_a^{\alpha b} \frac{1}{2}(q_a p_b + p_b q_a) P_\alpha \\ & + \sum_{\substack{abc \\ a\leq b\leq c}} {}_{(F)(1)}Y_{abc} q_a q_b q_c \\ & + \sum_{\substack{abc \\ (a,bc)\equiv(1,2\sigma\,2\sigma) \\ (2\sigma,1\,2\sigma)}} {}_{(F)(1)}Y_a^{bc} \frac{1}{2}(q_a p_b p_c + p_b p_c q_a) \end{aligned} \tag{V,2a} $$

$$ {}_{(F)}h_1' = {}^*h_1' + \sum_{\substack{abc \\ (a,bc)\equiv(1,2\sigma\,2\sigma) \\ (2\sigma,1\,2\sigma)}} {}_{(F)(1)}Y_a^{bc} \frac{1}{2}(q_a p_b p_c + p_b p_c q_a) \tag{V,2b} $$

where ${}^*h_1'$ is an operator defined by an equation identical to (II,25), in which, however, all the coefficients ${}_{(1)}Y$ have to be replaced by

TABLE XXV . Definitions of the coefficients ${}_{(F)}S^{abc}$ and ${}_{(F)}S^{c}_{ab}$

${}_{(F)}S^{abc} = S^{abc}$	$(abc) \neq (1\;2\sigma\;2\sigma)$
${}_{(F)}S^{1\,2\sigma\,2\sigma} = -\dfrac{2\pi c}{\hbar}\;{}^{*}A_{1\,2\sigma\,2\sigma}$	
${}^{*}A_{1\,2\sigma\,2\sigma} = -\dfrac{k_{122}}{4\hbar^{2}}\left(\dfrac{1}{\lambda_1^{\frac{1}{2}} + 2\lambda_2^{\frac{1}{2}}} - \dfrac{2}{\lambda_1^{\frac{1}{2}}}\right)$	
${}_{(F)}S^{c}_{ab} = S^{c}_{ab}$	$(ab,c) \neq (1\;2\sigma,\,2\sigma),(2\sigma\,2\sigma,1)$
${}_{(F)}S^{2\sigma}_{1\,2\sigma} = -\dfrac{4\pi c}{\hbar}\;{}^{*}B_{1\,2\sigma,2\sigma}$	
${}_{(F)}S^{1}_{2\sigma\,2\sigma} = -\dfrac{2\pi c}{\hbar}\;{}^{*}B_{2\sigma\,2\sigma,1}$	
${}^{*}B_{1\,2\sigma,\,2\sigma} = \dfrac{k_{122}}{4(\lambda_1^{\frac{1}{2}} + 2\lambda_2^{\frac{1}{2}})}$	
${}^{*}B_{2\sigma\,2\sigma,\,1} = \dfrac{k_{122}}{4}\left(\dfrac{1}{\lambda_1^{\frac{1}{2}} + 2\lambda_2^{\frac{1}{2}}} + \dfrac{2}{\lambda_1^{\frac{1}{2}}}\right)$	

new coefficients ${}_{(F)(1)}Y$.

The operators ${}_{(F)}h_1''$ and ${}_{(F)}h_1'''$ are given by relations similar to (V,2), in which the coefficients ${}_{(F)(1)}Y$ must be replaced by ${}_{(F)(1)}T$ and ${}_{(F)(1)}U$, respectively.

The various coefficients ${}^{\alpha\dots}_{(F)(1)}\mathcal{O}^{c\dots}_{a\dots}$ ($\mathcal{O}$ = Y , T , U) are in general equal to the coefficients ${}^{\alpha\dots}_{(1)}\mathcal{O}^{c\dots}_{a\dots}$ given in Table II (Chapter II). The only exceptions are the coefficients listed in Table XXVI.

OPERATOR ${}_{(F)}h_2'$

In computing the second-order effects, one must use the follow-

TABLE XXVI . Coefficients appearing in ${}_{(F)}h'_1$, ${}_{(F)}h''_1$, ${}_{(F)}h'''_1$

${}_{(F)(1)}Y_{1\,2\sigma\,2\sigma} = \dfrac{hc}{4}\,k_{122}$
${}_{(F)(1)}T_{1\,2\sigma\,2\sigma} = \dfrac{5hc}{8}\,k_{122}$
${}_{(F)(1)}U_{1\,2\sigma\,2\sigma} = \dfrac{3hc}{4}\,k_{122}$
${}_{(F)(1)}Y^{1\,2\sigma}_{2\sigma} = \dfrac{hc}{2\hbar^2}\,k_{122}$
${}_{(F)(1)}T^{1\,2\sigma}_{2\sigma} = \dfrac{hc}{4\hbar^2}\,k_{122}$
${}_{(F)(1)}U^{1\,2\sigma}_{2\sigma} = \dfrac{hc}{6\hbar^2}\,k_{122}$
${}_{(F)(1)}Y^{2\sigma\,2\sigma}_{1} = -\dfrac{hc}{4\hbar^2}\,k_{122}$
${}_{(F)(1)}T^{2\sigma\,2\sigma}_{1} = -\dfrac{hc}{8\hbar^2}\,k_{122}$
${}_{(F)(1)}U^{2\sigma\,2\sigma}_{1} = -\dfrac{hc}{12\hbar^2}\,k_{122}$

ing operator instead of the operator h'_2:

$$ {}_{(F)}h'_2 = {}^{*}h'_2 + \sum_{\substack{abcd \\ a\leq b\leq c\leq d}} {}_{(F)(2)}Y^{abcd}\, p_a p_b p_c p_d \qquad (V,3) $$

where ${}^{*}h'_2$ is an operator defined by an equation identical to that used for the definition of h'_2 (Equation II,28) in which, however all the coefficients ${}_{(2)}Y$ are replaced by ${}_{(F)(2)}Y$. It is convenient to write

$$ {}_{(F)(2)}{}^{\alpha\ldots}Y^{c\ldots}_{a\ldots} = {}_{*(2)}{}^{\alpha\ldots}Y^{c\ldots}_{a\ldots} + {}_{(T)}{}^{\alpha\ldots}f^{c\ldots}_{a\ldots} \qquad (V,4) $$

where ${}^{\alpha\ldots}_{*(2)}y^{c\ldots}_{a\ldots}$ is obtained by using Table III (Chapter II), after replacing all coefficients

$$_{(T)}\gamma,\ _{(T)}\varepsilon,\ _{(T)}\eta,\ \alpha \quad \text{by} \quad _{*(T)}\gamma,\ _{*(T)}\varepsilon,\ _{*(T)}\eta,\ {}_*\alpha$$

Actually,

$$^{\alpha\ldots}{}_*\alpha^{a\ldots}_{b\ldots} = {}^{\alpha\ldots}\alpha^{a\ldots}_{b\ldots} \tag{V,5}$$

and the nonvanishing coefficients ${}^{\alpha\ldots}_{(T)}\delta^{c\ldots}_{a\ldots}$ are given in Table XXVII.

The coefficients

$$^{\alpha\ldots}_{*(T)}\gamma^{c\ldots}_{a\ldots},\quad {}^{\alpha\ldots}_{*(T)}\varepsilon^{c\ldots}_{a\ldots},\quad {}^{\alpha\ldots}_{*(T)}\eta^{c\ldots}_{a\ldots}$$

are obtained by using Table VIII (Chapter II) after replacing

$$A_{1\,2\sigma\,2\sigma} \quad \text{by} \quad {}^*A_{1\,2\sigma\,2\sigma}$$

$$B_{1\,2\sigma,2\sigma} \quad \text{by} \quad {}^*B_{1\,2\sigma,2\sigma}$$

$$B_{2\sigma\,2\sigma,1} \quad \text{by} \quad {}^*B_{2\sigma\,2\sigma,1}$$

$$k_{122} \quad \text{by} \quad (5/4)\,k_{122}$$

OPERATOR $_{(F)}h''_2$

The operator $_{(F)}h''_2$ is given by a relation similar to (V,3) where $_{(F)(2)}y$ has to be replaced by $_{(F)(2)}T$. As before, it is convenient to write

$$^{\alpha\ldots}_{(F)(2)}T^{c\ldots}_{a\ldots} = {}^{\alpha\ldots}_{*(2)}T^{c\ldots}_{a\ldots} + \frac{1}{2}\,{}^{\alpha\ldots}_{(U)}\delta^{c\ldots}_{a\ldots} \tag{V,6}$$

Let us first observe that

$$^{\alpha\ldots}_{(U)}\delta^{c\ldots}_{a\ldots} = \frac{2}{3}\,{}^{\alpha\ldots}_{(T)}\delta^{c\ldots}_{a\ldots} \tag{V,7}$$

We wish now to compute the coefficients ${}^{\alpha\ldots}_{*(2)}T^{c\ldots}_{a\ldots}$ which are analogous to the coefficients ${}^{\alpha\ldots}_{(2)}T^{c\ldots}_{a\ldots}$, defined in Chapter II. The footnote below Table III (Chapter II) describes how to obtain ${}^{\alpha\ldots}_{(2)}T^{c\ldots}_{a\ldots}$ in terms of the coefficients α , ${}_{(U)}\gamma$, ${}_{(U)}\varepsilon$, ${}_{(U)}\eta$. This footnote may be taken over directly to compute ${}^{\alpha\ldots}_{*(2)}T^{c\ldots}_{a\ldots}$ in terms of a set of new coefficients, which we call ${}_{*}\alpha$, ${}_{*(U)}\gamma$, ${}_{*(U)}\varepsilon$, ${}_{*(U)}\eta$. Let us recall that ${}^{\alpha\ldots}{}_{*}\alpha^{a\ldots}_{b\ldots} = {}^{\alpha\ldots}\alpha^{a\ldots}_{b\ldots}$. The coefficients ${}^{\alpha\ldots}_{*(U)}\gamma^{c\ldots}_{a\ldots}$, ${}^{\alpha\ldots}_{*(U)}\varepsilon^{c\ldots}_{a\ldots}$, ${}^{\alpha\ldots}_{*(U)}\eta^{c\ldots}_{a\ldots}$ can be obtained by using Table VIII (Chapter II), in the following manner :

(a) change the Table as indicated in Chapter II(p.35 paragraph c)

(b) afterwards replace

$$A_{1\,2\sigma\,2\sigma} \text{ by } {}^{*}A_{1\,2\sigma\,2\sigma}$$

$$B_{1\,2\sigma,2\sigma} \text{ by } {}^{*}B_{1\,2\sigma,2\sigma}$$

$$B_{2\sigma\,2\sigma,1} \text{ by } {}^{*}B_{2\sigma\,2\sigma,1}$$

$$k_{122} \text{ by } (9/8)\,k_{122}$$

OPERATOR ${}_{(F)}h'_3$

As we know, a second contact transformation is necessary in order to compute energy to a third order. As a first step, however, we must compute the operator ${}_{(F)}h'_3$ to be used in the place of h'_3 . The operator ${}_{(F)}h'_3$ is defined by an equation similar to (II,30), in which the coefficients ${}_{(3)}Y$ are remplaced by new coefficients ${}_{(F)(3)}Y$

$${}^{\alpha\ldots}_{(F)(3)}Y^{c\ldots}_{a\ldots} = {}^{\alpha\ldots}_{*(3)}Y^{c\ldots}_{a\ldots} + {}^{a\ldots}_{(T)}g^{c\ldots}_{a\ldots} \qquad (V,8)$$

In the former, ${}^{\alpha\ldots}_{*(3)}Y^{c\ldots}_{a\ldots}$ is obtained by using Table IV (Chapter II), after replacing all the coefficients

$${}_{(T)}\theta \,,\, {}_{(T)}\mu \,,\, {}_{(T)}\nu \,,\, {}_{(T)}\xi \,,\, {}_{(T)}\pi \,,\, [\,{}_{(T)}\gamma \,,\, {}_{(T)}\varepsilon\,]$$

by

$${}_{*(T)}\theta \,,\, {}_{*(T)}\mu \,,\, {}_{*(T)}\nu \,,\, {}_{*(T)}\xi \,,\, {}_{*(T)}\pi \,,\, [\,{}_{*(T)}\gamma \,,\, {}_{*(T)}\varepsilon\,]$$

(actually , ${}^{\alpha\ldots}_{*(T)}\gamma^{c\ldots}_{a\ldots} \equiv {}^{\alpha\ldots}_{(T)}\gamma^{c\ldots}_{a\ldots}$ and ${}^{\alpha\ldots}_{*(T)}\varepsilon^{c\ldots}_{a\ldots} \equiv {}^{\alpha\ldots}_{(T)}\varepsilon^{c\ldots}_{a\ldots}$)

TABLE XXVII . Coefficients appearing in ${}_{(F)}h_2'$

$$ {}^{\alpha\beta}_{(T)}f^{1\,2\sigma} = \frac{\pi c}{4\hbar^{\frac{3}{2}}} \frac{k_{122}\, a^{\alpha\beta}_{2\sigma}}{I^e_{\alpha\alpha} I^e_{\beta\beta} \lambda_2^{\frac{3}{4}}} $$

$$ {}^{\alpha\beta}_{(T)}f^{2\sigma\,2\sigma} = -\frac{\pi c}{8\hbar^{\frac{3}{2}}} \frac{k_{122}\, a^{\alpha\beta}_{1}}{I^e_{\alpha\alpha} I^e_{\beta\beta} \lambda_1^{\frac{3}{4}}} $$

$$ {}^{\alpha}_{(T)}f^{a\,1\,2\sigma} \quad (s_a \neq 2) = \frac{\pi c}{\hbar^2} \frac{k_{122}\, \zeta^{\alpha}_{a\,2\sigma} (\lambda_a \lambda_2)^{\frac{1}{4}}}{(\lambda_a - \lambda_2) I^e_{\alpha\alpha}} $$

$$ {}^{\alpha}_{(T)}f^{a\,2\sigma\,2\sigma} \quad (a \neq 1) = -\frac{\pi c}{2\hbar^2} \frac{k_{122}\, \zeta^{\alpha}_{a1} (\lambda_a \lambda_1)^{\frac{1}{4}}}{(\lambda_a - \lambda_1) I^e_{\alpha\alpha}} $$

$$ {}^{\alpha}_{(T)}f^{1}_{a\,2\sigma} \quad (s_a \neq 2) = -\frac{\pi c}{2} \frac{k_{122}\, \zeta^{\alpha}_{a\,2\sigma} (\lambda_a + \lambda_2)}{2(\lambda_a - \lambda_2) I^e_{\alpha\alpha} (\lambda_a \lambda_2)^{\frac{1}{4}}} $$

$$ {}^{\alpha}_{(T)}f^{2\sigma}_{a\,1} \quad (s_a \neq 2) = \frac{\pi c}{2} \frac{k_{122}\, \zeta^{\alpha}_{a\,2\sigma} (\lambda_a + \lambda_2)}{(\lambda_a - \lambda_2) I^e_{\alpha\alpha} (\lambda_a \lambda_2)^{\frac{1}{4}}} $$

$$ {}^{\alpha}_{(T)}f^{2\sigma}_{a\,2\sigma} \quad (a \neq 1) = -\frac{\pi c}{2} \frac{k_{122}\, \zeta^{\alpha}_{a1} (\lambda_a + \lambda_1)}{(\lambda_a - \lambda_1) I^e_{\alpha\alpha} (\lambda_a \lambda_1)^{\frac{1}{4}}} $$

$$ {}_{(T)}f^{abcd} = \sideset{^*}{}\sum_{(lmno)\equiv(abcd)} \phi^{lm,no} $$

$$ \phi^{ab,1\,2\sigma} = -(\pi^2 c^2/\hbar)\, k_{122}\, {}^*A_{ab\,2\sigma} (1 + \delta_{a,2\sigma} + \delta_{b,2\sigma}) $$

$$ \phi^{ab,2\sigma\,2\sigma} = (\pi^2 c^2/2\hbar)\, k_{122}\, {}^*A_{ab\,1} (1 + \delta_{a1} + \delta_{b1}) $$

TABLE XXVII . Continued

$$ {}_{(T)}f^{cd}_{ab} = \phi^{cd}_{ab} + \frac{\psi^{c,d}_{a,b} + \psi^{d,c}_{a,b} + \psi^{c,d}_{b,a} + \psi^{d,c}_{b,a}}{(1+\delta_{ab})(1+\delta_{cd})} $$

$$ \begin{cases} \phi^{1\,2\sigma}_{ab} = -(\pi^2 c^2/\hbar)\, k_{122}\, {}^*B_{ab,2\sigma}(1+\delta_{a,2\sigma}+\delta_{b,2\sigma}) \\ \phi^{2\sigma\,2\sigma}_{ab} = (\pi^2 c^2/2\hbar)\, k_{122}\, {}^*B_{ab,1}(1+\delta_{a1}+\delta_{b1}) \end{cases} $$

$$ \begin{cases} \psi^{b,2\sigma}_{a,1} = -(\pi^2 c^2/\hbar) k_{122}\, {}^*B_{a\,2\sigma,b}(1+\delta_{a,b}+\delta_{b,2\sigma})(1+\delta_{a,2\sigma}) \\ \psi^{b,1}_{a,2\sigma} = (\pi^2 c^2/\hbar)\, k_{122}\, {}^*B_{a2\sigma,b}(1+\delta_{a,b}+\delta_{b,2\sigma})(1+\delta_{a,2\sigma}) \\ \psi^{b,2\sigma}_{a,2\sigma} = (\pi^2 c^2/\hbar)\, k_{122}\, {}^*B_{a\,1,b}(1+\delta_{a,b}+\delta_{1,b})(1+\delta_{a,1}) \end{cases} $$

the coefficients

$$ {}^{\alpha\ldots}_{*(T)}\theta^{c\ldots}_{a\ldots}, \quad {}^{\alpha\ldots}_{*(T)}\mu^{c\ldots}_{a\ldots}, \quad {}^{\alpha\ldots}_{*(T)}\nu^{c\ldots}_{a\ldots}, \quad {}^{\alpha\ldots}_{*(T)}\xi^{c\ldots}_{a\ldots}, \quad {}^{\alpha\ldots}_{*(T)}\pi^{c\ldots}_{a\ldots} $$

being obtained by using Table X (Chapter II) in the following manner :

(a) Replace S^{abc} and S^{c}_{ab} by ${}_{(F)}S^{abc}$ and ${}_{(F)}S^{c}_{ab}$ defined in Table XXV.

(b) Replace ${}^{\alpha\ldots}_{(2)}T^{c\ldots}_{a\ldots}$ by ${}^{\alpha\ldots}_{(F)(2)}T^{c\ldots}_{a\ldots}$ defined in the preceding paragraph of the present Chapter.

The quantities g occurring in the equation (V,8) are given in Table XXVIII. It should be noted that the following relation

$$ {}_{(F)(2)}T^{abcd} = \frac{1}{2}\,{}_{(U)}f^{abcd} = \frac{1}{3}\,{}_{(T)}f^{abcd} \qquad (V,9) $$

was taken into account in Table XXVIII.

TABLE XXVIII . Definitions of the quantities ${}^{\alpha\ldots}_{(T)}g^{b\ldots}_{a\ldots}$

$$ {}^{\alpha}_{(T)}g_a^{\;bcd} = -\frac{\hbar}{3} \sum_{\substack{m \\ s_m \neq s_a}} {}^{\alpha}S_{am} \; {}_{(T)}f^{mbcd}(1+\delta_{mb}+\delta_{mc}+\delta_{md}) $$

$$ {}_{(T)}g_a^{\;bcde} = \sideset{^*}{}\sum_{\substack{lmno \\ (lmno)\equiv(bcde)}} \phi_a^{\;l,mno} $$

$$ \phi_a^{\;b,cde} = -\frac{\hbar}{3} \sum_m {}_{(F)}S^{\;b}_{am}(1+\delta_{am}) \; {}_{(T)}f^{mcde}(1+\delta_{mc}+\delta_{md}+\delta_{me}) $$

FUNCTION ${}_{(F)}\mathcal{S}$

In the second contact transformation, one must use, instead of the function $\mathcal{S}$, a function ${}_{(F)}\mathcal{S}$ defined by an equation similar to (III,5), on which, however, the coefficients ${}^{\alpha\ldots}\mathcal{S}^{c\ldots}_{a\ldots}$ are replaced by new coefficients[2]

$$ {}^{\alpha\ldots}_{(F)}\mathcal{S}^{c\ldots}_{a\ldots} = {}^{*\alpha}\mathcal{S}^{c\ldots}_{a\ldots} + {}^{\alpha\ldots}\Sigma^{c\ldots}_{a\ldots} \tag{V,10} $$

The coefficients ${}^{*\alpha\ldots}\mathcal{S}^{c\ldots}_{a\ldots}$ are given by Table XII (Chapter III) after replacing

$$ {}^{\alpha\ldots}_{(2)}Y^{c\ldots}_{a\ldots} \quad \text{by} \quad {}^{\alpha\ldots}_{(F)(2)}Y^{c\ldots}_{a\ldots} $$

$\mathcal{A}_{1\,2\sigma\,2\sigma}$, $\mathcal{B}_{1\,2\sigma,2\sigma}$ and $\mathcal{B}_{2\sigma\,2\sigma,1}$ by ${}^*\mathcal{A}_{1\,2\sigma\,2\sigma}$, ${}^*\mathcal{B}_{1\,2\sigma,2\sigma}$ and ${}^*\mathcal{B}_{2\sigma\,2\sigma,1}$ respectively, these last coefficients being defined in Table XXIX. It can be seen that ${}^{*\alpha\beta\gamma}\mathcal{S}_a \equiv {}^{\alpha\beta\gamma}\mathcal{S}_a$.

The only non-vanishing ${}^{\alpha}\Sigma^{c\ldots}_{a\ldots}$ coefficients are of the types Σ^{abc}_d and Σ^{d}_{abc} ; then, ${}^{\alpha\beta\gamma}_{(F)}\mathcal{S}_a \equiv {}^{*\alpha\beta\gamma}\mathcal{S}_a$, ${}^{\alpha\beta}_{(F)}\mathcal{S}^b_a \equiv {}^{*\alpha\beta}\mathcal{S}^b_a$, ${}^{\alpha}_{(F)}\mathcal{S}^{ab}_c \equiv {}^{*\alpha}\mathcal{S}^{ab}_c$; ${}^{\alpha}_{(F)}\mathcal{S}_{abc} \equiv {}^{*\alpha}\mathcal{S}_{abc}$

The non-vanishing Σ^{abc}_d and Σ^{d}_{abc} coefficients can be written

TABLE XXIX . Definitions of $^{*}\mathcal{A}$ and $^{*}\mathcal{B}$

$$^{*}\mathcal{A}_{12\sigma 2\sigma} = -\frac{\hbar^2}{k_{122}} \, ^{*}A_{12\sigma 2\sigma}$$

$$^{*}\mathcal{B}_{12\sigma,2\sigma} = -\frac{1}{k_{122}} \, ^{*}B_{12\sigma,2\sigma}$$

$$^{*}\mathcal{B}_{2\sigma 2\sigma,1} = -\frac{1}{k_{122}} \, ^{*}B_{2\sigma 2\sigma,1}$$

The coefficients $^{*}\mathcal{A}$ and $^{*}\mathcal{B}$ are obtained by suppressing the fractions with a vanishing denominator in the corresponding coefficients $\mathcal{A}$ and $\mathcal{B}$ given in Table XIII (Chapter III).

as follows :

$$\Sigma_{d}^{abc} = -\frac{1}{\hbar^4}(1+\delta_{ad}+\delta_{bd}+\delta_{cd})\, {}_{(F)(2)}Y^{abcd}\, \mathcal{A}_{abc,d} \qquad (V,11)$$

$$\Sigma_{abc}^{d} = \frac{1}{\hbar^2}(1+\delta_{ad}+\delta_{bd}+\delta_{cd})\, {}_{(F)(2)}Y^{abcd}\, \mathcal{B}_{abc,d} \qquad (V,12)$$

where the non-vanishing coefficients ${}_{(F)(2)}Y^{abcd} \equiv {}_{(T)}f^{abcd}$ are defined in Table XXVII and where $\mathcal{A}_{abc,d}$ and $\mathcal{B}_{abc,d}$ are defined in Table XIII (Chapter III).

OPERATORS ${}_{(F)}h_0^{\dagger}$, ${}_{(F)}h_1^{\dagger}$, ${}_{(F)}h_2^{\dagger}$

The zeroth and first-order twice-transformed Hamiltonians are obviously given by

$${}_{(F)}h_0^{\dagger} = H_0 \qquad (V,13)$$

$${}_{(F)}h_1^{\dagger} = {}_{(F)}h_1' \qquad (V,14)$$

TABLE XXX . Coefficients occurring in ${}_{(F)}h_3^+$

$$ {}^{\alpha\beta\gamma}\mathcal{j}_a^{\cdot b} = -\hbar \sum_m {}^{\alpha\beta\gamma}\mathcal{S}_m \; {}_{(F)(1)}Y_a^{mb} $$

$$ {}^{\alpha\beta}\mathcal{j}_a^{\cdot bc} \quad {}^{\alpha\beta}\phi_a^{bc} + \frac{1}{(1+\delta_{bc})}\left({}^{\alpha\beta}\psi_a^{b,c} + {}^{\alpha\beta}\psi_a^{c,b} \right) $$

$$ \begin{cases} {}^{\alpha\beta}\phi_a^{bc} = \hbar \sum_m {}^{*\alpha\beta}\mathcal{S}_a^{m} \; {}_{(F)(1)}Y_m^{bc} \\ {}^{\alpha\beta}\psi_a^{b,c} = -\hbar \sum_m {}^{*\alpha\beta}\mathcal{S}_m^{b} \; {}_{(F)(1)}Y_a^{mc}(1+\delta_{mc}) \end{cases} $$

$$ {}^{\alpha\beta}\mathcal{j}_{abc}^{\cdot} = \sideset{^*}{}\sum_{\substack{\ell m n \\ (\ell m n)\equiv(abc)}} {}^{\alpha\beta}\phi_{\ell,mn} $$

$$ {}^{\alpha\beta}\phi_{a,bc} = \hbar \sum_m {}^{*\alpha\beta}\mathcal{S}_a^{m} \; {}_{(F)(1)}Y_{mbc}(1+\delta_{mb}+\delta_{mc}) $$

$$ {}^{\alpha}\mathcal{j}_a^{\cdot bcd} = \sideset{^*}{}\sum_{\substack{\ell m n \\ (\ell m n)\equiv(bcd)}} \left({}^{\alpha}\phi_a^{\ell m,n} + {}^{\alpha}\psi_a^{\ell m,n} \right) $$

$$ \begin{cases} {}^{\alpha}\phi_a^{bc,d} = -\hbar \sum_m {}^{*\alpha}\mathcal{S}_m^{bc} \; {}_{(F)(1)}Y_a^{md}(1+\delta_{md}) \\ {}^{\alpha}\psi_a^{b,cd} = \hbar \sum_m {}^{*\alpha}\mathcal{S}_a^{mb}(1+\delta_{mb}) \; {}_{(F)(1)}Y_m^{cd} \end{cases} $$

$$ {}^{\alpha}\mathcal{j}_{abc}^{\cdot d} = \sideset{^*}{}\sum_{\substack{\ell m n \\ (\ell m n)\equiv(abc)}} \left({}^{\alpha}\phi_{\ell m,n}^{d} + {}^{\alpha}\psi_{\ell,mn}^{d} \right) $$

$$ \begin{cases} {}^{\alpha}\phi_{ab,c}^{d} = -\hbar \sum_m {}^{*\alpha}\mathcal{S}_{mab}(1+\delta_{am}+\delta_{bm}) \; {}_{(F)(1)}Y_c^{md}(1+\delta_{md}) \\ {}^{\alpha}\psi_{a,bc}^{d} = \hbar \sum_m {}^{*\alpha}\mathcal{S}_a^{md}(1+\delta_{md}) \; {}_{(F)(1)}Y_{mbc}(1+\delta_{mb}+\delta_{mc}) \end{cases} $$

TABLE XXX . Continued

$$\dot{j}_a^{bcde} = \sum^*_{\substack{lmno \\ (lmno)\equiv(bcde)}} (\psi_a^{lmn,o} + \bar{\omega}_a^{lm,no})$$

$$\begin{cases} \psi_a^{bcd,e} = -\hbar \sum_m {}_{(F)}\varphi_m^{bcd} \; {}_{(F)(1)}Y_a^{me}(1+\delta_{me}) \\ \bar{\omega}_a^{bc,de} = \hbar \sum_m {}_{(F)}\varphi_a^{mbc}(1+\delta_{mb}+\delta_{mc}) \; {}_{(F)(1)}Y_m^{de} \end{cases}$$

$$\dot{j}_{abc}^{de} = \phi_{abc}^{de} + \frac{1}{1+\delta_{de}} \sum^*_{\substack{lmn \\ (lmn)\equiv(abc)}} (\psi_{lm,n}^{d,e} + \psi_{lm,n}^{c,d}) + \sum^*_{\substack{lmn \\ (lmn)\equiv(abc)}} \bar{\omega}_{l,mn}^{de}$$

$$\begin{cases} \phi_{abc}^{de} = \hbar \sum_m {}_{(F)}\varphi_{abc}^m \; {}_{(F)(1)}Y_m^{de} \\ \psi_{ab,c}^{d,e} = -\hbar \sum_m {}_{(F)}\varphi_{abm}^d (1+\delta_{am}+\delta_{bm}) \; {}_{(F)(1)}Y_c^{m,e}(1+\delta_{me}) \\ \bar{\omega}_{a,bc}^{de} = \hbar \sum_m {}_{(F)}\varphi_a^{mde}(1+\delta_{md}+\delta_{me}) \; {}_{(F)(1)}Y_{mbc}(1+\delta_{mb}+\delta_{mc}) \end{cases}$$

$$j_{abcde} = \sum^*_{\substack{klmno \\ (klmno)\equiv(abcde)}} \phi_{klm,no}$$

$$\phi_{abc,de} = \hbar \sum_m {}_{(F)}\varphi_{abc}^m \; {}_{(F)(1)}Y_{mde}(1+\delta_{md}+\delta_{me})$$

For the second-order operator,

$$_{(F)}h_2^{\dagger} = {}_{(F)}h_2' + \left[i\,{}_{(F)}\mathcal{S}, H_0 \right]_V \qquad (V,15)$$

a special computation is necessary. Considering that this operator is not of interest in the third-order calculation and that the second-order effects can as well be obtained from ${}_{(F)}h_2'$ we will not compute ${}_{(F)}h_2^{\dagger}$.

OPERATOR ${}_{(F)}h_3^{\dagger}$

In order to compute the third-order energy, one must use instead of $h_3^{\dagger}$ the operator ${}_{(F)}h_3^{\dagger}$ defined by an equation similar to (III,8) in which the coefficients ${}_{(3)}z$ are replaced by new coefficients ${}_{(F)(3)}z$.

$$ {}^{\alpha\ldots}_{(F)(3)}z^{c\ldots}_{a\ldots} = {}^{\alpha\ldots}_{*(3)}z^{c\ldots}_{a\ldots} + {}^{\alpha\ldots}\gamma^{c\ldots}_{a\ldots} \qquad (V,16) $$

${}^{\alpha\ldots}_{*(3)}z^{c\ldots}_{a\ldots}$ is obtained by using Table XV (Chapter III) after replacing the coefficients α, β, η by $*\alpha$, $*\beta$, $*\eta$

The coefficients ${}^{\alpha\ldots}{*\alpha}^{c\ldots}_{a\ldots}$ and ${}^{\alpha\ldots}{*\beta}^{c\ldots}_{a\ldots}$ are obtained by using Table XVII (Chapter III), and the coefficients ${}^{\alpha\ldots}{*\eta}^{c\ldots}_{a\ldots}$ by using Table XVI (Chapter III), after replacing in these two Tables ${}^{\alpha\ldots}\varphi^{b\ldots}_{a\ldots}$ by ${}^{\alpha\ldots}_{(F)}\varphi^{b\ldots}_{a\ldots}$ and ${}^{\alpha\ldots}_{(1)}y^{b\ldots}_{a\ldots}$ by ${}^{\alpha\ldots}_{(F)(1)}y^{b\ldots}_{a\ldots}$. The coefficients ${}^{\alpha\ldots}\gamma^{c\ldots}_{a\ldots}$ are given in Table XXX.

CONCLUSION

The formulation given in the present Chapter has been used successfully by S.Maes [20] in his theoretical study of the "Taylor-Benedict-Strong effect" (third-order corrections to Fermi resonance), and by M.H.Andrade e Silva and one of us [21] in their interpretation of the "Courtoy resonance", (simultaneous Fermi and second-order Coriolis resonances in CO_2).

Since the first publication of the computation described in the present Chapter, it has been extended in two different ways by various authors :

- Some fourth order operators appearing in ${}_{(F)}h_4^{\dagger}$ have been computed [22] (namely those involving $q_a q_b P_\alpha P_\beta P_\gamma P_\delta$ and $p_a p_b P_\alpha P_\beta P_\gamma P_\delta$)
- Other resonances have been considered : Fermi type resonance involving three vibrations [23] ($\lambda_1^{\frac{1}{2}} \simeq \lambda_2^{\frac{1}{2}} + \lambda_3^{\frac{1}{2}}$) and Coriolis type resonance [24] ($\lambda_1^{\frac{1}{2}} \simeq \lambda_2^{\frac{1}{2}}$).

-

NOTES TO CHAPTER V

1 . The wave numbers ω_s , associated with normal frequencies, are given by $\omega_s = \lambda_s^{1/2}/2\pi c$.

2 . Although the symbol is the same, the coefficients $^{*}\alpha\ldots\mathcal{S}^{c\ldots}_{a\ldots}$ defined here are not the same as those mentioned occasionaly in the reference [3].

PART 2

MATRIX ELEMENTS OF THE TRANSFORMED HAMILTONIAN

CHAPTER VI

NON-VANISHING MATRIX ELEMENTS FOR A MOLECULE OF GIVEN SYMMETRY

INTRODUCTION

As pointed out at the end of Chapter IV, the rotation-vibration energy levels, corrected to fourth order will generally be obtained by solving secular equations of the type (IV,7). When accidental resonances occur, we have seen in Chapter V that matrix elements, non-diagonal with respect to quantum numbers v_s need also be taken into account. In order to write down the secular determinant, we then need to compute matrix elements of the form :

$$(JKM\, v_s\, \ell_s\, m_s \ldots \,|\, h \,|\, J, K+\Delta K, M, v_s+\Delta v_s, \ell_s+\Delta \ell_s, m_s+\Delta m_s \ldots) \quad (VI,1)$$

where h is one of the operators $h_0^\dagger$, $h_1^\dagger$, $h_2^\dagger$, $h_3^\dagger$ or $h_4^\dagger$ (possibly H_1 or h_2' in the case of first or second order accidental resonances). In any event the operator h is totally symmetric (with respect to all symmetry operations belonging to the point group of the equilibrium configuration of the molecule). It can then be shown that the matrix element (VI,1) will be non-vanishing, in a molecule of given symmetry, only if the quantum jumps ΔK, Δv_s, $\Delta \ell_s$, Δm_s, ... satisfy a relation which has been derived by one of us [25].

The aim of the present chapter is to indicate, for a molecule of given symmetry, which are the non-vanishing elements which can possibly appear in the secular determinant. In the next chapter, the true order of magnitude of these non-vanishing matrix elements will be discussed allowing a selection among them of those which give a significant contribution to the rotation-vibration energy in a fourth order computation. These two kinds of considerations (symmetry properties and discussion on orders of magnitude) prove to be very useful in avoiding unnecessary computations when writing down the secular determinant.

AXIALLY SYMMETRIC MOLECULES

Since axially symmetric molecules have non-degenerate and twofold degenerate normal vibrations, we are interested in matrix elements of the type :

$$(JKM\, v_s\, \ell_s \ldots |\bar{h}| J, K+\Delta K, M, v_s+\Delta v_s, \ell_s+\Delta \ell_s \ldots) \tag{VI,2}$$

The point group of the equilibrium configuration can be one of the following :

a) C_N, C_{Nh}, C_{Nv}, D_N, D_{Nh}, D_{Nd} (N odd), S_{2N} (N odd)

b) $D_{\frac{N}{2}d}$ ($\frac{N}{2}$ even), S_N ($\frac{N}{2}$ even)

where N is the foldness of the principal symmetry axis z. The symmetry species $A\, B\, E_1\, E_2 \ldots$ are defined according to the nature of the transformation law associated with the symmetry operation $\mathcal{R}$ which is a proper rotation by $\frac{2\pi}{N}$ about the z axis for molecules belonging to one of the groups listed under a) above and an improper rotation by $\frac{2\pi}{N}$ about the z axis for molecules belonging to one of the groups listed under b) above.

A non-degenerate normal coordinate q_s can be either symmetric or antisymmetric with respect to operation $\mathcal{R}$. The corresponding symmetry species are designated by a symbol A or B respectively[1].

We shall assume that each pair of twofold degenerate normal coordinates q_{s1}, q_{s2} is "oriented" in such a manner that, when operation $\mathcal{R}$ is performed it transforms according to the matrix [26]

$$\begin{Bmatrix} \cos \dfrac{2\pi a_s}{N} & \sin \dfrac{2\pi a_s}{N} \\ -\sin \dfrac{2\pi a_s}{N} & \cos \dfrac{2\pi a_s}{N} \end{Bmatrix} \qquad \text{(VI,3)}$$

where a_s is an integer $(1 \leqslant a_s < \frac{N}{2})$. When $a_s = 1, 2, 3 \ldots$ the symmetry species of vibration s is designated by a symbol E_1, E_2, $E_3 \ldots$ respectively[1]. If $N = 3$ or $N = 4$, the only possible value of a_s is 1 and a symbol E is used instead of E_1.

In what follows the subscript s which defines a normal vibration will be replaced by n for a non-degenerate normal vibration belonging to species A, by n' for a non-degenerate normal vibration belonging to species B, by t for a twofold degenerate normal vibration, so that the matrix element (VI,2) can be written as follows :

$$(JKM\, v_n\, v_{n'}\, v_t\, \ell_t \ldots |h| J, K+\Delta K, M, v_n+\Delta v_n, v_{n'}+\Delta v_{n'}, v_t+\Delta v_t, \ell_t+\Delta\ell_t \ldots) \qquad \text{(VI,4)}$$

It can be shown that this matrix element is non-vanishing only if

$$\lambda\, \Delta K + \sum_t a_t\, \Delta \ell_t + \frac{N}{2} \sum_{n'} \Delta v_{n'} = p N \qquad \text{(VI,5)}$$

where p is an arbitrary integer which is positive, negative or equal to zero. Depending on whether the molecule belongs to one of the groups listed under a) or b) above, λ is equal to -1 or $\frac{N}{2} - 1$ respectively.

Relation (VI,5) is obviously verified for matrix elements which are diagonal with respect to all quantum numbers ; it indicates furthermore which non-diagonal matrix elements are non-vanishing for an axially symmetric molecule of given symmetry. If one is interested in matrix elements which are diagonal with respect to all quantum numbers v_s (i.e.all quantum numbers v_n $v_{n'}$ and v_t), it can be shown [25] that relation (VI,5) can be simplified and written as follows :

$$-\Delta K + \sum_t a_t \Delta \ell_t = p N \qquad \text{(VI,6)}$$

As long as no accidental resonances occur, the secular equation contains only matrix elements which are diagonal with respect to all the quantum numbers v_s and relation (VI,6) can be used in order to find the non-vanishing matrix elements. It is only when accidental resonances occur that relation (VI,5) must be used to find the non-vanishing matrix elements which are non-diagonal with respect to the quantum numbers v_s.

LINEAR MOLECULES

In the case of a linear molecule belonging to group $C_{\infty v}$ or $D_{\infty h}$ relation (VI,5) can still be used, however :

- N is infinite, so that p must be taken equal to zero in the right hand side of the relation (VI,5)
- Operation $\mathcal{R}$ used in the definition of symmetry species is a proper rotation (in that respect groups $C_{\infty v}$ and $D_{\infty h}$ are similar to groups listed under a) in second paragraph above). Then λ must be taken equal to -1.
- If $\mathcal{R}$ is a rotation by an angle $d\varphi$ about the z axis, all two-fold degenerate normal coordinates (they belong to species Π in group $C_{\infty v}$, to species Π_g or Π_u in group $D_{\infty h}$) transform under $\mathcal{R}$ according to the matrix

$$\begin{Bmatrix} \cos d\varphi & \sin d\varphi \\ -\sin d\varphi & \cos d\varphi \end{Bmatrix} \qquad \text{(VI,7)}$$

so that all the coefficients a_t are equal to 1 .

- There are no non-degenerate vibrations antisymmetric with

respect to $\mathfrak{R}$ so that all the $\Delta v_{n'}$ must be taken equal to zero.

Then both equations (VI,5) and (VI,6) reduce to :

$$-\Delta K + \sum_t \Delta \ell_t = 0 \qquad \text{(VI,8)}$$

It is obvious that in the case of linear molecules, this relation can be derived much more easily as a consequence of the Sayvetz condition [27]

$$P_z = p_z \qquad \text{(VI,9)}$$

or

$$K = \sum_t \ell_t \qquad \text{(VI,10)}$$

ASYMMETRIC MOLECULES

Asymmetric molecules belonging to one of the symmetry groups C_2, C_{2v}, D_2 or D_{2h} have at least one twofold proper-rotation axis which will be assumed to be the z axis. In this event, relation (VI,5) reduces to

$$-\Delta K + \sum_{n'} \Delta v_{n'} = 2p \qquad \text{(VI,11)}$$

where n' is associated with a normal vibration antisymmetric with respect to a rotation by π about the z axis.

Relation (VI,6) reduces to

$$-\Delta K = 2p \qquad \text{(VI,12)}$$

showing that, in matrix elements diagonal with respect to all the v_s quantum numbers, ΔK must be even.

For asymmetric molecules belonging to a symmetry group C_s or C_i, no restriction on matrix elements arises from the argument used in the present chapter. In the case of C_i however, the remark made in footnote 1 holds.

SPHERICAL TOP MOLECULES

The study of symmetry properties in spherical top molecules is beyond the scope of this Monograph (see references [28-39]).

NOTE TO CHAPTER VI

1 . Symbols $A\ B\ E_1\ E_2\ E_3$ .... need for certain groups to be equipped with superscripts ' , " or with subscripts 1, 2, g or u. Further restrictions on non-vanishing matrix elements may of course arise from the corresponding symmetry properties.

CHAPTER VII

ORDERS OF MAGNITUDE OF MATRIX ELEMENTS AND OF THEIR CONTRIBUTIONS TO THE ENERGY

In Chapter I, the expansion of the Hamiltonian was carried out on the basis that the zero-order Hamiltonian consisted of the terms

$$H_0 = \frac{1}{2}\sum_{s,\sigma} hc\,\omega_s\left[(p_{s\sigma}/\hbar)^2 + q_{s\sigma}^2\right] + \frac{1}{2}\sum_{\alpha}(P_\alpha^2/I_{\alpha\alpha}^e) \qquad \text{(VII,1)}$$

where ω_s are the normal frequencies of the molecule ; σ is an index which takes the values 1;1 and 2, or 1,2, and 3 as ω_s is respectively nondegenerate, twofold degenerate or threefold degenerate. $q_{s\sigma}$ is the normal coordinate used to describe the normal oscillation ω_s ; $p_{s\sigma}$ is the linear momentum conjugate to $q_{s\sigma}$; P_α are the components of the total angular momentum directed along the x, y and z axes fixed in the molecule ; and $I_{\alpha\alpha}^e$ are the three principal moments of inertia. Equation (VII,1) is tantamount to stating that

$$\left[J(J+1) - K^2\right]B_e^{xx}$$

and, indeed, $K^2 B_e^{zz}$ where $B_e^{\alpha\alpha} = h/8\pi^2 I_{\alpha\alpha}^e c$ are of the same order of magnitude as $\sum_s \omega_s(v_s + d_s/2)$,where d_s designates the degree of degeneracy of vibration s . $B_e^{\alpha\alpha}$ is of the order of magnitude of from 0.1 to 10 cm^{-1} in the case of the molecules which are commonly studied, while ω_s is of the order of magnitude of from 300 to 3000 cm^{-1}, so that the above assumption is, indeed, true only for relatively large values of J and K , say $J \approx 10$ to 30. While this constitutes an important range of frequencies of particular interest in the infrared, there is another range of frequencies of especial interest to the microwave spectroscopists where $J \approx K$ is of the order of from 0 to 10. The quantities

$$\left[J(J+1)-K^2\right]B_e^{xx}$$

and $K^2 B_e^{zz}$ are, in general, not comparable to

$$\sum_s \omega_s \left(v_s + \frac{d_s}{2}\right)$$

for the latter values of J and K and they might more justifiably be regarded as a part of H_1 or H_2. Other investigators have, for this reason, expanded the Hamiltonian in other ways [40] and the results obtained by the several methods vary slightly in appearance, the difference being entirely one of identifying certain terms obtained by one method as belonging to a given order of magnitude which, by another method, are regarded as of another order of magnitude.

The problem which exists may be seen by considering the term $-p_z^* P_z / I_{zz}^e$ which stems from the Coriolis interaction between rotation and vibration due to the twofold degenerate perpendicular vibrations in a symmetric top molecule. The term p_z^* has the form

$$p_z^* = \sum_t \zeta_t^z p_{xt}, \qquad \text{(VII,2)}$$

where $\zeta_t^z p_{xt}$ is the angular momentum associated with the degenerate vibration ω_t, ζ_t^z being a constant which depends on the nature of the normal frequency. The term $-p_z^* P_z / I_{zz}^e$ contributes a term to the energy equal to $-2\sum_t \zeta_t^z l_t B_e^{zz} K hc$, where $|l_t|$ is the quantum number of angular momentum of vibration which takes the values $v_t, v_t - 2, \ldots$ 0 or 1. The zero-order rotational energy will be equal to

$$\left\{J(J+1)B_e^{xx} + K^2\left[B_e^{zz} - B_e^{xx}\right]\right\} hc$$

and for small values of J and K (i.e., $J \approx K \approx 1$) it is of the same order of magnitude as

$$-2hc\sum_t \zeta_t^z l_t B_e^{zz} K$$

For large values of J (i.e., $J \approx 30$) the zero-order energy is clearly of a different order of magnitude than

$$-2hc \sum_t \zeta_t^{z} l_t B_e^{zz} K.$$

The importance of assigning a contribution to the energy to the right order of magnitude may be seen from the following example. Consider a transition from the normal vibration state where all $v_s = 0$ (and therefore all $l_t = 0$ also) to the first excited vibration state where $v_t = 1$ (and therefore also $|l_t| = 1$). The term $-p_z^* P_z / I_{zz}^e$ contributes an amount equal to zero to the energy of the normal state but yields an amount $-2\zeta_t^z B_e^{zz} Khc$ to the energy of the first excited state. For large values of K the energy term $-2\zeta_t^z B_e^{zz} Khc$ is certainly small compared to the energy term

$$\left\{J(J+1) B_e^{xx} + K^2 (B_e^{zz} - B_e^{xx})\right\} hc.$$

The selection rule states, however, that $\Delta J = \pm 1, 0$; $\Delta K = \Delta l = \pm 1$. Consider the case where $\Delta J = 0$ and $\Delta K = \Delta l = \pm 1$. The frequency condition gives for the positions of a line,

$$\nu = \nu_0 \pm 2K \left\{\left[(1 - \zeta_t^z) B_e^{zz} - B_e^{xx}\right]\right\}, \tag{VII,3}$$

where ν_0 is the vibration frequency. In other words the term $-2\zeta_t^z B_e^{zz} Khc$, which energywise is small compared to

$$\left\{J(J+1) B_e^{xx} + K^2 (B_e^{zz} - B_e^{xx})\right\} hc$$

has contributed an amount to the frequency entirely comparable to the rotational energy itself.

We shall study the contributions to the energy of the various terms comprising the Hamiltonian in the following sections, for various values of J and K, and consider the orders of magnitude to which they must be assigned in each case.

We shall be concerned, in the following, with operators O which make up the various terms in $h_m^\dagger$, $h_m^\dagger$ being an operator of order m in the twice-transformed Hamiltonian $H^\dagger$ so that

$$H^\dagger = h_0^\dagger + h_1^\dagger + h_2^\dagger + \ldots + h_m^\dagger + \ldots, \qquad \text{(VII,4)}$$

with $h_0^\dagger = H_0$ and $h_1^\dagger = h_1'$. The operators O are of the form

$$O = {}^{\alpha_1 \ldots \alpha_j}_{(m)}Z^{c\ldots}_{a\ldots} \, \tfrac{1}{2}(q_a \ldots p_c \ldots + p_c \ldots q_a \ldots)P_{\alpha_1} \ldots P_{\alpha_j} \qquad \text{(VII,5)}$$

and have both diagonal matrix elements, $(\Gamma|O|\Gamma)$ and nondiagonal matrix elements $(\Gamma|O|\Gamma')$, Γ denoting the collection of quantum numbers v_s, l_s, m_s ..., J, K, M which characterize the basic wave functions ψ_0[1]. Since the operators O are diagonal with respect to J and M, the symbol Γ' denotes a collection of quantum numbers $v_s + \Delta v_s$, $l_s + \Delta l_s$, $m_s + \Delta m_s$ ..., J, K + ΔK, M.

The matrix elements of an operator O are of an order of magnitude which decreases, on the average, in direct proportion to the increase in the order of the expanded Hamiltonian in which the operator occurs. To remove any ambiguity we shall designate m as the <u>index of magnitude</u>, where m is the order of a term in the expansion (VII,4) or in the corresponding expansion of the rotation-vibration energy

$$E_{v,R} = E_0 + E_1 + E_2 + \ldots + E_m + \ldots \qquad \text{(VII,6)}$$

A large index of magnitude thus corresponds to a small order of magnitude[2].

The index of magnitude n of a matrix element, $(\Gamma|O|\Gamma)$ or $(\Gamma|O|\Gamma')$ of an operator O appearing in $h_m^\dagger$ is not necessarily[3] equal to m. In addition, the contribution of such a matrix element to the rotation-vibration energy may be of an order of magnitude E_k, that is to say, it may have an index of magnitude k different from n or m.

We now propose to study systematically

(a) the index of magnitude n of the matrix elements $(\Gamma|O|\Gamma)$ and $(\Gamma|O|\Gamma')$

(b) the index of magnitude k of their contribution to the rotation-vibration energy.

We shall consider in succession the diagonal and non-diagonal contributions for the following three particular cases :

(1) $J \approx 30$, $K \approx 30$;

(2) $J \approx 30$, $K \approx 1$;

(3) $J \approx 1$, $K \approx 1$.

If we consider a vibration-rotation band observed by infrared or Raman spectroscopic techniques, then for a molecule of axial symmetry the three cases referred to correspond respectively to the energy levels which are involved (1) in the sub-bands of large K values ; (2) in the wings of the sub-bands with small K values ; (3) in the central region of sub-bands with small K values. Moreover, the third case where J and K are small corresponds in general to the energy levels involved in the pure rotational spectra observed in the microwave region.

It is perhaps worthwhile to point out that the value 30 for J and K is a value suitable for a rather representative group of molecules. Actually, of course, values may vary some from molecule to molecule.

DIAGONAL CONTRIBUTIONS

A . Index of Magnitude n of the Element $(\Gamma|O|\Gamma)$

The validity of the series expansion of the vibration-rotation Hamiltonian

$$H = H_0 + H_1 + H_2 + \dots + H_m + \dots \qquad \text{(VII,7)}$$

rests on the comparison of the orders of magnitude of the various diagonal elements $(\Gamma|H_0|\Gamma)$, $(\Gamma|H_1|\Gamma)$, $(\Gamma|H_2|\Gamma)$, etc. Considering, therefore, an operator O defined by the relation (VII,5), the index of magnitude n of $(\Gamma|O|\Gamma)$ will then be equal to m to the same extent as the expansion (VII,7) is valid[4].

The form of the expansion (VII,7) used by us in Chapter I leads to an expression for H indicated schematically in Table XXXI. The operators which appear in $h_0^\dagger$, $h_1^\dagger$, $h_2^\dagger$, etc. are given in the first, second, third, etc. rows, respectively, of Table XXXI. Now it is legitimate to write $\sum_\alpha P_\alpha^2/2I_{\alpha\alpha}^e$ and $\sum_{s\sigma} hc\omega_s[(p_{s\sigma}/\hbar)^2 + q_{s\sigma}^2]$ in the same Hamiltonian H_0 only if the diagonal elements of these two operators are of the same order of magnitude. This condition is fulfilled when $(J/v^{\frac{1}{2}}) \approx (\omega/B_e)^{\frac{1}{2}}$. Explicitly this states that, if v_s is taken small, J must be of the order of 30. If J is of the order of 1 the relations stated in Note 2 to the present Chapter demonstrate that $\sum_\alpha P_\alpha^2 / I_{\alpha\alpha}^e$ should, in fact, be included in H_2.

Consider the expansion (VII,4) of the transformed Hamiltonian $H^\dagger$ and particularly the operator O appearing in $h_m^\dagger$, which is defined by the relation (VII,5). The operator O has been included with $h_m^\dagger$, taking account of the fact that the matrix elements of $(P_\alpha/\hbar)$ have a value which is of the order[5] of 30 (i.e., $J \approx 30$). If these elements are of the order of 1 (i.e., $J \approx 1$) the operator O, which contains j operators P_α, must, in fact, be classified with $h_{m+j}^\dagger$. One obtains then for $H^\dagger$ the expansion schematically indicated in Table XXXI (reading this time vertically) the operators appearing in the first, second, third column, etc., corresponding respectively to $h_0^\dagger$, $h_2^\dagger$, $h_4^\dagger$, ...

At this point, two possible courses of action may be pursued.

(a) One may utilize different expansions for H according to whether $J \approx K \approx 30$, $J \approx K \approx 1$ or $J \approx 30$, $K \approx 1$, in which case the index of magnitude of $(\Gamma|O|\Gamma)$ shall always be equal to the order m of the operator $h_m^\dagger$ to which O belongs ;

(b) one may utilize, whatever the values of the quantum numbers may be, the same expansion for H , accepting then as a matter of fact that n is not necessarily equal to m. We have adopted the second

TABLE XXXI . Orders of magnitude of operators appearing in expansion of Hamiltonian $H^{\dagger}$

	0	2	4	6	8	10	12
0	${}_{(0)}Zr^2$	${}_{(0)}ZP^2$					
1		${}_{(1)}Zr^2P$					
2		${}_{(2)}Zr^4$	${}_{(2)}Zr^2P^2$	${}_{(2)}ZP^4$			
3			${}_{(3)}Zr^4P$	${}_{(3)}Zr^2P^3$			
4			${}_{(4)}Zr^6 + {}_{(4)}Zr^2$	${}_{(4)}Zr^4P^2 + {}_{(4)}ZP^2$	${}_{(4)}Zr^2P^4$	${}_{(4)}ZP^6$	
5				${}_{(5)}Zr^6P + {}_{(5)}Zr^2P$	${}_{(5)}Zr^4P^3$	${}_{(5)}Zr^2P^5$	
6				${}_{(6)}Zr^8 + {}_{(6)}Zr^4$	${}_{(6)}Zr^6P^2 + {}_{(6)}Zr^2P^2$	${}_{(6)}Zr^4P^4 + {}_{(6)}ZP^4$	...
7					${}_{(7)}Zr^8P + {}_{(7)}Zr^4P$	...	
8					...		

In this table, we use the following simplified notation : P^{j} instead of $P_{\alpha_1} P_{\alpha_2} \dots P_{\alpha_j}$; r^2 for the product of two vibrational operators $q_a q_b$, $p_a p_b$, $q_a p_b$, $p_a q_b$; r^4 for the product of four vibrational operators etc.; ${}_{(m)}Z$ instead of ${}^{\alpha_1 \alpha_2 \dots}_{(m)}Z^{cd\dots}_{ab\dots}$. In addition, the sums $\sum$ over rotational ($\alpha_1 \alpha_2 \dots$) and vibrational ($abc\dots$) indices are omitted.

alternative in choosing for H the expansion[6] used in Chapter I .

Table XXXII gives the index of magnitude of the elements $(\Gamma|O|\Gamma)$. We may note that when $m=1,3,5,\ldots$, the operators O having elements diagonal in v_s contain only odd powers of P_α . In order to have matrix elements diagonal in K they must contain an operator P_z (or an odd number of operators P_z). The element $(\Gamma|O|\Gamma)$, thus, necessarily contains K as a factor. Its index of magnitude n cannot be less than $m+1$ when $J \approx 30$, $K \approx 1$.

TABLE XXXII . Diagonal contributions .

$J \approx K \approx 30$	$n = m$		$k = n$
$J \approx 30$, $K \approx 1$	$n = m$	if m is even	
	$n = m+1$	if m is odd	
$J \approx K \approx 1$	$n = m + j$		

B . Index of Magnitude k of the Contribution of the Element $(\Gamma|O|\Gamma)$ to the Energy.

For the diagonal contribution of $(\Gamma|O|\Gamma)$ to the energy we have evidently: $k = n$.

NONDIAGONAL CONTRIBUTIONS

A . Index of Magnitude n of the Element $(\Gamma|O|\Gamma')$

A matrix element $(\Gamma|O|\Gamma')$ nondiagonal in K contains the following factor :

$$(J,K|P_{\alpha_1}\cdots P_{\alpha_j}|J,K\pm|\Delta K|)=\hbar^j f(J,K)[J(J+1)-K(K\pm1)]^{\frac{1}{2}}$$
$$\times[J(J+1)-(K\pm1)(K\pm2)]^{\frac{1}{2}}\cdots[J(J+1)-(K\pm|\Delta K|\mp1)(K\pm|\Delta K|)]^{\frac{1}{2}} \qquad \text{(VII,8)}$$

where the number of radicals is equal to $|\Delta K|$ and where $f(J,K)$

designates a nonhomogeneous polynomial in J and K of degree $j - |\Delta K|$. Under these circumstances the index of magnitude n of an element $(\Gamma | O | \Gamma')$ is obtained without difficulty. The results are summarized in Table XXXIII and are obtained, in particular, from the following observations :

(1) When $J \approx K \approx 30$ the $|\Delta K|$ radicals in Equation (VII,8) are proportional to $J^{\frac{1}{2}}$ or $K^{\frac{1}{2}}$ and not to J or K .

(2) The terms of highest power in $f(J, K)$ are always of even degree in J . When $j - |\Delta K|$ is odd, these terms are thus

$$\left(J^{j-|\Delta K|-1}\right) K, \left(J^{j-|\Delta K|-3}\right) K^3, \ldots$$

Thus for $J \approx 30$, $K \approx 1$ the polynomial is equivalent to $J^{j-|\Delta K|-1}$ and the factor (VII,8) is proportional to J^{j-1} . The index of magnitude n of $(\Gamma | O | \Gamma')$ is, therefore, equal to $m + 1$.

It may be noted that the index of magnitude of the diagonal element $(\Gamma | O | \Gamma)$ could have been obtained from Table XXXIII by setting $\Delta K = 0$. One finds, in fact, that the results obtained for n are identical to those given in Table XXXII (j and m being necessarily of the same parity in order for the operator O to have elements diagonal in v_s).

TABLE XXXIII . Nondiagonal contributions .

$J \approx K \approx 30$	$n = m + \frac{\lvert\Delta K\rvert}{2}$	
$J \approx 30$, $K \approx 1$	$n = m$	if $j - \lvert\Delta K\rvert$ is even
	$n = m + 1$	if $j - \lvert\Delta K\rvert$ is odd
$J \approx K \approx 1$	$n = m + j$	

B. Index of Magnitude k of the Contribution of the Element $(\Gamma|O|\Gamma')$ to the Energy

1. Linear and Axially Symmetric and Spherically Symmetric Molecules

To determine to which order a nondiagonal element $(\Gamma|O|\Gamma')$ contributes in the energy one must, in general, compare the index of magnitude n of this element with the index of magnitude p of the difference $(\Gamma|H^{\dagger}|\Gamma) - (\Gamma'|H^{\dagger}|\Gamma')$.

If $n \leqslant p$ one has, for all practical purposes, returned to a degenerate problem where it is necessary to solve a secular equation. In this instance the element $(\Gamma|O|\Gamma')$ contributes to the energy in order n; i.e., the contribution to the energy has an index of magnitude $k = n$.

If $n > p$ the principal contributions of $(\Gamma|O|\Gamma')$ to the energy may be obtained by a perturbation calculation. They are of the form

$$\frac{(\Gamma|O|\Gamma')(\Gamma'|O|\Gamma)}{(\Gamma|H^{\dagger}|\Gamma) - (\Gamma'|H^{\dagger}|\Gamma')}$$

and have an index of magnitude $k = 2n - p$. The preceding results are valid, however, only if there are no elements of the type $(\Gamma|\Gamma')$ having an index of magnitude n' less than the index of magnitude n of the element $(\Gamma|O|\Gamma')$. In order to find the value of k in all cases let us suppose that there exist operators O' which have an element $(\Gamma|O'|\Gamma')$ with an index of magnitude less than the index of magnitude n of $(\Gamma|O|\Gamma')$. We shall designate by n' the minimum index of magnitude of the various elements $(\Gamma|O'|\Gamma')$. Under these conditions, if $n' > p$, the element $(\Gamma|O|\Gamma')$ makes a contribution to the energy of the form

$$\frac{(\Gamma|O|\Gamma')(\Gamma'|O'|\Gamma)}{(\Gamma|H^{\dagger}|\Gamma) - (\Gamma'|H^{\dagger}|\Gamma')}$$

with[7] an index of magnitude $k = n + n' - p$. When $n' \leqslant p$ it is not possible to use the usual non-degenerate perturbation method, but it is necessary to resort to the degenerate form and solve a secular equation. One has in this case that $k = n$. Table XXXIV summarizes the results we have just stated.

TABLE XXXIV . Nondiagonal contributions (for linear, axially symmetric and spherically symmetric molecules).

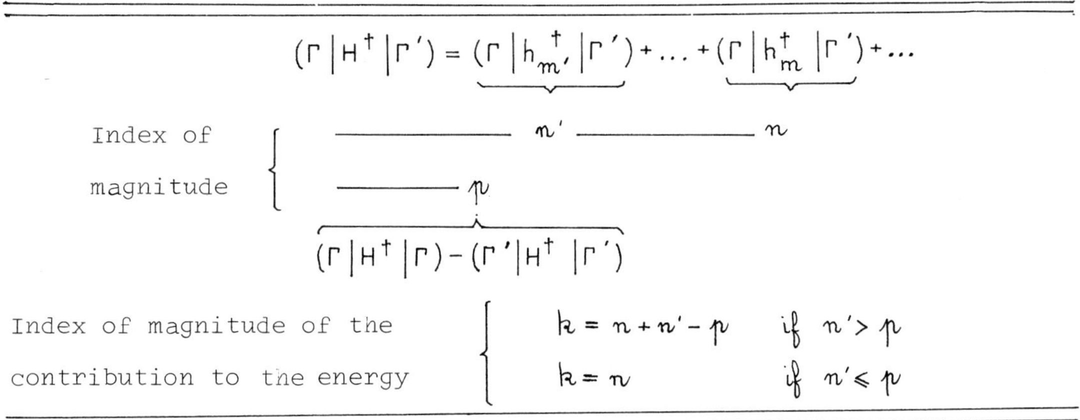

$$(\Gamma|H^{\dagger}|\Gamma') = (\Gamma|h^{\dagger}_{m'}|\Gamma') + \ldots + (\Gamma|h^{\dagger}_{m}|\Gamma') + \ldots$$

Index of magnitude: n' ; n ; p for $(\Gamma|H^{\dagger}|\Gamma) - (\Gamma'|H^{\dagger}|\Gamma')$

Index of magnitude of the contribution to the energy:

$k = n + n' - p$ if $n' > p$

$k = n$ if $n' \leqslant p$

2 . Asymmetric Molecules

Asymmetric molecules represent a special case in which the basic functions utilized to calculate the matrix elements are not eigenfunctions of H_0 . The discussion of the preceding paragraph remains valid, however, if one uses the linear combinations of the basic wave functions which diagonalize H_0 to calculate the matrix elements.

C . Resonances

A result of the preceding discussion was that the element $(\Gamma|O|\Gamma')$ contributes to the energy in order $n+n'$ ($k = n + n'$) when the different levels are well separated from one another[8], so that $p = 0$. When the two levels characterized by Γ and Γ' have nearly identical values, the contribution to the energy of $(\Gamma|O|\Gamma')$ has an index of magnitude k lying between $n + n'$ and n. This condition is referred to as resonance between the levels characterized by Γ and Γ'. The resonance is "weak" if $0 < p < n'$, i.e., $k = n + n' - p$. In this case it is still possible to get the contribution of $(\Gamma | O | \Gamma')$ to the energy by means of a perturbation calculation. On the other hand, if $p \geqslant n'$, $k = n$, the resonance is "strong". The problem is now quasi-degenerate and the contribution of $(\Gamma | O | \Gamma')$ to the energy is obtained by solving a secular equation.

One can distinguish two types of resonance : accidental and essential. A resonance is called accidental if the two resonating levels Γ and Γ' corresponding to different values of the vibrational quantum numbers v_s accidentally have the same energies by virtue of the particular values of the force constants and masses of the nuclei. Examples of these would be Fermi resonance and Coriolis resonance. Essential resonance comes about in the following manner. Two levels corresponding to the same values of the quantum numbers v_s and J may, because of the essential degeneracies and quasi-degeneracies which characterize the zero-order vibration-rotation problem, nevertheless have almost identical energies for different values of the quantum numbers l_s, m_s and K, ΔK being small. Resonances between such sublevels are called essential resonances. Examples of these would be l-type resonance and K-type resonance.

D . Essential Resonance in Linear and Axially Symmetric Molecules

The levels which participate in an essential resonance in a molecule of axial symmetry are described by the quantum numbers $\Gamma \equiv v_s$, l_s ..., J, K, M and $\Gamma' \equiv v_s, l_s + \Delta_s$ ..., J, K + ΔK, M. The calculation of the index of magnitude p of $(\Gamma | H^{\dagger} | \Gamma) - (\Gamma' | H^{\dagger} | \Gamma')$ is given in the Appendix. The results are as follows :

(1) $$(\Gamma | H^{\dagger} | \Gamma) - (\Gamma' | H^{\dagger} | \Gamma') = 0$$

if $\Gamma' \equiv v_s, -l_s$ .., J, $-$K, M. The index of magnitude p is infinite. This case constitutes the extreme case of strong resonance[9] and is referred to as <u>doubling</u>.

(2) One has, in general, in the absence of doubling, $p = 1$ if $K \approx 30$ and $p = 2$ if $K \approx 1$.

(3) It may happen, as a result of particular values of the molecular constants, that the index of magnitude p will have, for certain pairs of levels, a value greater than 1 or 2 as suggested under (2) above. It can even happen that p may accidentally be greater than n' in which case it is said that the essential resonance is <u>accidentally strong</u>.

For a molecule of axial symmetry the essential resonances can be classed in the following manner :

- K is small ($K \approx 1$)
  - strong resonance ($k = n$)
    - doubling where $p = \infty$
    - accidentally strong resonance ($p > n'$)
    - resonance for which $n' = 2$
  - weak resonance ($k = n + n' - p > n$ with $p = 2$ generally) : all other cases
- K is large ($K \approx 30$)
  - strong resonance ($k = n$) : accidentally strong resonance ($p > n'$)
  - weak resonance ($k = n + n' - p > n$ with $p = 1$ generally) : all other cases

The following two remarks are pertinent to the foregoing.

(a) The doubling of the level $K=30$, which can occur from the elements $(K|K \pm 60)$, has not been considered.

(b) The preceding considerations are applicable also to linear molecules where, however, one has only the small values of K in accordance with the Sayvetz condition [27]:

$$K = \sum_{s} \ell_{s} \tag{VII,9}$$

APPENDIX TO CHAPTER VII

Herein will be found the calculation of

$$\begin{aligned}\Delta\varepsilon = {} & \left(K+\Delta K, \ell_t + \Delta\ell_t, \ell_{t'} + \Delta\ell_{t'} \dots \left| \frac{H^\dagger}{hc} \right| K+\Delta K, \ell_t + \Delta\ell_t, \ell_{t'} + \Delta\ell_{t'} \dots\right) \\ & - \left(K \ell_t \ell_{t'} \dots \left| \frac{H^\dagger}{hc} \right| K \ell_t \ell_{t'} \dots\right)\end{aligned} \tag{VII,A1}$$

It is assumed in this calculation that the molecule has axial symmetry, that $\Delta J = 0$ and $\Delta v_s = 0$, that ΔK, $\Delta\ell_t$ and $\Delta\ell_{t'}$ are not all equal to zero, but that ΔK, $\Delta\ell_t$, $\Delta\ell_{t'} \dots$ are zero or of the order of 1. By limiting the expansion of $H^\dagger$ to $h_0^\dagger + h_1^\dagger + h_2^\dagger$ we obtain[10]

$$\begin{aligned}\varepsilon = {} & \left(K \ell_t \ell_{t'} \dots \left| H^\dagger/hc \right| K \ell_t \ell_{t'} \dots\right) \\ = {} & \sum_s \omega_s \left(v_s + \frac{d_s}{2}\right) + B_e^{xx} J(J+1) + (B_e^{zz} - B_e^{xx}) K^2 - 2B_e^{zz} \sum_t \zeta_t^z \ell_t K \\ & + \sum_{s} \sum_{\substack{s' \\ s \leq s'}} x_{ss'} \left(v_s + \frac{d_s}{2}\right)\left(v_{s'} + \frac{d_{s'}}{2}\right) + \sum_{t} \sum_{\substack{t' \\ t \leq t'}} x_{\ell_t \ell_{t'}} \ell_t \ell_{t'} \\ & - \sum_s \alpha_s^x \left(v_s + \frac{d_s}{2}\right) J(J+1) - \sum_s (\alpha_s^z - \alpha_s^x)\left(v_s + \frac{d_s}{2}\right) K^2 \\ & - D_e^J J^2 (J+1)^2 - D_e^{JK} J(J+1) K^2 - D_e^K K^4,\end{aligned} \tag{VII,A2}$$

from which one derives

$$\Delta\varepsilon = \Delta\varepsilon^{(0)} + \Delta\varepsilon^{(1)} + \Delta\varepsilon_a^{(2)} + \Delta\varepsilon_b^{(2)} + \Delta\varepsilon_c^{(2)} \tag{VII,A3}$$

where

$$\Delta \varepsilon^{(0)} = (B_e^{zz} - B_e^{xx}) \left[2K\Delta K + (\Delta K)^2 \right] \tag{VII,A4}$$

$$\Delta \varepsilon^{(1)} = -2 \sum_t B_e^{zz} \zeta_t^z (K\Delta l_t + l_t \Delta K + \Delta K \Delta l_t) \tag{VII,A5}$$

$$\Delta \varepsilon_a^{(2)} = \sum_t x_{l_t l_t} \left[2 l_t \Delta l_t + (\Delta l_t)^2 \right]$$

$$+ \sum_t \sum_{\substack{t' \\ t<t'}} x_{l_t l_{t'}} (l_t \Delta l_{t'} + l_{t'} \Delta l_t + \Delta l_t \Delta l_{t'}) \tag{VII,A6}$$

$$\Delta \varepsilon_b^{(2)} = -\sum_s (\alpha_s^z - \alpha_s^x)(v_s + \frac{d_s}{2}) \left[2K\Delta K + (\Delta K)^2 \right] \tag{VII,A7}$$

$$\Delta \varepsilon_c^{(2)} = -D_e^{JK} J(J+1) \left[2K\Delta K + (\Delta K)^2 \right]$$

$$- D_e^K \left[4K^3 \Delta K + 6K^2 (\Delta K)^2 + 4K(\Delta K)^3 + (\Delta K)^4 \right] \tag{VII,A8}$$

One obtains thus for the index of magnitude p of these various terms the following values :

Index of magnitude of $\Delta\varepsilon^{(0)}$:

$\Delta K \neq 0$, $K \approx 30$; $p = 1$

$\Delta K \neq 0$, $K \approx 1$; $p = 2$

$\Delta K = 0$, or $\Delta K = -2K$; $p = \infty$ (that is to say, $\Delta\varepsilon^{(0)} = 0$).

Index of magnitude of $\Delta\varepsilon^{(1)}$:

$\Delta l_t \neq 0$, $K \approx 30$; $p = 1$

$\Delta l_t \neq 0$, $K \approx 1$;
or
$\Delta l_t = 0$ $\Delta K \neq 0$; } $p = 2$

$K\Delta l_t + l_t \Delta K + \Delta K \Delta l_t = 0$; $p = \infty$

Index of magnitude of $\Delta\varepsilon_a^{(2)}$:

At least one $\Delta l_t \neq 0$; $\quad p = 2$

$$\left.\begin{array}{l}\Delta l_t = \Delta l_{t'} = \ldots = 0 \\ \text{or} \\ \Delta l_t = -2l_t,\ \Delta l_{t'} = -2l_{t'}, \ldots\end{array}\right\} \quad p = \infty$$

Index of magnitude of $\Delta\varepsilon_b^{(2)}$:

$\Delta K \neq 0 \qquad K \approx 30; \qquad p = 3$

$\Delta K \neq 0 \qquad K \approx 1; \qquad p = 4$

$\Delta K = 0$ or $\Delta K = -2K; \qquad p = \infty$

Index of magnitude of $\Delta\varepsilon_c^{(2)}$

$\Delta K \neq 0, \qquad K \approx 30; \qquad p = 3$

$\Delta K \neq 0, \qquad K \approx 1, \qquad J \approx 30; \qquad p = 4$

$\Delta K \neq 0, \qquad K \approx 1, \qquad J \approx 1; \qquad p = 6$

$\Delta K = 0$ or $\Delta K = -2K; \qquad p = \infty$

Combining the foregoing one may draw the following conclusions concerning the index of magnitude p of $\Delta\varepsilon$ in its entirety :

(a) In the general case since at least ΔK or one of the Δl_t are different from zero the following result is obtained. When $K \approx 30$, then $p = 1$, and $\Delta\varepsilon$ becomes

$$\Delta\varepsilon \approx \left(B_e^{zz} - B_e^{xx}\right)\left(2K + \Delta K\right)\Delta K - 2\sum_t B_e^{zz}\zeta_t^z\left(K\Delta l_t + l_t\Delta K + \Delta l_t\Delta K\right) \qquad \text{(VII,A9)}$$

When $K \approx 1$, then $p = 2$, and $\Delta\varepsilon$ will become

$$\Delta\varepsilon \approx \left(B_e^{zz} - B_e^{xx}\right)\left(2K + \Delta K\right)\Delta K$$
$$- 2\sum_t B_e^{zz} \zeta_t^z \left(K\Delta l_t + l_t \Delta K + \Delta K \Delta l_t\right)$$
$$+ \sum_t x_{l_t l_t} \left(\Delta l_t + 2 l_t\right)\Delta l_t$$
$$+ \sum_{\substack{t \\ t<t'}} \sum_{t'} x_{l_t l_{t'}} \left(l_t \Delta l_{t'} + l_{t'} \Delta l_t + \Delta l_t \Delta l_{t'}\right) \qquad \text{(VII,A10)}$$

(b) If

$$\Delta K = -2K \ , \quad \Delta l_t = -2 l_t \ , \quad \Delta l_{t'} = -2 l_{t'} \quad \ldots, \qquad \text{(VII, A11)}$$

which can also be written as

$$K + \Delta K = -K \ , \quad l_t + \Delta l_t = -l_t \ , \quad l_{t'} + \Delta l_{t'} = -l_{t'}, \ldots, \qquad \text{(VII, A12)}$$

one sees that[11]

$$\Delta\varepsilon^{(0)} = \Delta\varepsilon^{(1)} = \Delta\varepsilon_a^{(2)} = \Delta\varepsilon_b^{(2)} = \Delta\varepsilon_c^{(2)} = 0 \qquad \text{(VII,A13)}$$

ε being a function of K^2, l_t^2 , $K l_t$ and $l_t\, l_{t'}$, the $\Delta\varepsilon^{(m)}$ associated with the Hamiltonian $h_m^{\dagger}$ of higher order than the second are identically zero. One thus obtains the following result for the index of magnitude p of $\Delta\varepsilon$: if $K + \Delta K = -K$, $l_t + \Delta l_t = -l_t$, $l_{t'} + \Delta l_{t'} = -l_{t'}$ $\ldots, p = \infty$. This result shows that the diagonal elements of $H^{\dagger}$ are the same for the values of the quantum numbers K, l_t , $l_{t'}$ ... as for the negative values, $-K$, $-l_t$, $-l_{t'}$ ... . This is not surprising if one recalls that changing all the signs of K , l_t , $l_{t'}$ ... is equivalent to changing the sense of the z axis attached to the molecule and the sense of the triangle of reference used in the space configuration of the degenerate normal vibrations.

The situation which we have just described corresponds to the case of doubling.

(c) The form of the relations (VII,A 9) and (VII,A 10):

$$\Delta\varepsilon = CK + D \qquad \text{(VII,A14)}$$

shows that for given values of the constants $B_e^{xx}, B_e^{zz}, \zeta_t^z, x_{l_t l_t}, x_{l_t l_{t'}}$ and for given values of $l_t, l_{t'}, \ldots, \Delta K, \Delta l_t, \Delta l_{t'}, \ldots,$ there exist, as a general rule, a particular value of the quantum number K for which the difference $\Delta\varepsilon$ has an index of magnitude greater than the value 1 or 2 predicted in the general case. This particular value of K will evidently be obtained by solving the equation $\Delta\varepsilon = 0$, $\Delta\varepsilon$ being defined by (VII,A 9) or (VII,A 10) according as K is respectively of the order 30 or 1. This situation gives rise to accidentally strong essential resonances. One sees, in principle, that such resonances are always possible, but that it is only accidental that they will arise from values of K such that the attendant effects will, in practice, be observed in the spectrum of a molecule. On the other hand if both C and D happen to be very small in Equation (VII, A 14), accidentally strong essential resonance will occur for all values of K. This situation has been recently observed in the microwave spectrum of fluoroform [41,42].

NOTES TO CHAPTER VII

1. These functions are, except in the case of asymmetric molecules, the eigenfunctions of the operator H_0.

2. Let us set $E_m \approx E_0\gamma^m$ or, which is the equivalent, $E_{m+1}/E_m \approx \gamma$. Recalling now that $\gamma \approx (B/\omega)^{\frac{1}{2}}$ and that $B \approx 1\ \text{cm}^{-1}$ whereas $\omega \approx 1000\ \text{cm}^{-1}$ we obtain $\gamma \approx 1/30$.

3. As we shall see from what follows, n designates the true index of magnitude of a matrix element while m designates a conventional index of magnitude linked to a particular choice of expansion

of the Hamiltonian.

4. The same criteria are evidently utilized in writing the expansion (VII,7) of H and that of the transformed Hamiltonians H' and $H^{\dagger}$.

5. When $J \approx 30$ the matrix elements $(K|P_x/\hbar\;|K \pm 1)$ and $(K|\;P_y/\hbar\;|K \pm 1)$ are of the order of 30 if $K \approx 1$ and the matrix element $(K|P_z/\hbar|K)$ is of the order of 30 if $K \approx 30$.

6. As indicated, we have used for the Hamiltonian an expansion where J is taken to be approximately 30. It is important to note that the same convention has been adopted in a coherent way for the two contact transformations [see for example Chapter II, Equation (II,10)].

7. However it can be shown that $k = n$ (even if $n' > p$) when the two levels Γ and Γ' are involved in a strong resonance (see paragraph C). This situation is discussed in reference [8].

8. $(\Gamma|H^{\dagger}|\Gamma)-(\Gamma'|H^{\dagger}|\Gamma')$ is then equivalent to $(\Gamma|H_0|\Gamma)-(\Gamma'|H_0|\Gamma')$ whose index of magnitude is zero.

9. The secular equation is written

$$\begin{vmatrix} A-x & B \\ B^* & A-x \end{vmatrix} = 0$$

with $A=(\Gamma|H^{\dagger}|\Gamma) = (\Gamma'|H^{\dagger}|\Gamma')$, $B=(\Gamma|H^{\dagger}|\Gamma')$ whence $x = A \pm |B|$. The degenerate levels Γ and Γ' are thus doubled by the resonance.

10. See Table XXXV, page 112.

11. It is possible to verify that the relations (VII,A 11) constitute a necessary condition for Equation (VII,A 13) to be satisfied. For exemple, when $\Delta K = 0$, one has certainly $\Delta\varepsilon^{(0)} = 0$; but $\Delta\varepsilon^{(1)}$ will only be equal to zero if $K = 0$. The condition $\Delta K = 0$, $K = 0$ is indeed a particular case of (VII,A 11).

CHAPTER VIII

MATRIX ELEMENTS OF THE TRANSFORMED HAMILTONIAN IN TERMS OF QUANTUM NUMBERS (TO THIRD ORDER)

INTRODUCTION

In the present chapter we shall tabulate the matrix elements of $h_0^\dagger$, $h_1^\dagger$, $h_2^\dagger$ and $h_3^\dagger$ and give their expression in terms of quantum numbers. We shall not consider the case of accidental resonances so that only those matrix elements which are diagonal with respect to all quantum numbers v_s will be given. To compute them one may, instead of the operators $h_0^\dagger, h_1^\dagger, h_2^\dagger, h_3^\dagger$, use the operators H_0, H_1, h_2' and h_3' respectively ; it is only when matrix elements of the fourth order Hamiltonian are to be considered (Chapter IX) that the second contact transformation will need to be taken into account.

Since all matrix elements studied in this chapter will be diagonal with respect to $J, M, v_s ..$, these quantum numbers will be omitted in writing the matrix elements. Only the quantum numbers $K, \ell_t ...$, with respect to which the matrix elements are possibly non-diagonal, will be explicitly written [1].

All the results given in this chapter have already been published in other papers. They are rewritten together here for the convenience of the reader.

AXIALLY SYMMETRIC MOLECULES

The matrix elements diagonal with respect to all quantum numbers are given in Table XXXV. The matrix elements of h_2' were first obtained by Nielsen [10], those of h_3' by Maes [20]. Expressions of the

TABLE XXXV . Diagonal matrix elements for axially symmetric molecules

$$(K\ell_t\ldots|h_0^\dagger|K\ell_t\ldots) = hc\left\{\sum_s \omega_s\left(v_s+\frac{d_s}{2}\right)+B_e^{xx}J(J+1)+(B_e^{zz}-B_e^{xx})K^2\right\}$$

$$(K\ell_t\ldots|h_1^\dagger|K\ell_t\ldots) = -2hc\,B_e^{zz}\sum_t \zeta_t^z \ell_t K$$

$$(K\ell_t\ldots|h_2^\dagger|K\ell_t\ldots) = hc\left\{\sum_{\substack{ss'\\ s\leq s'}} x_{ss'}\left(v_s+\frac{d_s}{2}\right)\left(v_{s'}+\frac{d_{s'}}{2}\right)+\sum_{\substack{tt'\\ t\leq t'}} x_{\ell_t\ell_{t'}}\ell_t\ell_{t'}\right.$$

$$-\sum_s \alpha_s^x\left(v_s+\frac{d_s}{2}\right)J(J+1)-\sum_s(\alpha_s^z-\alpha_s^x)\left(v_s+\frac{d_s}{2}\right)K^2$$

$$\left.-D_e^J J^2(J+1)^2-D_e^{JK}J(J+1)K^2-D_e^K K^4-b_0^{xx}J(J+1)-(b_0^{zz}-b_0^{xx})K^2\right\}$$

$$(K\ell_t\ldots|h_3^\dagger|K\ell_t\ldots) = hc\sum_t \ell_t K\left[\eta_t^J J(J+1)+\eta_t^K K^2+\sum_s \eta_{ts}\left(v_s+\frac{d_s}{2}\right)+\eta_t\right]$$

Taking into account their dependance upon quantum numbers and their order of magnitude, one sees that the term $-hc[b_0^{xx}J(J+1)+(b_0^{zz}-b_0^{xx})K^2]$ may conveniently be written together with matrix elements of $h_4^\dagger$ while the term $hc\sum_t \ell_t K\eta_t$ may conveniently be written with matrix elements of $h_5^\dagger$ (see Table XLIX in Chapter X)

---

coefficients x_{ss}, $x_{ss'}$, $x_{\ell_t\ell_t}$, $x_{\ell_t\ell_{t'}}$, α_s^x, α_s^z, D_e^J, D_e^{JK}, D_e^K, b_0^{xx}, b_0^{zz}, in terms of molecular constants can be found[2] in reference [43].

If we now consider matrix elements diagonal in v_s but nondiagonal in K or/and ℓ_t, it can be shown, by direct computation or by using arguments developed in Chapter VI [44 , 45 , 46] that only the matrix elements listed in Table XXXVI can possibly be non-vanishing for an axially symmetric molecule. Furthermore, while elements a and d are non-vanishing for any molecule with axial symmetry, the other matrix elements are non-vanishing only for particular values of the foldness N of the molecular axis. This is shown in Table XXXVII where the operator responsible for each matrix element is also indicated

TABLE XXXVI . Nondiagonal matrix elements for axially symmetric molecules

a. $(\ell_t \ell_{t'} \ldots | h_2^{\dagger} | \ell_t \pm 2, \ell_{t'} \mp 2 \ldots) = hc R_{\pm}^{tt'} \left[(v_t + \ell_t + 1 \pm 1)(v_t - \ell_t + 1 \mp 1)(v_{t'} + \ell_{t'} + 1 \mp 1)(v_{t'} - \ell_{t'} + 1 \pm 1) \right]^{\frac{1}{2}}$

b. $(\ell_t \ell_{t'} \ldots | h_2^{\dagger} | \ell_t \pm 2, \ell_{t'} \pm 2 \ldots) = hc S_{\pm}^{tt'} \left[(v_t + \ell_t + 1 \pm 1)(v_t - \ell_t + 1 \mp 1)(v_{t'} + \ell_{t'} + 1 \pm 1)(v_{t'} - \ell_{t'} + 1 \mp 1) \right]^{\frac{1}{2}}$

c. $(\ell_t \ldots | h_2^{\dagger} | \ell_t \pm 4 \ldots) = hc U_{\pm}^{t} \left[(v_t - \ell_t + 1 \mp 3)(v_t - \ell_t + 1 \mp 1)(v_t + \ell_t + 1 \pm 1)(v_t + \ell_t + 1 \pm 3) \right]^{\frac{1}{2}}$

d. $(\ell_t K \ldots | h_2^{\dagger} | \ell_t \pm 2, K \pm 2 \ldots) = hc F_{\pm}^{t} \left\{ [J(J+1) - K(K \pm 1)][J(J+1) - (K \pm 1)(K \pm 2)](v_t \pm \ell_t + 2)(v_t \mp \ell_t) \right\}^{\frac{1}{2}}$

e. $(\ell_t K \ldots | h_2^{\dagger} | \ell_t \pm 2, K \mp 1 \ldots) = hc E_{\pm}^{t} (2K \mp 1) \left\{ [J(J+1) - K(K \mp 1)](v_t \pm \ell_t + 2)(v_t \mp \ell_t) \right\}^{\frac{1}{2}}$

f. $(\ell_t K \ldots | h_2^{\dagger} | \ell_t \pm 2, K \mp 2 \ldots) = hc G_{\pm}^{t} \left\{ [J(J+1) - K(K \mp 1)][J(J+1) - (K \mp 1)(K \mp 2)](v_t \pm \ell_t + 2)(v_t \mp \ell_t) \right\}^{\frac{1}{2}}$

g. $(K \ldots | h_2^{\dagger} | K \pm 3 \ldots) = hc (R_7^{x} \pm i R_7^{y}) \left\{ [J(J+1) - K(K \pm 1)][J(J+1) - (K \pm 1)(K \pm 2)][J(J+1) - (K \pm 2)(K \pm 3)] \right\}^{\frac{1}{2}} (2K \pm 3)$

h. $(K \ldots | h_2^{\dagger} | K \pm 4 \ldots) = hc (R_6 \pm i R_6') \left\{ [J(J+1) - K(K \pm 1)][J(J+1) - (K \pm 1)(K \pm 2)][J(J+1) - (K \pm 2)(K \pm 3)][J(J+1) - (K \pm 3)(K \pm 4)] \right\}^{\frac{1}{2}}$

i. $(K \ldots | h_3^{\dagger} | K \pm 3 \ldots) = hc \sum_t a_t'' \ell_t \left\{ [J(J+1) - K(K \pm 1)][J(J+1) - (K \pm 1)(K \pm 2)][J(J+1) - (K \pm 2)(K \pm 3)] \right\}^{\frac{1}{2}}$

j. $(\ell_t K \ldots | h_3^{\dagger} | \ell_t \pm 2, K \mp 1 \ldots) = \frac{hc}{2} \left\{ [{}^{x}\mathcal{R}_t - {}^{y}\mathcal{T}_t \pm i({}^{x}\mathcal{T}_t + {}^{y}\mathcal{R}_t)](\ell_t \pm 1) + \sum_{\substack{t' \\ t' \neq t}} [{}^{x}\mathcal{R}_{t't} - {}^{y}\mathcal{T}_{t't} \pm i({}^{x}\mathcal{T}_{t't} + {}^{y}\mathcal{R}_{t't})] \ell_{t'} \right\}$

$$\times \left\{ (v_t \mp \ell_t)(v_t \pm \ell_t + 2)[J(J+1) - K(K \mp 1)] \right\}^{\frac{1}{2}}$$

(in the simplified notation of Table XXXI, Chapter VII). From Table XXXVII we see that for a molecule with a threefold axis, elements a, d, e, g, i and j are non-vanishing ; for a molecule with a fourfold axis, a, b, c, d, f and h are non-vanishing. The expressions of coefficients appearing in Table XXXVI in terms of molecular constants can be found in reference [44] $(R_{\pm}^{tt'}, S_{\pm}^{tt'}, U_{\pm}^{t})$, in reference [45] $(F_{\pm}^{t}, E_{\pm}^{t}, G_{\pm}^{t})$ and in reference [46] $(R_7^x, R_7^y, R_6, R_6')$ for an axially symmetric molecule belonging to any given symmetry. Matrix elements $a \dots j$ are responsible for "essential resonances", namely "rotational ℓ type" (d, e, f, j),

TABLE XXXVII . Non-vanishing matrix elements for an axially symmetric molecule of given symmetry

Non-vanishing matrix element	Operator		Foldness of molecular axis
a	$h_2^{\dagger}$	r^4	$N \geqslant 3$
b	$h_2^{\dagger}$	r^4	$N = 4, 6 \dots, 2p$
c	$h_2^{\dagger}$	r^4	$N = 4, 8 \dots, 4p$
d	$h_2^{\dagger}$	$r^2 P^2$	$N \geqslant 3$
e	$h_2^{\dagger}$	$r^2 P^2$	$N = 3, 5 \dots, 2p+1$
f	$h_2^{\dagger}$	$r^2 P^2$	$N = 4, 6 \dots, 2p$
g	$h_2^{\dagger}$	P^4	$N = 3$
h	$h_2^{\dagger}$	P^4	$N = 4$
i	$h_3^{\dagger}$	$r^2 P^3$	$N = 3$
j	$h_3^{\dagger}$	$r^4 P$	$N = 3, 5 \dots, 2p+1$

"vibrational ℓ type" (a, b, c) and " K type" (g, h, i) resonances.

The orders of magnitude of matrix elements listed in Tables XXXV and XXXVI can be easily obtained using the results of Chapter VII (see for example reference [8]).

LINEAR MOLECULES

Matrix elements of $h_0^\dagger$, $h_2^\dagger$ and $h_3^\dagger$ (matrix elements of $h_1^\dagger$ are zero) are given for a linear molecule in Table XXXVIII. The symbol ℓ is written for $\sum_t \ell_t$ and is therefore equal to K according to the Sayvetz condition (equation (VI,10)) [27]. The only nondiagonal matrix elements correspond to the elements a and d in Table XXXVI, which had been shown to be non-vanishing for any molecule with axial symmetry. The expressions of the coefficients appearing in Table XXXVIII can be found in terms of molecular constants[3] in reference [43] (x_{ss}, $x_{ss'}$, $x_{\ell_t\ell_t}$, $x_{\ell_t\ell_{t'}}$, α_s, D_e), or [20] (η_{tn}), or [43,47] ($F_\pm^t$), or [49] ($R_\pm^{tt'}$).

ASYMMETRIC MOLECULES

Matrix elements of $h_0^\dagger$ and $h_2^\dagger$ (matrix elements of $h_1^\dagger$ and $h_3^\dagger$ are zero) are given in Table XXXIX for asymmetric molecules. Expressions of coefficients B_v^{xx}, B_v^{yy}, B_v^{zz}, B_v^{xy}, B_v^{xz}, B_v^{yz}, D_J, D_{JK}, D_K, x_{ss} and $x_{ss'}$ are given in references [10] or [43] in terms of molecular constants, while expressions for R_4, R_4', R_5, R_5', R_6, R_6', R_7^x, R_7^y, R_8^x, R_8^y, R_9^x, R_9^y, R_{10} R_{10}', R_{11}^x, R_{11}^y can be found[4] in reference [46].

SPHERICAL TOP MOLECULES

The case of these molecules will not be studied here. Appropriate discussions can be found in references [28] to [39].

NOTES TO CHAPTER VIII

1 . The case of spherical top molecules will not be discussed so that we need to consider only nondegenerate and twofold degenerate normal vibrations. For this reason, no reference is made in

TABLE XXXVIII . Matrix elements for linear molecules

$$(\ell_t \dots | h_0^\dagger | \ell_t \dots) = hc \left\{ \sum_s \omega_s \left(v_s + \frac{d_s}{2}\right) + B_e \left[J(J+1) - \ell^2\right] \right\}$$

$$(\ell_t \dots | h_2^\dagger | \ell_t \dots) = hc \left\{ \sum_{\substack{ss' \\ s \leq s'}} x_{ss'} \left(v_s + \frac{d_s}{2}\right)\left(v_{s'} + \frac{d_{s'}}{2}\right) + \sum_{\substack{tt' \\ t \leq t'}} x_{\ell_t \ell_{t'}} \ell_t \ell_{t'} - \sum_s \alpha_s \left(v_s + \frac{d_s}{2}\right)\left[J(J+1) - \ell^2\right] - D_e^J \left[J(J+1) - \ell^2\right]^2 \right\}$$

$$(\ell_t \dots | h_3^\dagger | \ell_t \dots) = hc \sum_{tt'} \sum_n \eta_{tn} \left(v_n + \frac{1}{2}\right) \ell_t \ell_{t'}$$

$$(\ell_t \ell_{t'} \dots | h_2^\dagger | \ell_t \pm 2, \ell_{t'} \mp 2 \dots) = hc\, R_\pm^{tt'} \left[(v_t + \ell_t + 1 \pm 1)(v_t - \ell_t + 1 \mp 1)(v_t' + \ell_{t'} + 1 \mp 1)(v_{t'} - \ell_{t'} + 1 \pm 1)\right]^{\frac{1}{2}}$$

$$(\ell_t \dots | h_2^\dagger | \ell_t \pm 2 \dots) = hc\, F_\pm^t \left\{ \left[J(J+1) - \ell(\ell \pm 1)\right]\left[J(J+1) - (\ell \pm 1)(\ell \pm 2)\right](v_t \pm \ell_t + 2)(v_t \mp \ell_t) \right\}^{\frac{1}{2}}$$

TABLE XXXIX . Matrix elements for asymmetric molecules

$$(K\ldots|h_0^{\dagger}+h_2^{\dagger}|K\ldots)=hc\Big\{\sum_{s}\omega_s(v_s+\tfrac{1}{2})+\sum_{\substack{ss'\\ s\leqslant s'}}x_{ss'}(v_s+\tfrac{1}{2})(v_{s'}+\tfrac{1}{2})+\tfrac{1}{2}(B_v^{xx}+B_v^{yy})\big[J(J+1)-K^2\big]+B_v^{zz}K^2 -D_J J^2(J+1)^2-D_{JK}J(J+1)K^2-D_K K^4\Big\}$$

$$(K\ldots|h_0^{\dagger}+h_2^{\dagger}|K\pm 2\ldots)=hc\Big\{\big[J(J+1)-K(K\pm 1)\big]\big[J(J+1)-(K\pm 1)(K\pm 2)\big]\Big\}^{\frac{1}{2}}\Big\{\tfrac{1}{4}(B_v^{xx}-B_v^{yy}\pm 2iB_v^{xy}) +(R_4\pm iR'_4)J(J+1)+(R_5\pm iR'_5)\big[K^2+(K\pm 2)^2\big]+R_{10}\pm iR'_{10}\Big\}$$

$$(K\ldots|h_2^{\dagger}|K\pm 4\ldots)=hc\Big\{\big[J(J+1)-K(K\pm 1)\big]\big[J(J+1)-(K\pm 1)(K\pm 2)\big]\big[J(J+1)-(K\pm 2)(K\pm 3)\big] \times\big[J(J+1)-(K\pm 3)(K\pm 4)\big]\Big\}^{\frac{1}{2}}(R_6\pm iR'_6)$$

$$(K\ldots|h_2^{\dagger}|K\pm 1\ldots)=hc\big[J(J+1)-K(K\pm 1)\big]^{\frac{1}{2}}(2K\pm 1)\Big\{\tfrac{1}{2}(B_v^{xz}\mp iB_v^{yz})+(R_8^{y}\mp iR_8^{x})J(J+1) +(R_9^{y}\mp iR_9^{x})\big[K^2+(K\pm 1)^2\big]+R_{11}^{y}\mp iR_{11}^{x}\Big\}$$

$$(K\ldots|h_2^{\dagger}|K\pm 3)=hc\Big\{\big[J(J+1)-K(K\pm 1)\big]\big[J(J+1)-(K\pm 1)(K\pm 2)\big]\big[J(J+1)-(K\pm 2)(K\pm 3)\big]\Big\}^{\frac{1}{2}}(2K\pm 3)(R_7^{x}\pm iR_7^{y}).$$

the present chapter to quantum numbers m_s associated with threefold degenerate vibrations.

2. The coefficients D_e^J, D_e^{JK} and D_e^K are respectively equal to coefficients D_J, D_{JK} and D_K given in reference [43].

3. For linear molecules, D_e and α_s are written for D_e^J and α_s^x respectively. Moreover, the coefficient $F_\pm^t$ is equal to q_t^o given in references[43] and [47]; it is equal to $q/2$ if notation of Herzberg [48] is used. $R_\pm^{tt'}$ is equal to $r^{tt'}/4$ of reference [49]

4. Most of these coefficients have been firstly computed in reference [10], but the expressions given there contain some misprints

CHAPTER IX

MATRIX ELEMENTS OF THE TRANSFORMED HAMILTONIAN IN TERMS OF QUANTUM NUMBERS (FOURTH ORDER)

INTRODUCTION

The present chapter concerns itself with arriving at values for the non-vanishing matrix elements of $h_4^\dagger$

In doing so we have confined ourselves to linear and axially symmetric molecules. We further restrict ourselves in the case of axially symmetric molecules to those having a three-fold or a four-fold axis of symmetry. Should matrix elements be required in the case of molecules of other symmetries, they can be obtained using the same techniques [1].

NON-VANISHING MATRIX ELEMENTS OF $h_4^\dagger$ AND THEIR ORDER OF MAGNITUDE

The expression for the twice transformed Hamiltonian $h_4^\dagger$ is given by equation (IV,2). It is convenient to designate by $h_{41}^\dagger$, $h_{42}^\dagger$, $h_{43}^\dagger$ $h_{44}^\dagger$, $h_{45}^\dagger$ and $h_{46}^\dagger$ the terms which have respectively the following dependence upon the vibrational ($\mathfrak{r}$) and rotational (P) operators : $P^6, \mathfrak{r}^2 P^4, \mathfrak{r}^4 P^2, P^2, \mathfrak{r}^6$, and $\mathfrak{r}^2$.

The problem of calculating the matrix elements of $h_4^\dagger$ for various symmetry classes of molecules is a formidable one because of the extremely large number of high order products of rotational and vibrational operators. While calculating the many possible matrix elements of $h_2^\dagger$ and $h_3^\dagger$ it was found that large numbers were identically zero for the symmetry classes of interest. This observation led to a rule given by Amat [25] which states that the matrix element

$$(J\,K\,M\; v_n\, v_t\, \ell_t \ldots | h^{\dagger}_m | J, K+\Delta K, M, v_n, v_t, \ell_t + \Delta \ell_t \ldots)$$

is identically zero unless

$$-\Delta K + \sum_t a_t \Delta \ell_t = 0\;, \pm N\;, \pm 2N\;, \pm 3N \ldots \qquad \text{(IX,1)}$$

where N is the foldness of the principal symmetry axis and a_t defines the symmetry species E_{a_t} $(1 \leq a_t < \frac{N}{2})$ of the degenerate vibration t (see Chapter VI). This rule reduces the matrix element calculation for $h^{\dagger}_3$ for axially symmetric molecules having $N = 3$ and 4, to matrix elements of the type $(K | K \pm 3)$ and $(K \ell_t \,|\, K \pm 1\,,\, \ell_t \mp 2)$ (see Chapter VIII). When the rule is applied to $h^{\dagger}_4$,the non-vanishing matrix elements that must be calculated explicitly are listed in Table XL. Included in Table XL are the indices of magnitude k, n, and n' defined in Chapter VII. The index n denotes the order of magnitude of a particular matrix element $(\Gamma | 0 | \Gamma')$ in terms of the smallness paramater $\gamma \approx \left(\frac{B}{\omega}\right)^{1/2} \approx \frac{1}{30}$ raised to the power of the index. Similarly $n'(n' \leq n)$ is the index of magnitude of the matrix element $(\Gamma | 0' | \Gamma')$ of an operator $0'$ which occurs in a <u>lower</u> order of the expanded Hamiltonian (i.e. either h'_2 or h'_3) and has the same type of matrix element as the operator 0 belonging to $h^{\dagger}_4$. If the rotation-vibration energy obtained by solving the secular equation (IV,7) is expanded as

$$E = E_0 + E_1 + E_2 + \ldots + E_k + \ldots \qquad \text{(IX,2)}$$

with

$$E_k \simeq E_0\, \gamma^k. \qquad \text{(IX,3)}$$

then k is called the "energy index of magnitude". In Table XL, k indicates to which of the above orders in the expansion of the energy, the matrix element $(\Gamma \,|\, 0 \,|\, \Gamma')$ contributes. Further details of the order of magnitude analysis adopted here may be found in Chapter VII as well as a detailed discussion of the conditions for "doubling", "strong resonance" and "weak resonance".

OPERATOR TECHNIQUES FOR THE CALCULATION OF THE MATRIX ELEMENTS $(\Gamma|h_4^{\dagger}|\Gamma')$

The method to be described here for the evaluation of matrix elements of $h_4^{\dagger}$ can be divided into three steps. First the symmetry properties of the coefficients ${}_{(4)}Z$ occurring in $h_4^{\dagger}$ are used to recast $h_4^{\dagger}$ in explicit Hermitian form. In the second step, the operators occurring in the explicitly Hermitian form of $h_4^{\dagger}$ are expanded in terms of complete sets of operators having known matrix elements. Finally in the third step the matrix elements of $h_4^{\dagger}$ are classified according to quantum number dependence and the resulting coefficients defined in terms of the molecular coefficients ${}^{\alpha\beta\ldots}_{(4)}Z^{b\ldots}_{a\ldots}$.

For the sake of brevity the first two steps will be illustrated by giving examples of the techniques employed. A complete discussion of these techniques is given elsewhere [7]. The third step is given in its entirety so that the matrix elements can be used without recourse to any previous work.

SYMMETRY PROPERTIES OF THE COEFFICIENTS ${}^{\alpha\ldots}_{(4)}Z^{b\ldots}_{a\ldots}$

While the form of $h_4^{\dagger}$ given in Chapter IV is convenient for the purposes of compactness, it is a considerable aid in the calculation of matrix elements to have $h_4^{\dagger}$ written in terms of operator combinations which are explicitly Hermitian. By making use of certain symmetry properties of the coefficients ${}^{\alpha\beta\ldots}_{(4)}Z^{b\ldots}_{a\ldots}$ two techniques have been developed which transform $h_4^{\dagger}$ to the desired Hermitian form. We shall designate these two techniques as the method of inspection and the method of nested anticommutators. While both methods have been useful, since one provides an independent check on the other, the method of inspection requires the least additional notation and is used on all parts of $h_4^{\dagger}$, except $h_{41}^{\dagger}$ where the method of nested anticommutators proves more convenient.

As an example of the inspection method consider the coefficients ${}^{\alpha\beta\gamma\delta}_{(4)}Z^{ab}$ which occur in $h_{42}^{\dagger}$. If one writes out the individual terms in the summation over $\alpha\beta\gamma\delta$ one finds :

$$\sum_{\alpha\beta\gamma\delta}\left(\sum_{ab} {}^{\alpha\beta\gamma\delta}_{(4)}Z^{ab}\, p_a p_b P_\alpha P_\beta P_\gamma P_\delta\right)$$

$$=\sum_{ab}\left({}^{xxxy}_{(4)}Z^{ab}\, p_a p_b P_x^3 P_y + {}^{yxxx}_{(4)}Z^{ab}\, p_a p_b P_y P_x^3\right) + \text{others terms} \tag{IX,4}$$

By inspection of the defining relations for ${}^{\alpha\beta\gamma\delta}_{(4)}Z^{ab}$ in Chapter IV we find that

$${}^{\alpha\alpha\alpha\beta}_{(4)}Z^{ab} = {}^{\beta\alpha\alpha\alpha}_{(4)}Z^{ab} \tag{IX,5}$$

Thus the sum on the right in (IX,4) can be written in symmetric Hermitian form

$$\sum_{ab}\frac{1}{2}\left({}^{xxxy}_{(4)}Z^{ab} + {}^{yxxx}_{(4)}Z^{ab}\right) p_a p_b \left(P_x^3 P_y + P_y P_x^3\right)$$

By inspection seven other identities such as that in equation (IX,5) can be found which reduce the 81 rotational operator products in $h_{42}^{\dagger}$ to 36 Hermitian terms.

Because $h_{41}^{\dagger}$ consists of 729 terms, a more systematic approach than the method of inspection was found necessary. It is well known that if we have a set of Hermitian operators A, B, C... which do not commute which each other, then products of these operators are, in general, not Hermitian. However, the particular combinations $[A,B]^+ \equiv AB + BA$ and $i[A,B]^- = i(AB - BA)$ are Hermitian. Using the first of these forms it is readily seen that nested anticommutator forms such as,

$$\left[[A,[B,C]^+]^+ ,\ [D,[E,F]^+]^+\right]^+$$

$$\left[[A,B]^+ ,\ [[C,D]^+ ,[E,F]^+]^+\right]^+$$

$$\left[[A,B]^+ ,\ [C\,[D,[E,F]^+]^+]^+\right]^+ \qquad \text{(IX,6)}$$

are also Hermitian. The nested anticommutator method consists of breaking $h^{\dagger}_{41}$ into three sums; one for each of the three nested forms above. In order to accomplish this, each of the coefficients ${}^{\alpha\beta\gamma\delta\varepsilon\eta}_{(4)}Z$ has to be broken into three parts. Each of these parts has the requisite symmetry to interchange of indices in order to contract the sum in $h^{\dagger}_{41}$ to one of the nested anticommutator forms in (IX,6).

<u>THE OPERATOR FORMALISM</u>

It is the purpose of this section to describe and illustrate the operator formalism which has been used to calculate the matrix elements of $h^{\dagger}_{4}$. Briefly the formalism consists of constructing sets of shift operators [51] which span the appropriate higher order manifolds in which matrix elements are desired. For instance, the set P_+, P_-, P_z spans the manifold $\{P^n\}$ for $n=1$. In like fashion, the set P^2, P_z^2, $P^+P_z, P^-P_z, (P^+)^2, (P^-)^2$ spans the manifold $\{P^2\}$. Any operator product $P_\alpha P_\beta \ldots$ can then be expanded in terms of the set of symmetrized shift operators that span the manifold to which it belongs. Since each symmetrized shift operator has only one non-vanishing matrix element which can be written down in simple closed form, the problem is solved.

There are several simplifying features in the procedure outlined above which are worthy of mention.

(a) The sets of shift operators are easily constructed even for large n including their matrix elements.

(b) General expressions for the expansion coefficients of an operator form $P_\alpha P_\beta \ldots$ in terms of the shift operators have been found, from which tables have been constructed.

(c) The technique is equally well adapted to treating the vibrational portions of the Hamiltonian as it is to treating the purely rotational portions.

A complete discussion of the mechanics of this formalism and its relation to the irreductible tensor techniques of Fano and Racah [52] would go beyond the scope of this chapter. For the interested reader some of the details are contained in reference [32].

As a rather simple illustration of the method let us consider the problem of $\{r^2\}$ manifold of the twofold oscillator. Louck [32] has given the shift operators in unsymmetrized form for $\{r\}$ as

$$\begin{aligned}
\xi &= \frac{p_+}{\hbar} + iq_+ \\
\eta &= \frac{p_-}{\hbar} + iq_+ \\
\bar{\xi} &= -\frac{p_-}{\hbar} - iq_- \\
\bar{\eta} &= -\frac{p_-}{\hbar} + iq_+
\end{aligned} \qquad \text{(IX,7)}$$

where $p_\pm = p_{t_1} \pm ip_{t_2}$ and $q_\pm = q_{t_1} \pm iq_{t_2}$.

These operators shift eigenstates as follows :

$$\begin{aligned}
\xi\psi(v_t \ell_t) &= -i\left[2(v_t+\ell_t+2)\right]^{\frac{1}{2}}\psi(v_t+1, \ell_t+1) \\
\eta\psi(v_t \ell_t) &= i\left[2(v_t+\ell_t)\right]^{\frac{1}{2}}\psi(v_t-1, \ell_t-1) \\
\bar{\xi}\psi(v_t \ell_t) &= -i\left[2(v_t-\ell_t+2)\right]^{\frac{1}{2}}\psi(v_t+1, \ell_t-1) \\
\bar{\eta}\psi(v_t \ell_t) &= i\left[2(v_t-\ell_t)\right]^{\frac{1}{2}}\psi(v_t-1, \ell_t+1) ,
\end{aligned} \qquad \text{(IX,8)}$$

and obey the commutation relations $[\xi,\eta]^- = [\bar{\xi},\bar{\eta}]^- = 4$. The set of symmetrized shift operators spanning the manifold $\{r^2\}$ is shown in Table XLI along with their matrix elements. The expansion coefficients for any operator in the manifold $\{r^2\}$ are listed in Table XLII.

THE NON-VANISHING MATRIX ELEMENTS OF $h_4^\dagger$

Having briefly illustrated the technique, we present the results in the form of tables of the non-vanishing matrix elements of $h_4^\dagger$. The quantum number dependence for diagonal and off-diagonal matrix elements is given in Tables XLIII and XLIV respectively (by off-diagonal matrix elements, we mean elements which are diagonal in v_s and off-diagonal in ℓ_t or K). The coefficients appearing in these two Tables are further defined in Tables XLV and XLVI in terms of the coefficients [2] derived in Chapter IV. Moreover, in these last two tables, appear symbols ${}^{\alpha\beta\gamma\delta}u^{\pm}$, ${}^{\alpha\beta\gamma\delta}v(n,n)$, $a_1^{(2)}$, $a_2^{(2)}$,... which are defined in Tables XLVII ; the letters a, b, c, d, e, f are written for $t_1, t_2, t_1', t_2', t_1'', t_2''$ respectively when no other explicit definition is given in the tables for these letters.

Finally, the Tables XLV and XLVI are numbered from 1 to 6, according as they contain respectively the coefficients occurring in matrix elements of $h_{41}^\dagger$, $h_{42}^\dagger$... to $h_{46}^\dagger$; let us point out that the Tables XLVI, 4 and 6, are missing because the operators $h_{44}^\dagger$ and $h_{46}^\dagger$ have no off-diagonal matrix element. In Tables XLV,1 ; XLV,2a ; XLV,2b ; XLV,3a ; XLVII,b XLVII,c ; XLVII,d and XLVII,e , each row corresponds to a coefficient. These coefficients are obtained by adding up the expressions given at the head of the various columns, each of these expressions being multiplied by the numerical factor written at the intersection of the row and the column under consideration.

Further details on matters discussed in this Chapter can be found in reference [7].

TABLE XL . Index of magnitude analysis of $(\Gamma \mid h_4^\dagger \mid \Gamma')$ for linear and axially symmetric molecules with $N = 3,4$

$h_4^\dagger$	$J \simeq K \simeq 30$				$J \simeq 30, K \simeq 1$						$J \simeq K \simeq 1$							
	n	n'	k		n	n'	k				n	n'	k					
			weak resonance (w.r.)	accidentally strong resonance (a.s.r.)			doubling (d.)	strong resonance, n'=2 (s.r.)	weak resonance (w.r.)	accidentally strong resonance (a.s.r.)			doubling (d.)	strong resonance, n'=2 (s.r.)	weak resonance (w.r.)	accidentally strong resonance (a.s.r.)	origin of n'	N
$P^6\ (K\|K \pm 3)$	5.5	3.5	8	5.5	5	3	–	–	6	5	10	6	–	–	14	10	P^4	3
$P^6\ (K\|K \pm 4)$	6	4	9	6	4	2	4	4	–	–	10	6	10	–	14	10	P^4	4
$P^6\ (K\|K \pm 6)$	7	–	13	7	4	–	4	–	6	4	10	–	10	–	18	10		3
$r^2 P^4\ (K\|K \pm 3)$	5.5	3.5	8	5.5	5	3	–	–	6	5	8	6	–	–	12	8	P^4	3
$r^2 P^4\ (K\|K \pm 4)$	6	4	9	6	4	2	4	4	–	–	8	6	8	–	12	8	P^4	4

TABLE XL . Continued

$h_4^\dagger$	J≃K≃30 n	n'	k (w.r.)	k (a.s.r.)	J≃30, K≃1 n	n'	k (d.)	k (s.r.)	k (w.r.)	k (a.s.r.)	J≃K≃1 n	n'	k (d.)	k (s.r.)	k (w.r.)	k (a.s.r.)	origin of n'	N
$r^2 P^4$ $(K\ell_t \vert K\pm 1, \ell_t \mp 2)$	4.5	2.5	6	4.5	5	3	-	-	6	5	8	4	-	-	10	8	$r^2 P^2$	3
$r^2 P^4$ $(K\ell_t \vert K\pm 2, \ell_t \pm 2)$	5	3	7	5	4	2	4	4	-	-	8	4	8	-	10	8	$r^2 P^2$	3, 4, ∞
$r^2 P^4$ $(K\ell_t \vert K\pm 2, \ell_t \mp 2)$	5	3	7	5	4	2	4	4	-	-	8	4	8	-	10	8	$r^2 P^2$	4
$r^2 P^4$ $(K\ell_t \vert K\pm 4, \ell_t \mp 2)$	6	-	11	6	4	-	4	-	6	4	8	-	8	-	14	8		3
$r^4 P^2$ $(K\ell_t \vert K\pm 1, \ell_t \mp 2)$	4.5	2.5	6	4.5	5	3	-	-	6	5	6	4	-	-	8	6	$r^2 P^2$	3
$r^4 P^2$ $(K\ell_t \vert K\pm 1, \ell_t \pm 4)$	4.5	-	8	4.5	5	-	-	-	8	5	6	-	-	-	10	6		3
$r^4 P^2$ $(K\ell_t \vert K\pm 2, \ell_t \pm 2)$	5	3	7	5	4	2	4	4	-	-	6	4	6	-	8	6	$r^2 P^2$	3, 4, ∞
$r^4 P^2$ $(K\ell_t \vert K\pm 2, \ell_t \mp 2)$	5	3	7	5	4	2	4	4	-	-	6	4	6	-	8	6	$r^2 P^2$	4
$r^4 P^2$ $(K\ell_t \vert K\pm 2\ \ell_t \mp 4)$	5	-	9	5	4	-	4	-	6	4	6	-	6	-	10	6		3
$r^4 P^2$ $(K\ell_t\ell_{t'} \vert K\pm 1, \ell_t \pm 2, \ell_{t'} \pm 2)$	4.5	-	8	4.5	5	-	-	-	8	5	6	-	-	-	10	6		3
$r^4 P^2$ $(K\ell_t\ell_{t'} \vert K\pm 2, \ell_t \mp 2, \ell_{t'} \mp 2)$	5	-	9	5	4	-	4	-	6	4	6	-	6	-	10	6		3

TABLE XL . Continued

$h_4^{\dagger}$	$J \simeq K \simeq 30$ n	n'	k (w.r.)	k (a.s.r.)	$J \simeq 30, K \simeq 1$ n	n'	k (d.)	k (s.r.)	k (w.r.)	k (a.s.r.)	$J \simeq K \simeq 1$ n	n'	k (d.)	k (s.r.)	k (w.r.)	k (a.s.r.)	origin of n'	N
$r^4 P^2$ $(\ell_t \ell_{t'} \| \ell_t \pm 2, \ell_{t'} \mp 2)$	4	2	5	4	4	2	4	4	–	–	6	2	6	6	–	–	r^4	3, 4, ∞
$r^4 P^2$ $(\ell_t \ell_{t'} \| \ell_t \pm 2, \ell_{t'} \pm 2)$	4	2	5	4	4	2	4	4	–	–	6	2	6	6	–	–	r^4	4
$r^4 P^2$ $(\ell_t \| \ell_t \pm 4)$	4	2	5	4	4	2	4	4	–	–	6	2	6	6	–	–	r^4	4
r^6 $(\ell_t \| \ell_t \pm 4)$	4	2	5	4	4	2	4	4	–	–	4	2	4	4	–	–	r^4	4
r^6 $(\ell_t \| \ell_t \pm 6)$	4	–	7	–	4	–	4	–	6	4	4	–	4	–	6	4		3
r^6 $(\ell_t \ell_{t'} \| \ell_t \pm 2, \ell_{t'} \mp 2)$	4	2	5	4	4	2	4	4	–	–	4	2	4	4	–	–	r^4	3, 4, ∞
r^6 $(\ell_t \ell_{t'} \| \ell_t \pm 2, \ell_{t'} \pm 2)$	4	2	5	4	4	2	4	4	–	–	4	2	4	4	–	–	r^4	4
r^6 $(\ell_t \ell_{t'} \| \ell_t \pm 2, \ell_{t'} \pm 4)$	4	–	7	4	4	–	4	–	6	4	4	–	4	–	6	4		3
r^6 $(\ell_t \ell_{t'} \ell_{t''} \| \ell_t \pm 2, \ell_{t'} \pm 2, \ell_{t''} \pm 2)$	4	–	7	4	4	–	4	–	6	4	4	–	4	–	6	4		3

TABLE XLI . The shift operators spanning the manifold $\{r^2\}$ of the twofold oscillator

Shift Operator		Matrix Element	
Designation	Form	Type	Quantum Number Dependence
$\psi_1^{(2)}$	$\frac{1}{4}(\xi\eta - \bar{\xi}\bar{\eta})$	$(v_t \ell_t \mid v_t \ell_t)$	ℓ_t
$\psi_2^{(2)}$	$\frac{1}{4}(\xi\eta + \bar{\xi}\bar{\eta}) + 1$	$(v_t \ell_t \mid v_t \ell_t)$	$v_t + 1$
$\psi_3^{(2)}$	$\frac{1}{4}(\bar{\xi}\eta + \xi\bar{\eta})$	$(v_t \ell_t \mid v_t, \ell_t \pm 2)$	$\frac{1}{2}\left[(v_t + \ell_t + 1 \pm 1)(v_t - \ell_t + 1 \pm 1)\right]^{\frac{1}{2}}$
$\psi_4^{(2)}$	$\frac{i}{4}(\bar{\xi}\eta - \xi\bar{\eta})$	$(v_t \ell_t \mid v_t, \ell_t \pm 2)$	$\pm \frac{i}{2}\left[(v_t + \ell_t + 1 \pm 1)(v_t - \ell_t + 1 \pm 1)\right]^{\frac{1}{2}}$
$\psi_5^{(2)}$	$\frac{1}{4}(\xi\bar{\xi} + \eta\bar{\eta})$	$(v_t \ell_t \mid v_t \pm 2, \ell_t)$	$-\frac{1}{2}\left[(v_t - \ell_t + 1 \pm 1)(v_t + \ell_t + 1 \pm 1)\right]^{\frac{1}{2}}$
$\psi_6^{(2)}$	$\frac{i}{4}(\xi\bar{\xi} - \eta\bar{\eta})$	$(v_t \ell_t \mid v_t \pm 2, \ell_t)$	$\pm \frac{i}{2}\left[(v_t - \ell_t + 1 \pm 1)(v_t + \ell_t + 1 \pm 1)\right]^{\frac{1}{2}}$
$\psi_7^{(2)}$	$\frac{1}{4}(\xi^2 + \eta^2)$	$(v_t \ell_t \mid v_t \pm 2, \ell_t \pm 2)$	$-\frac{1}{2}\left[(v_t + \ell_t + 1 \pm 1)(v_t + \ell_t + 1 \pm 3)\right]^{\frac{1}{2}}$
$\psi_8^{(2)}$	$\frac{i}{4}(\xi^2 - \eta^2)$	$(v_t \ell_t \mid v_t \pm 2, \ell_t \pm 2)$	$\pm \frac{i}{2}\left[(v_t + \ell_t + 1 \pm 1)(v_t + \ell_t + 1 \pm 3)\right]^{\frac{1}{2}}$
$\psi_9^{(2)}$	$\frac{1}{4}(\bar{\xi}^2 - \bar{\eta}^2)$	$(v_t \ell_t \mid v_t \pm 2, \ell_t \mp 2)$	$-\frac{1}{2}\left[(v_t - \ell_t + 1 \pm 1)(v_t - \ell_t + 1 \pm 3)\right]^{\frac{1}{2}}$
$\psi_{10}^{(2)}$	$\frac{i}{4}(\bar{\xi}^2 - \bar{\eta}^2)$	$(v_t \ell_t \mid v_t \pm 2, \ell_t \mp 2)$	$\pm \frac{i}{2}\left[(v_t - \ell_t + 1 \pm 1)(v_t - \ell_t + 1 \pm 3)\right]^{\frac{1}{2}}$

TABLE XLII . The expansion coefficients for the twofold oscillator manifold $\{r^2\}$

	$\psi_1^{(2)}$	$\psi_2^{(2)}$	$\psi_3^{(2)}$	$\psi_4^{(2)}$	$\psi_5^{(2)}$	$\psi_6^{(2)}$	$\psi_7^{(2)}$	$\psi_8^{(2)}$	$\psi_9^{(2)}$	$\psi_{10}^{(2)}$
$p_{t_1}^2/\hbar^2$	0	$+\frac{1}{2}$	$-\frac{1}{2}$	0	$-\frac{1}{2}$	0	$+\frac{1}{4}$	0	$+\frac{1}{4}$	0
$p_{t_2}^2/\hbar^2$	0	$+\frac{1}{2}$	$+\frac{1}{2}$	0	$-\frac{1}{2}$	0	$-\frac{1}{4}$	0	$-\frac{1}{4}$	0
$p_{t_1} p_{t_2}/\hbar^2$	0	0	0	$-\frac{1}{2}$	0	0	0	$-\frac{1}{4}$	0	$+\frac{1}{4}$
$q_{t_1}^2$	0	$+\frac{1}{2}$	$-\frac{1}{2}$	0	$+\frac{1}{2}$	0	$-\frac{1}{4}$	0	$-\frac{1}{4}$	0
$q_{t_2}^2$	0	$+\frac{1}{2}$	$+\frac{1}{2}$	0	$+\frac{1}{2}$	0	$+\frac{1}{4}$	0	$+\frac{1}{4}$	0
$q_{t_1} q_{t_2}$	0	0	0	$-\frac{1}{2}$	0	0	0	$+\frac{1}{4}$	0	$-\frac{1}{4}$
$[q_{t_1}, p_{t_1}/\hbar]^+$	0	0	0	0	0	$+1$	0	$-\frac{1}{2}$	0	$-\frac{1}{2}$
$[q_{t_2}, p_{t_2}/\hbar]^+$	0	0	0	0	0	$+1$	0	$+\frac{1}{2}$	0	$+\frac{1}{2}$
$q_{t_1} p_{t_2}/\hbar$	$+\frac{1}{2}$	0	0	0	0	0	$-\frac{1}{4}$	0	$+\frac{1}{4}$	0
$q_{t_2} p_{t_1}/\hbar$	$-\frac{1}{2}$	0	0	0	0	0	$-\frac{1}{4}$	0	$+\frac{1}{4}$	0

TABLE XLIII . The diagonal matrix elements of $h_4^\dagger$

$$(K|h_{41}^\dagger|K) = hc\Big[\mu_1^J J^3(J+1)^3 + \mu_2^J J^2(J+1)^2 + \mu_3^J J(J+1) + \mu_1^{JK} K^2 J^2(J+1)^2 + \mu_2^{JK} K^2 J(J+1) + \mu_3^{JK} K^4 J(J+1) + \mu_1^K K^2 + \mu_2^K K^4 + \mu_3^K K^6\Big]$$

$$(K|h_{42}^\dagger|K) = hc\Big[\sum_t \mu_1^{tJ}(v_t+1)J^2(J+1)^2 + \sum_t \mu_2^{tJ}(v_t+1)J(J+1) + \sum_t \mu^{tJK}(v_t+1)J(J+1)K^2 + \sum_t \mu_1^{tK}(v_t+1)K^2 + \sum_t \mu_2^{tK}(v_t+1)K^4 + \sum_n \mu_1^{nJ}(v_n+\tfrac{1}{2})J^2(J+1)^2 + \sum_n \mu_2^{nJ}(v_n+\tfrac{1}{2})J(J+1) + \sum_n \mu^{nJK}(v_n+\tfrac{1}{2})J(J+1)K^2 + \sum_n \mu_1^{nK}(v_n+\tfrac{1}{2})K^2 + \sum_n \mu_2^{nK}(v_n+\tfrac{1}{2})K^4\Big]$$

$$(K|h_{43}^\dagger|K) = hc\Big[\sum_t \mu_3^{tJ}(v_t+1)^2 J(J+1) + \sum_t \mu_4^{tJ} \ell_t^2 J(J+1) + \sum_t \mu_3^{tK}(v_t+1)^2 K^2 + \sum_t \mu_4^{tK}\ell_t^2 K^2 + \sum_n \mu_3^{nJ}(v_n+\tfrac{1}{2})^2 J(J+1) + \sum_n \mu_3^{nK}(v_n+\tfrac{1}{2})^2 K^2 + \sum_{\substack{nn' \\ n<n'}} \mu^{nn'J}(v_n+\tfrac{1}{2})(v_{n'}+\tfrac{1}{2})J(J+1) + \sum_{\substack{nn' \\ n<n'}} \mu^{nn'K}(v_n+\tfrac{1}{2})(v_{n'}+\tfrac{1}{2})K^2 + \sum_{nt}\mu^{ntJ}(v_n+\tfrac{1}{2})(v_t+1)J(J+1) + \sum_{nt}\mu^{ntK}(v_n+\tfrac{1}{2})(v_t+1)K^2 + \sum_{\substack{tt' \\ t<t'}} \mu_1^{tt'J}(v_t+1)(v_{t'}+1)J(J+1) + \sum_{\substack{tt' \\ t<t'}} \mu_2^{tt'J}\ell_t\ell_{t'}J(J+1) + \sum_{\substack{tt' \\ t<t'}} \mu_1^{tt'K}(v_t+1)(v_{t'}+1)K^2 + \sum_{\substack{tt' \\ t<t'}} \mu_2^{tt'K}\ell_t\ell_{t'}K^2 + \mu_4^J J(J+1) + \mu_4^K K^2\Big]$$

TABLE XLIII . Continued

$$(K|h^{\dagger}_{44}|K) = hc\left[\mu^{J}_{5}\, J(J+1) + \mu^{K}_{5} K^{2}\right]$$

$$\begin{aligned}(K|h^{\dagger}_{45}|K) = hc\Big[&\sum_{t}\mu^{t}_{1}(v_t+1)^3 + \sum_{t}\mu^{t}_{2}\,\ell^2_t(v_t+1) + \sum_{t}\mu^{t}_{3}(v_t+1) + \sum_{n}\mu^{n}_{1}(v_n+\tfrac{1}{2})^3 + \sum_{n}\mu^{n}_{2}(v_n+\tfrac{1}{2})\\
&+\sum_{\substack{tt'\\ t\neq t'}}\mu^{tt'}_{1}(v_t+1)^2(v_{t'}+1) + \sum_{\substack{tt'\\ t\neq t'}}\mu^{tt'}_{2}\,\ell^2_t(v_{t'}+1) + \sum_{\substack{tt'\\ t\neq t'}}\mu^{tt'}_{3}\,\ell_t\ell_{t'}(v_t+1) + \sum_{nt}\mu^{nt}_{1}(v_t+1)^2(v_n+\tfrac{1}{2})\\
&+\sum_{nt}\mu^{nt}_{2}\,\ell^2_t(v_n+\tfrac{1}{2}) + \sum_{nt}\mu^{nt}_{3}(v_t+1)(v_n+\tfrac{1}{2})^2 + \sum_{\substack{nn'\\ n\neq n'}}\mu^{nn'}(v_n+\tfrac{1}{2})^2(v_{n'}+\tfrac{1}{2})\\
&+\sum_{\substack{tt't''\\ t<t'<t''}}\mu^{tt't''}_{1}(v_t+1)(v_{t'}+1)(v_{t''}+1) + \sum_{\substack{tt't''\\ t\neq t'\neq t''\\ t<t'}}\mu^{tt't''}_{2}\,\ell_t\ell_{t'}(v_{t''}+1) + \sum_{\substack{nn'n''\\ n\leq n'<n''}}\mu^{nn'n''}(v_n+\tfrac{1}{2})(v_{n'}+\tfrac{1}{2})(v_{n''}+\tfrac{1}{2})\\
&+\sum_{\substack{tt'n\\ t<t'}}\mu^{tt'n}_{1}(v_t+1)(v_{t'}+1)(v_n+\tfrac{1}{2}) + \sum_{\substack{tt'n\\ t<t'}}\mu^{tt'n}_{2}\,\ell_t\ell_{t'}(v_n+\tfrac{1}{2}) + \sum_{\substack{nn't\\ n<n'}}\mu^{nn't}(v_n+\tfrac{1}{2})(v_{n'}+\tfrac{1}{2})(v_t+1)\Big]\end{aligned}$$

$$(K|h^{\dagger}_{46}|K) = hc\left[\sum_{t}\mu^{t}_{4}(v_t+1) + \sum_{n}\mu^{n}_{3}(v_n+\tfrac{1}{2})\right]$$

TABLE XLIV . Non-vanishing off-diagonal matrix elements in K and ℓ_t of $h_4^\dagger$ for linear and axially symmetric molecules with N = 3,4

$$(K|h_{41}^\dagger|K\pm 6) = hc \quad f_6^{\pm}\, a^{\pm}(6,JK)$$

$$(K|h_{41}^\dagger + h_{42}^\dagger|K\pm 4) = hc\left[f_4^{J\pm}J(J+1) + f_4^{K\pm}(2K\pm 4)^2 + f_4^{\pm} + \sum_n f_4^{n\pm}(v_n+\tfrac{1}{2}) + \sum_t f_4^{t\pm}(v_t+1)\right] a^{\pm}(4,JK)$$

$$(K|h_{41}^\dagger + h_{42}^\dagger|K\pm 3) = hc\left[f_3^{J\pm}J(J+1) + f_3^{K\pm}(2K\pm 3)^2 + f_3^{\pm} + \sum_n f_3^{n\pm}(v_n+\tfrac{1}{2}) + \sum_t f_3^{t\pm}(v_t+1)\right](2K\pm 3)\, a^{\pm}(3,JK)$$

$$(K\ell_t|h_{42}^\dagger + h_{43}^\dagger|K\pm 1, \ell_t\mp 2) = hc\left[f_{12}^{t,J\pm}J(J+1) + f_{12}^{t,K\pm}(2K\pm 1)^2 + f_{12}^{t\pm} + f_{12}^{t,t\pm}(v_t+1) + \sum_{\substack{t'\\(t'\neq t)}} f_{12}^{t,t'\pm}(v_{t'}+1) + \sum_n f_{12}^{t,n\pm}(v_n+\tfrac{1}{2})\right](2K\pm 1)\; a^{\pm}(1,JK)\left[(v_t+\ell_t+1\mp 1)(v_t-\ell_t+1\pm 1)\right]^{\frac{1}{2}}$$

$$(K\ell_t|h_{42}^\dagger + h_{43}^\dagger|K\pm 2, \ell_t\pm 2) = hc\left[f_{22+}^{t,J\pm}J(J+1) + f_{22+}^{t,K\pm}(2K\pm 2)^2 + f_{22+}^{t\pm} + f_{22+}^{t,t\pm}(v_t+1) + \sum_{\substack{t'\\(t'\neq t)}} f_{22+}^{t,t'\pm}(v_{t'}+1) + \sum_n f_{22+}^{t,n\pm}(v_n+\tfrac{1}{2})\right] a^{\pm}(2,JK)\left[(v_t+\ell_t+1\pm 1)(v_t-\ell_t+1\mp 1)\right]^{\frac{1}{2}}$$

$$(K\ell_t|h_{42}^\dagger + h_{43}^\dagger|K\pm 2, \ell_t\mp 2) = hc\left[f_{22-}^{t,J\pm}J(J+1) + f_{22-}^{t,K\pm}(2K\pm 2)^2 + f_{22-}^{t\pm} + f_{22-}^{t,t\pm}(v_t+1) + \sum_{\substack{t'\\(t'\neq t)}} f_{22-}^{t,t'\pm}(v_{t'}+1) + \sum_n f_{22-}^{t,n\pm}(v_n+\tfrac{1}{2})\right] a^{\pm}(2,JK)\left[(v_t+\ell_t+1\mp 1)(v_t-\ell_t+1\pm 1)\right]^{\frac{1}{2}}$$

TABLE XLIV . Continued

$$(K\ell_t|h_{42}^{\dagger}|K\pm 4,\ell_t\mp 2) = hc\ f_{42}^{t\pm}\ a^{\pm}(4,JK)\left[(v_t+\ell_t+1\mp 1)(v_t-\ell_t+1\pm 1)\right]^{\frac{1}{2}}$$

$$(K\ell_t|h_{43}^{\dagger}|K\pm 1,\ell_t\pm 4) = hc\ f_{14}^{t\pm}\ (2K\pm 1)\,a^{\pm}(1,JK)\left[(v_t-\ell_t+1\mp 3)(v_t-\ell_t+1\mp 1)(v_t+\ell_t+1\pm 1)(v_t+\ell_t+1\pm 3)\right]^{\frac{1}{2}}$$

$$(K\ell_t|h_{43}^{\dagger}|K\pm 2,\ell_t\mp 4) = hc\ f_{24}^{t\pm}\ a^{\pm}(2,JK)\left[(v_t-\ell_t+1\pm 3)(v_t-\ell_t+1\pm 1)(v_t+\ell_t+1\mp 1)(v_t+\ell_t+1\mp 3)\right]^{\frac{1}{2}}$$

$$(K\ell_t\ell_{t'}|h_{43}^{\dagger}|K\pm 1,\ell_t\pm 2,\ell_{t'}\pm 2) = hc\ f_{122}^{t,t'\pm}(2K\pm 1)\,a^{\pm}(1,JK)\left[(v_t+\ell_t+1\pm 1)(v_t-\ell_t+1\mp 1)(v_{t'}+\ell_{t'}+1\pm 1)(v_{t'}-\ell_{t'}+1\mp 1)\right]^{\frac{1}{2}}$$

$$(K\ell_t\ell_{t'}|h_{43}^{\dagger}|K\pm 2,\ell_t\mp 2,\ell_{t'}\mp 2) = hc\ f_{222}^{tt'\pm}\ a^{\pm}(2,JK)\left[(v_t+\ell_t+1\mp 1)(v_t-\ell_t+1\pm 1)(v_{t'}+\ell_{t'}+1\mp 1)(v_{t'}-\ell_{t'}+1\pm 1)\right]^{\frac{1}{2}}$$

$$(\ell_t\ell_{t'}|h_{43}^{\dagger}+h_{45}^{\dagger}|\ell_t\pm 2,\ell_{t'}\pm 2) = hc\left[g_{22+}^{tt',J\pm}J(J+1)+g_{22+}^{tt',K\pm}K^2+g_{22+}^{tt',t\pm}(v_t+1)+g_{22+}^{tt',t'\pm}(v_{t'}+1)+\sum_{\substack{t''\\(t''\neq tt')}}g_{22+}^{tt',t''\pm}(v_{t''}+1)\right.$$
$$\left.+\sum_{n}g_{22+}^{tt',n\pm}(v_n+\tfrac{1}{2})\right]\left[(v_t+\ell_t+1\pm 1)(v_t-\ell_t+1\mp 1)(v_{t'}+\ell_{t'}+1\pm 1)(v_{t'}-\ell_{t'}+1\mp 1)\right]^{\frac{1}{2}}$$

$$(\ell_t\ell_{t'}|h_{43}^{\dagger}+h_{45}^{\dagger}|\ell_t\pm 2,\ell_{t'}\mp 2) = hc\left[g_{22-}^{tt',J\pm}J(J+1)+g_{22-}^{tt',K\pm}K^2+g_{22-}^{tt',t\pm}(v_t+1)+g_{22-}^{tt',t'\pm}(v_{t'}+1)+\sum_{\substack{t''\\(t''\neq t,t')}}g_{22-}^{tt',t''\pm}(v_{t''}+1)\right.$$
$$\left.+\sum_{n}g_{22-}^{tt'n\pm}(v_n+\tfrac{1}{2})\right]\left[(v_t+\ell_t+1\pm 1)(v_t-\ell_t+1\mp 1)(v_{t'}+\ell_{t'}+1\mp 1)(v_{t'}-\ell_{t'}+1\pm 1)\right]^{\frac{1}{2}}$$

TABLE XLIV . Continued

$$(\ell_t|h_{43}^{\dagger}+h_{45}^{\dagger}|\ell_t\pm 4) = hc\left[g_4^{t,J\pm}J(J+1)+g_4^{t,K\pm}K^2+g_4^{t,t\pm}(v_t+1)+\sum_{\substack{t'\\(t'\neq t)}}g_4^{t,t'\pm}(v_{t'}+1)+\sum_n g_4^{t,n\pm}(v_n+\tfrac{1}{2})\right]$$

$$\times\left[(v_t-\ell_t+1\mp 3)(v_t-\ell_t+1\mp 1)(v_t+\ell_t+1\pm 1)(v_t+\ell_t+1\pm 3)\right]^{\frac{1}{2}}$$

$$(\ell_t\ell_{t'}|h_{45}^{\dagger}|\ell_t\pm 2,\ell_{t'}\pm 4) = hc\ g_{24}^{tt'\pm}\left[(v_{t'}-\ell_{t'}+1\mp 3)(v_{t'}-\ell_{t'}+1\mp 1)(v_{t'}+\ell_{t'}+1\pm 1)(v_{t'}+\ell_{t'}+1\pm 3)(v_t+\ell_t+1\pm 1)(v_t-\ell_t+1\mp 1)\right]^{\frac{1}{2}}$$

$$(\ell_t\ell_{t'}\ell_{t''}|h_{45}^{\dagger}|\ell_t\pm 2,\ell_{t'}\pm 2,\ell_{t''}\pm 2) = hc\ g_{222}^{tt't''\pm}\left[(v_t+\ell_t+1\pm 1)(v_t-\ell_t+1\mp 1)(v_{t'}+\ell_{t'}+1\pm 1)(v_{t'}-\ell_{t'}+1\mp 1)(v_{t''}+\ell_{t''}+1\pm 1)(v_{t''}-\ell_{t''}+1\mp 1)\right]^{\frac{1}{2}}$$

$$(\ell_t|h_{45}^{\dagger}|\ell_t\pm 6) = hc\ g_6^{t\pm}\left[(v_t-\ell_t+1\mp 5)(v_t-\ell_t+1\mp 3)(v_t-\ell_t+1\mp 1)(v_t+\ell_t+1\pm 1)(v_t+\ell_t+1\pm 3)(v_t+\ell_t+1\pm 5)\right]^{\frac{1}{2}}$$

The symbols $a^{\pm}(6,JK)$, $a^{\pm}(4,JK)$, $a^{\pm}(3,JK)$, $a^{\pm}(2,JK)$ and $a^{\pm}(1,JK)$ are obtained from the following general expression :

$$a^{\pm}(i,JK) = \left[J(J+1)-K(K\pm 1)\right]^{\frac{1}{2}}\left[J(J+1)-(K\pm 1)(K\pm 2)\right]^{\frac{1}{2}}\ \ldots\ldots\ \left\{J(J+1)-\left[K\pm(i-1)\right](K\pm i)\right\}^{\frac{1}{2}}$$

TABLE XLV-1 . The coefficients μ occurring in the diagonal matrix elements of h_{41}^{+}

	$[\overline{\omega}_{hhhh'hh}+\overline{\omega}_{xxxx'xx}]\frac{\hbar^6}{hc}$	$[\overline{\omega}_{xxxx'hh}+\overline{\omega}_{hhhh'xx}]\frac{\hbar^6}{hc}$	$[\overline{\omega}_{hxxh'hh}+\overline{\omega}_{xhhx'xx}]\frac{\hbar^6}{hc}$	$[\overline{\omega}_{xhhx'hh}+\overline{\omega}_{hxxh'xx}]\frac{\hbar^6}{hc}$	$[\overline{\omega}_{hhxx'hh}+\overline{\omega}_{hhxx'xx}]\frac{\hbar^6}{hc}$	$[\overline{\omega}_{hzzh'hh}+\overline{\omega}_{xzzx'xx}]\frac{\hbar^6}{hc}$	$[\overline{\omega}_{xzzx'hh}+\overline{\omega}_{hzzh'xx}]\frac{\hbar^6}{hc}$	$[\overline{\omega}_{zzhh'hh}+\overline{\omega}_{zzxx'xx}]\frac{\hbar^6}{hc}$	$[\overline{\omega}_{zzxx'hh}+\overline{\omega}_{zzhh'xx}]\frac{\hbar^6}{hc}$	$[\overline{\omega}_{zhhz'hh}+\overline{\omega}_{zxxz'xx}]\frac{\hbar^6}{hc}$	$[\overline{\omega}_{zxxz'hh}+\overline{\omega}_{zhhz'xx}]\frac{\hbar^6}{hc}$	$[\overline{\omega}_{zhzh'hh}+\overline{\omega}_{zxzx'xx}]\frac{\hbar^6}{hc}$	$[\overline{\omega}_{zxzx'hh}+\overline{\omega}_{zhzh'xx}]\frac{\hbar^6}{hc}$	$[\overline{\omega}_{hxhx'hh}+\overline{\omega}_{hxhx'xx}]\frac{\hbar^6}{hc}$	
μ_1^J	$+\frac{5}{16}$	$+\frac{1}{16}$	$+\frac{1}{16}$	$+\frac{1}{16}$	$+\frac{1}{8}$	0	0	0	0	0	0	0	0	$+\frac{1}{8}$	.....
μ_2^J	$-\frac{5}{8}$	$+\frac{3}{8}$	0	$-\frac{1}{4}$	$+\frac{1}{4}$	$+\frac{3}{8}$	$+\frac{1}{8}$	$+\frac{1}{2}$	$-\frac{1}{2}$	0	0	$+\frac{1}{4}$	-1	$-\frac{1}{4}$	.....
μ_3^J	$-\frac{15}{16}$	$-\frac{3}{16}$	$-\frac{3}{16}$	$-\frac{3}{16}$	$-\frac{3}{8}$	$+\frac{3}{8}$	$+\frac{1}{8}$	$+\frac{3}{4}$	$+\frac{1}{4}$	$+\frac{3}{8}$	$+\frac{1}{8}$	$+\frac{3}{4}$	$+\frac{1}{4}$	$-\frac{3}{8}$	.....
μ_1^{JK}	$+\frac{1}{2}$	$-\frac{1}{2}$	$-\frac{1}{4}$	$+\frac{1}{4}$	0	$-\frac{1}{4}$	$+\frac{1}{4}$	-1	$+1$	0	0	$-\frac{1}{2}$	$+\frac{1}{2}$	0	.....
μ_2^{JK}	$+\frac{45}{16}$	$-\frac{31}{16}$	$+\frac{5}{16}$	$+\frac{13}{16}$	$-\frac{7}{8}$	$-\frac{5}{2}$	$+\frac{1}{2}$	$-\frac{7}{2}$	$+\frac{7}{2}$	$-\frac{5}{4}$	$+\frac{5}{4}$	$-\frac{7}{2}$	$+\frac{5}{2}$	$+\frac{5}{8}$	.....
μ_3^{JK}	$+\frac{15}{16}$	$+\frac{3}{16}$	$+\frac{3}{16}$	$+\frac{3}{16}$	$+\frac{3}{8}$	$-\frac{3}{4}$	$-\frac{1}{4}$	-1	$-\frac{1}{2}$	$-\frac{3}{4}$	$-\frac{1}{4}$	$-\frac{3}{2}$	$-\frac{1}{2}$	$+\frac{3}{8}$	.....
μ_1^K	$-\frac{7}{4}$	$+\frac{7}{4}$	$+\frac{5}{8}$	$-\frac{5}{8}$	0	$+\frac{9}{8}$	$-\frac{9}{8}$	$+\frac{7}{2}$	$-\frac{7}{2}$	$+\frac{1}{2}$	$-\frac{1}{2}$	$+\frac{9}{4}$	$-\frac{9}{4}$	0	.....
μ_2^K	$-\frac{35}{16}$	$+\frac{25}{16}$	$-\frac{5}{16}$	$-\frac{9}{16}$	$+\frac{5}{8}$	$+\frac{5}{2}$	-1	$+\frac{15}{4}$	$-\frac{15}{4}$	$+\frac{13}{8}$	$-\frac{13}{8}$	$+4$	-3	$-\frac{3}{8}$	.....
μ_3^K	$-\frac{5}{16}$	$-\frac{1}{16}$	$-\frac{1}{16}$	$-\frac{1}{16}$	$-\frac{1}{8}$	$+\frac{3}{8}$	$+\frac{1}{8}$	$+\frac{3}{4}$	$+\frac{1}{4}$	$+\frac{3}{8}$	$+\frac{1}{8}$	$+\frac{3}{4}$	$+\frac{1}{4}$	$-\frac{1}{8}$	.....

TABLE XLV-1 . Continued

	$\frac{\hbar^6}{hc}[\varpi_{zzzz,yy}+\varpi_{zzzz,xx}]$	$\frac{\hbar^6}{hc}[\varpi_{yyyy,zz}+\varpi_{xxxx,zz}]$	$\frac{\hbar^6}{hc}\varpi_{yyxx,zz}$	$\frac{\hbar^6}{hc}[\varpi_{yxxy,zz}+\varpi_{xyyx,zz}]$	$\frac{\hbar^6}{hc}\varpi_{yxyx,zz}$	$\frac{\hbar^6}{hc}[\varpi_{yzzy,zz}+\varpi_{xzzx,zz}]$	$\frac{\hbar^6}{hc}[\varpi_{zzyy,zz}+\varpi_{zzxx,zz}]$
(μ_1^J)	0	0	0	0	0	0	0
(μ_2^J)	0	0	0	0	0	0	0
(μ_3^J)	0	$+\frac{3}{8}$	$+\frac{1}{4}$	$+\frac{1}{8}$	$+\frac{1}{4}$	0	0
(μ_1^{JK})	0	0	0	0	0	0	0
(μ_2^{JK})	0	$-\frac{1}{4}$	$+\frac{1}{2}$	$-\frac{1}{4}$	$-\frac{1}{2}$	$+\frac{1}{2}$	0
(μ_3^{JK})	$+\frac{1}{2}$	$-\frac{3}{4}$	$-\frac{1}{2}$	$-\frac{1}{4}$	$-\frac{1}{2}$	$+\frac{1}{2}$	+1
(μ_1^K)	0	0	0	0	0	0	0
(μ_2^K)	0	$+\frac{5}{8}$	$-\frac{5}{4}$	$+\frac{7}{8}$	$+\frac{3}{4}$	$-\frac{3}{2}$	0
(μ_3^K)	$-\frac{1}{2}$	$+\frac{3}{8}$	$+\frac{1}{4}$	$+\frac{1}{8}$	$+\frac{1}{4}$	$-\frac{1}{2}$	-1

	$\frac{\hbar^6}{hc}[\varpi_{zyyz,zz}+\varpi_{zxxz,zz}]$	$\frac{\hbar^6}{hc}[\varpi_{zyzy,zz}+\varpi_{zxzx,zz}]$	$\frac{\hbar^6}{hc}\varpi_{zzzz,zz}$	$\frac{\hbar^6}{hc}[\varpi_{zzzy,zy}+\varpi_{zzzx,zx}]$	$\frac{\hbar^6}{hc}[\varpi_{zyyy,zy}+\varpi_{zxxx,zx}]$	$\frac{\hbar^6}{hc}[\varpi_{yzyy,zy}+\varpi_{xzxx,zx}]$	$\frac{\hbar^6}{hc}[\varpi_{zyzz,zy}+\varpi_{zxzz,zx}]$	
(μ_1^J)	0	0	0	0	0	0	0	.....
(μ_2^J)	0	0	0	0	0	0	0	.....
(μ_3^J)	0	0	0	0	$+\frac{3}{2}$	$+\frac{3}{2}$	0	.....
(μ_1^{JK})	0	0	0	$+\frac{5}{8}$	$-\frac{1}{4}$	$-\frac{3}{4}$	0	.....
(μ_2^{JK})	0	0	0	$+\frac{9}{2}$	$-\frac{23}{8}$	$-\frac{27}{4}$	$+\frac{5}{2}$	.....
(μ_3^{JK})	$+\frac{1}{2}$	+1	0	0	$-\frac{3}{2}$	$-\frac{3}{2}$	0	.....
(μ_1^K)	0	0	0	-3	$+\frac{13}{8}$	$+\frac{33}{8}$	$-\frac{1}{2}$	.....
(μ_2^K)	0	-1	0	$-\frac{17}{2}$	$+\frac{47}{8}$	$+\frac{63}{8}$	$-\frac{13}{2}$	.....
(μ_3^K)	$-\frac{1}{2}$	-1	+1	-2	$+\frac{3}{2}$	$+\frac{3}{2}$	-2	.....

TABLE XLV-1 . Continued

	$\frac{\hbar^6}{hc}\left[\bar{\omega}_{xz,yxzy}+\bar{\omega}_{yz,xyzx}\right]$	$\frac{\hbar^6}{hc}\left[\bar{\omega}_{xz,xzyy}+\bar{\omega}_{yz,yzxx}\right]$	$\frac{\hbar^6}{hc}\left[\bar{\omega}_{xz,zxyy}+\bar{\omega}_{yz,zyxx}\right]$	$\frac{\hbar^6}{hc}\left[\bar{\omega}_{xz,xyyz}+\bar{\omega}_{yz,yxxz}\right]$	$\frac{\hbar^6}{hc}\left[\bar{\omega}_{xz,xyzy}+\bar{\omega}_{yz,yxzx}\right]$	$\frac{\hbar^6}{hc}\left[\bar{\omega}_{xz,xyxy}+\bar{\omega}_{yz,zxyx}\right]$	$\frac{\hbar^6}{hc}\left[\bar{\omega}_{xy,xyyy}+\bar{\omega}_{xy,yxxx}\right]$	$\frac{\hbar^6}{hc}\left[\bar{\omega}_{xy,xxyx}+\bar{\omega}_{xy,yyxy}\right]$	$\frac{\hbar^6}{hc}\bar{\omega}_{xy,zxyz}$	$\frac{\hbar^6}{hc}\bar{\omega}_{xy,yxzz}$	$\frac{\hbar^6}{hc}\bar{\omega}_{xy,xzzy}$	$\frac{\hbar^6}{hc}\bar{\omega}_{xy,zyzx}$	$\frac{\hbar^6}{hc}\bar{\omega}_{xy,yzxz}$	$\frac{\hbar^6}{hc}\left[\gamma_{\{xxx,xxx\}}+\gamma_{\{yyy,yyy\}}\right]$	
(μ_1^J)....	0	0	0	0	0	0	$+\frac{1}{4}$	$+\frac{1}{4}$	0	0	0	0	0	$+\frac{5}{32}$	.....
(μ_2^J)....	0	0	0	0	0	0	$-\frac{3}{4}$	$-\frac{5}{4}$	0	+1	$+\frac{1}{2}$	$+\frac{1}{2}$	$+\frac{1}{2}$	$-\frac{5}{16}$	.....
(μ_3^J)....	$+\frac{1}{2}$	$+\frac{1}{2}$	$+\frac{1}{2}$	$+\frac{1}{2}$	$+\frac{1}{2}$	$+\frac{1}{2}$	$-\frac{3}{4}$	$-\frac{3}{4}$	$+\frac{1}{2}$	$+\frac{1}{2}$	$+\frac{1}{2}$	$+\frac{1}{2}$	$+\frac{1}{2}$	$-\frac{15}{32}$	.....
(μ_1^{JK})....	$-\frac{3}{4}$	$+\frac{1}{4}$	$+\frac{1}{4}$	$+\frac{1}{4}$	$-\frac{1}{4}$	$-\frac{1}{4}$	$+\frac{1}{2}$	$+\frac{3}{2}$	0	−2	−1	−1	−1	$+\frac{1}{4}$	.....
(μ_2^{JK})....	$-\frac{17}{4}$	$-\frac{1}{4}$	$-\frac{1}{4}$	$-\frac{9}{4}$	$-\frac{9}{4}$	$-\frac{9}{4}$	$+\frac{17}{4}$	$+\frac{21}{4}$	−5	−7	−6	−6	−6	$+\frac{45}{32}$	.....
(μ_3^{JK})....	$-\frac{1}{2}$	$-\frac{1}{2}$	$-\frac{1}{2}$	$-\frac{1}{2}$	$-\frac{1}{2}$	$-\frac{1}{2}$	$+\frac{3}{4}$	$+\frac{3}{4}$	−1	−1	−1	−1	−1	$+\frac{15}{32}$	.....
(μ_1^K)....	$+\frac{31}{8}$	$-\frac{9}{8}$	$-\frac{9}{8}$	$+\frac{11}{8}$	$+\frac{11}{8}$	$+\frac{11}{8}$	$-\frac{9}{4}$	$-\frac{19}{4}$	+2	+7	$+\frac{9}{2}$	$+\frac{9}{2}$	$+\frac{9}{2}$	$+\frac{7}{8}$	.....
(μ_2^K)....	$+\frac{37}{8}$	$+\frac{5}{8}$	$+\frac{5}{8}$	$+\frac{21}{8}$	$+\frac{21}{8}$	$+\frac{21}{8}$	$-\frac{7}{2}$	−4	$+\frac{13}{2}$	$+\frac{15}{2}$	+7	+7	+7	$-\frac{35}{32}$	.....
(μ_3^K)....	$+\frac{1}{2}$	$+\frac{1}{2}$	$+\frac{1}{2}$	$+\frac{1}{2}$	$+\frac{1}{2}$	$+\frac{1}{2}$	$-\frac{1}{4}$	$-\frac{1}{4}$	$+\frac{1}{2}$	$+\frac{1}{2}$	$+\frac{1}{2}$	$+\frac{1}{2}$	$+\frac{1}{2}$	$-\frac{5}{32}$	.....

TABLE XLV-1 . Continued

	$[\ell_{\{yxx,yyy\}}+\ell_{\{yyx,xxx\}}]\frac{\hbar^6}{hc}$	$[\ell_{\{xyx,yyy\}}+\ell_{\{yxy,xxx\}}]\frac{\hbar^6}{hc}$	$[\ell_{\{zzy,yyy\}}+\ell_{\{zzx,xxx\}}]\frac{\hbar^6}{hc}$	$[\ell_{\{zyz,yyy\}}+\ell_{\{zxz,xxx\}}]\frac{\hbar^6}{hc}$	$[\ell_{\{yxx,yxx\}}+\ell_{\{yyx,yyx\}}]\frac{\hbar^6}{hc}$	$[\ell_{\{xyx,yxx\}}+\ell_{\{yxy,yyx\}}]\frac{\hbar^6}{hc}$	$[\ell_{\{zzy,xxy\}}+\ell_{\{zzx,yyx\}}]\frac{\hbar^6}{hc}$	$[\ell_{\{zyz,yxx\}}+\ell_{\{zxz,yyx\}}]\frac{\hbar^6}{hc}$	$[\ell_{\{xyx,xyx\}}+\ell_{\{yxy,yxy\}}]\frac{\hbar^6}{hc}$	$[\ell_{\{zzy,xyx\}}+\ell_{\{zzx,yxy\}}]\frac{\hbar^6}{hc}$	$[\ell_{\{zyz,xyx\}}+\ell_{\{zxz,yxy\}}]\frac{\hbar^6}{hc}$	$[\ell_{\{zzy,zzy\}}+\ell_{\{zzx,zzx\}}]\frac{\hbar^6}{hc}$	$[\ell_{\{zyz,zzy\}}+\ell_{\{zxz,zzx\}}]\frac{\hbar^6}{hc}$	$[\ell_{\{zyz,zyz\}}+\ell_{\{zxz,zxz\}}]\frac{\hbar^6}{hc}$	
(μ_1^J)	$+\frac{1}{8}$	$+\frac{1}{16}$	0	0	$+\frac{1}{8}$	$+\frac{1}{8}$	0	0	$+\frac{1}{32}$	0	0	0	0	0	.....
(μ_2^J)	$+\frac{3}{8}$	0	$+\frac{3}{8}$	0	$-\frac{1}{2}$	$-\frac{5}{8}$	$+\frac{1}{4}$	0	$-\frac{3}{16}$	$+\frac{1}{8}$	0	0	0	0	.....
(μ_3^J)	$-\frac{3}{8}$	$-\frac{3}{16}$	$+\frac{3}{4}$	$+\frac{3}{8}$	$-\frac{3}{8}$	$-\frac{3}{8}$	$+\frac{1}{2}$	$+\frac{1}{4}$	$-\frac{3}{32}$	$+\frac{1}{4}$	$+\frac{1}{8}$	0	0	0	.....
(μ_1^{JK})....	$-\frac{3}{4}$	$-\frac{1}{4}$	$-\frac{1}{4}$	0	$+\frac{3}{4}$	$+\frac{3}{4}$	0	0	$+\frac{1}{4}$	$-\frac{1}{4}$	0	$+\frac{1}{4}$	0	0	.....
(μ_2^{JK})....	$-\frac{13}{8}$	$-\frac{7}{16}$	$-\frac{11}{4}$	-1	$+\frac{21}{8}$	$+\frac{23}{8}$	$-\frac{3}{2}$	$-\frac{1}{2}$	$+\frac{25}{32}$	$-\frac{5}{4}$	$-\frac{1}{2}$	$+2$	$+\frac{3}{2}$	$+\frac{1}{4}$	.....
(μ_3^{JK})....	$+\frac{3}{8}$	$+\frac{3}{16}$	$-\frac{3}{2}$	$-\frac{3}{4}$	$+\frac{3}{8}$	$+\frac{3}{8}$	-1	$-\frac{1}{2}$	$+\frac{3}{32}$	$-\frac{1}{2}$	$-\frac{1}{4}$	$+1$	$+1$	$+\frac{1}{4}$	.....
(μ_1^K)....	$+\frac{19}{8}$	$+\frac{7}{8}$	$+\frac{9}{8}$	$+\frac{1}{4}$	$-\frac{5}{2}$	$-\frac{21}{8}$	$+\frac{1}{4}$	0	$-\frac{3}{4}$	$+\frac{7}{4}$	$+\frac{1}{4}$	$-\frac{5}{4}$	$-\frac{1}{2}$	0	.....
(μ_2^K)	$+\frac{5}{4}$	$+\frac{7}{16}$	$+\frac{25}{8}$	$+\frac{11}{8}$	$-\frac{17}{8}$	$-\frac{9}{4}$	$+\frac{7}{4}$	$+\frac{3}{4}$	$-\frac{19}{32}$	$+\frac{11}{8}$	$+\frac{5}{8}$	-4	$-\frac{7}{2}$	$-\frac{3}{4}$	.....
(μ_3^K)	$-\frac{1}{8}$	$-\frac{1}{16}$	$+\frac{3}{4}$	$+\frac{3}{8}$	$-\frac{1}{8}$	$-\frac{1}{8}$	$+\frac{1}{2}$	$+\frac{1}{4}$	$-\frac{1}{32}$	$+\frac{1}{4}$	$+\frac{1}{8}$	-1	-1	$-\frac{1}{4}$	.....

TABLE XLV-1 . Continued

	(μ_1^J)	(μ_2^J)	(μ_3^J)	(μ_1^{JK})	(μ_2^{JK})	(μ_3^{JK})	(μ_1^K)	(μ_2^K)	(μ_3^K)
$\left[\lambda_{\{yzx,yzx\}} + \lambda_{\{zyx,zyx\}} + \lambda_{\{zxy,zxy\}}\right]\frac{\hbar^6}{hc}$	0	$+\frac{1}{4}$	$+\frac{1}{4}$	$-\frac{1}{2}$	-3	$-\frac{1}{2}$	$+\frac{9}{4}$	$+\frac{7}{2}$	$+\frac{1}{4}$
$\left[\lambda_{\{yzx,zxy\}} + \lambda_{\{yzx,zyx\}} + \lambda_{\{zxy,zyx\}}\right]\frac{\hbar^6}{hc}$	0	$+\frac{1}{2}$	$+\frac{1}{2}$	-1	-6	-1	$+\frac{9}{2}$	$+7$	$+\frac{1}{2}$
$\lambda_{\{yzy,xzx\}}\frac{\hbar^6}{hc}$	0	$-\frac{1}{8}$	$+\frac{1}{8}$	$+\frac{1}{4}$	$+1$	$-\frac{1}{4}$	$-\frac{7}{8}$	$-\frac{5}{4}$	$+\frac{1}{8}$
$\left[\lambda_{\{yzy,yzy\}} + \lambda_{\{xzx,xzx\}}\right]\frac{\hbar^6}{hc}$	0	$+\frac{1}{16}$	$+\frac{3}{16}$	$-\frac{1}{8}$	-1	$-\frac{3}{8}$	$+\frac{11}{16}$	$+\frac{9}{8}$	$+\frac{3}{16}$
$\left[\lambda_{\{xzx,zyy\}} + \lambda_{\{yzy,zxx\}}\right]\frac{\hbar^6}{hc}$	0	$-\frac{1}{4}$	$+\frac{1}{4}$	$+\frac{1}{2}$	$+\frac{5}{2}$	$-\frac{1}{2}$	$-\frac{9}{4}$	-3	$+\frac{1}{4}$
$\left[\lambda_{\{yzy,zyy\}} + \lambda_{\{xzx,zxx\}}\right]\frac{\hbar^6}{hc}$	0	$+\frac{1}{4}$	$+\frac{3}{4}$	$-\frac{1}{2}$	$-\frac{7}{2}$	$-\frac{3}{2}$	$+\frac{9}{4}$	$+4$	$+\frac{3}{4}$
$\lambda_{\{zyy,zxx\}}\frac{\hbar^6}{hc}$	0	$-\frac{1}{2}$	$+\frac{1}{2}$	$+1$	$+6$	-1	$-\frac{9}{2}$	-7	$+\frac{1}{2}$
$\left[\lambda_{\{zyy,zyy\}} + \lambda_{\{zxx,zxx\}}\right]\frac{\hbar^6}{hc}$	0	$+\frac{1}{4}$	$+\frac{3}{4}$	$-\frac{1}{2}$	-3	$-\frac{3}{2}$	$+\frac{9}{4}$	$+\frac{7}{2}$	$+\frac{3}{4}$
$\left[\lambda_{\{yzy,zzz\}} + \lambda_{\{xzx,zzz\}}\right]\frac{\hbar^6}{hc}$	0	0	0	0	0	$+\frac{1}{2}$	0	$-\frac{1}{2}$	$-\frac{1}{2}$
$\left[\lambda_{\{zyy,zzz\}} + \lambda_{\{zxx,zzz\}}\right]\frac{\hbar^6}{hc}$	0	0	0	0	0	$+1$	0	0	-1
$\lambda_{\{zzz,zzz\}}\frac{\hbar^6}{hc}$	0	0	0	0	0	0	0	0	$+\frac{1}{2}$

TABLE XLV-2 a . The coefficients μ^{tJ}, μ^{tJK}, μ^{tK} occurring in the diagonal matrix elements of h^{+}_{42}

	μ^{tJ}_1	μ^{tJ}_2	μ^{tJK}	μ^{tK}_1	μ^{tK}_2
$\frac{\hbar^4}{2hc}[\tau^{+}_{xxxx} + \tau^{+}_{yyyy}]$	$+\frac{3}{8}$	$-\frac{1}{4}$	$-\frac{3}{4}$	$+\frac{5}{8}$	$+\frac{3}{8}$
$\frac{\hbar^4}{2hc}\tau^{+}_{xxyy}$	$+\frac{1}{4}$	$+\frac{1}{2}$	$-\frac{1}{2}$	$-\frac{5}{4}$	$+\frac{1}{4}$
$\frac{\hbar^4}{2hc}[\tau^{+}_{xyyx} + \tau^{+}_{yxxy}]$	$+\frac{1}{8}$	$-\frac{1}{4}$	$-\frac{1}{4}$	$+\frac{7}{8}$	$+\frac{1}{8}$
$\frac{\hbar^4}{2hc}\tau^{+}_{xyxy}$	$+\frac{1}{4}$	$-\frac{1}{2}$	$-\frac{1}{2}$	$+\frac{3}{4}$	$+\frac{1}{4}$
$\frac{\hbar^4}{2hc}[\tau^{+}_{xzzx} + \tau^{+}_{yzzy}]$	0	$+\frac{1}{2}$	$+\frac{1}{2}$	$-\frac{3}{2}$	$-\frac{1}{2}$
$\frac{\hbar^4}{2hc}[\tau^{+}_{xxzz} + \tau^{+}_{yyzz}]$	0	0	+1	0	−1
$\frac{\hbar^4}{2hc}[\tau^{+}_{zxxz} + \tau^{+}_{zyyz}]$	0	0	$+\frac{1}{2}$	0	$-\frac{1}{2}$
$\frac{\hbar^4}{2hc}[\tau^{+}_{xzxz} + \tau^{+}_{yzyz}]$	0	0	+1	−1	−1
$\frac{\hbar^4}{2hc}\tau^{+}_{zzzz}$	0	0	0	0	+1

TABLE XLV-2b . The coefficients $\mu^{nJ}, \mu^{nJK}, \mu^{nK}$ occurring in the diagonal matrix elements of $h^{\dagger}_{42}$

	μ_1^{nJ}	μ_2^{nJ}	μ^{nJK}	μ_1^{nK}	μ_2^{nK}
$\frac{\hbar^4}{2hc}\left[{}^{xxxx}v(nn) + {}^{yyyy}v(nn)\right]$	$+\frac{3}{8}$	$-\frac{1}{4}$	$-\frac{3}{4}$	$+\frac{5}{8}$	$+\frac{3}{8}$
$\frac{\hbar^4}{2hc}\,{}^{xxyy}v(nn)$	$+\frac{1}{4}$	$+\frac{1}{2}$	$-\frac{1}{2}$	$-\frac{5}{4}$	$+\frac{1}{4}$
$\frac{\hbar^4}{2hc}\left[{}^{xyyx}v(nn) + {}^{yxxy}v(nn)\right]$	$+\frac{1}{8}$	$-\frac{1}{4}$	$-\frac{1}{4}$	$+\frac{7}{8}$	$+\frac{1}{8}$
$\frac{\hbar^4}{2hc}\,{}^{xyxy}v(nn)$	$+\frac{1}{4}$	$-\frac{1}{2}$	$-\frac{1}{2}$	$+\frac{3}{4}$	$+\frac{1}{4}$
$\frac{\hbar^4}{2hc}\left[{}^{xzzx}v(nn) + {}^{yzzy}v(nn)\right]$	0	$+\frac{1}{2}$	$+\frac{1}{2}$	$-\frac{3}{2}$	$-\frac{1}{2}$
$\frac{\hbar^4}{2hc}\left[{}^{xxzz}v(nn) + {}^{yyzz}v(nn)\right]$	0	0	+ 1	0	− 1
$\frac{\hbar^4}{2hc}\left[{}^{zxxz}v(nn) + {}^{zyyz}v(nn)\right]$	0	0	$+\frac{1}{2}$	0	$-\frac{1}{2}$
$\frac{\hbar^4}{2hc}\left[{}^{xzxz}v(nn) + {}^{yzyz}v(nn)\right]$	0	0	+ 1	− 1	− 1
$\frac{\hbar^4}{2hc}\,{}^{zzzz}v(nn)$	0	0	0	0	+ 1

TABLE XLV-3a . The coefficients μ^{tJ} and μ^{tK} occurring in the diagonal matrix elements of $h^{\dagger}_{43}$

	$\frac{\hbar^2}{hc} a_1^{(2)}\left(\hbar^4\, {}^{\alpha\beta}_{(4)}Z^{aaaa} + {}^{\alpha\beta}_{(4)}Z_{aaaa} + \hbar^4\, {}^{\alpha\beta}_{(4)}Z^{bbbb} + {}^{\alpha\beta}_{(4)}Z_{bbbb}\right)$	$\frac{\hbar^2}{hc} a_2^{(2)}\left(\hbar^4\, {}^{\alpha\beta}_{(4)}Z^{aaaa} + {}^{\alpha\beta}_{(4)}Z_{aaaa} + \hbar^4\, {}^{\alpha\beta}_{(4)}Z^{bbbb} + {}^{\alpha\beta}_{(4)}Z_{bbbb}\right)$	$\frac{\hbar^2}{hc} a_1^{(2)}\left(\hbar^4\, {}^{\alpha\beta}_{(4)}Z^{aabb} + {}^{\alpha\beta}_{(4)}Z_{aabb} + \hbar^2\, {}^{\alpha\beta}_{(4)}Z^{aa}_{aa} + \hbar^2\, {}^{\alpha\beta}_{(4)}Z^{bb}_{bb}\right)$	$\frac{\hbar^2}{hc} a_2^{(2)}\left(\hbar^4\, {}^{\alpha\beta}_{(4)}Z^{aabb} + {}^{\alpha\beta}_{(4)}Z_{aabb} + \hbar^2\, {}^{\alpha\beta}_{(4)}Z^{aa}_{aa} + \hbar^2\, {}^{\alpha\beta}_{(4)}Z^{bb}_{bb}\right)$	$\frac{\hbar^2}{hc} a_1^{(2)}\left(\hbar^2\, {}^{\alpha\beta}_{(4)}Z^{bb}_{aa} + \hbar^2\, {}^{\alpha\beta}_{(4)}Z^{aa}_{bb}\right)$	$\frac{\hbar^2}{hc} a_2^{(2)}\left(\hbar^2\, {}^{\alpha\beta}_{(4)}Z^{bb}_{aa} + \hbar^2\, {}^{\alpha\beta}_{(4)}Z^{aa}_{bb}\right)$	$\frac{\hbar^2}{hc} a_1^{(2)}\left(\hbar^2\, {}^{\alpha\beta}_{(4)}Z^{ab}_{ab}\right)$	$\frac{\hbar^2}{hc} a_2^{(2)}\left(\hbar^2\, {}^{\alpha\beta}_{(4)}Z^{ab}_{ab}\right)$
μ_3^{tJ}	$+\frac{9}{16}$	0	$+\frac{3}{16}$	0	$+\frac{1}{16}$	0	$+\frac{1}{16}$	0
μ_4^{tJ}	$-\frac{3}{16}$	0	$-\frac{1}{16}$	0	$+\frac{5}{16}$	0	$-\frac{3}{16}$	0
μ_3^{tK}	0	$+\frac{9}{16}$	0	$+\frac{3}{16}$	0	$+\frac{1}{16}$	0	$+\frac{1}{16}$
μ_4^{tK}	0	$-\frac{3}{16}$	0	$-\frac{1}{16}$	0	$+\frac{5}{16}$	0	$-\frac{3}{16}$

TABLE XLV-3b The coefficients $\mu^{nJ}, \mu^{nK}, \mu^{ss'J}, \mu^{ss'K}, \mu_J^4$ and μ_K^4 occurring in the diagonal matrix elements of $h_{43}^{\dagger}$.

$$\mu_3^{nJ} = \frac{\hbar^2}{2hc} a_1^{(2)} \left(3\hbar^4 \, {}^{\alpha\beta}_{(4)}Z^{nnnn} + 3 \, {}^{\alpha\beta}_{(4)}Z_{nnnn} + \hbar^2 \, {}^{\alpha\beta}_{(4)}Z^{nn}_{nn}\right)$$

$$\mu_3^{nK} = \frac{\hbar^2}{2hc} a_2^{(2)} \left(3\hbar^4 \, {}^{\alpha\beta}_{(4)}Z^{nnnn} + 3 \, {}^{\alpha\beta}_{(4)}Z_{nnnn} + \hbar^2 \, {}^{\alpha\beta}_{(4)}Z^{nn}_{nn}\right)$$

$$\mu^{nn'J} = \frac{\hbar^2}{4hc} a_1^{(2)} \left(4\hbar^4 \, {}^{\alpha\beta}_{(4)}Z^{nnn'n'} + 4 \, {}^{\alpha\beta}_{(4)}Z_{nnn'n'} + \hbar^2 \, {}^{\alpha\beta}_{(4)}Z^{n'n'}_{nn} + \hbar^2 \, {}^{\alpha\beta}_{(4)}Z^{nn}_{n'n'}\right)$$

$$\mu^{nn'K} = \frac{\hbar^2}{4hc} a_2^{(2)} \left(4\hbar^2 \, {}^{\alpha\beta}_{(4)}Z^{nnn'n'} + 4 \, {}^{\alpha\beta}_{(4)}Z_{nnn'n'} + \hbar^2 \, {}^{\alpha\beta}_{(4)}Z^{n'n'}_{nn} + \hbar^2 \, {}^{\alpha\beta}_{(4)}Z^{nn}_{n'n'}\right)$$

$$\mu^{ntJ} = \frac{\hbar^2}{2hc} a_1^{(2)} \left(\hbar^4 \, {}^{\alpha\beta}_{(4)}Z^{aann} + \hbar^4 \, {}^{\alpha\beta}_{(4)}Z^{bbnn} + {}^{\alpha\beta}_{(4)}Z_{aann} + {}^{\alpha\beta}_{(4)}Z_{bbnn} + \hbar^2 \, {}^{\alpha\beta}_{(4)}Z^{aa}_{nn} + \hbar^2 \, {}^{\alpha\beta}_{(4)}Z^{nn}_{aa} + \hbar^2 \, {}^{\alpha\beta}_{(4)}Z^{bb}_{nn} + \hbar^2 \, {}^{\alpha\beta}_{(4)}Z^{nn}_{bb}\right)$$

$$\mu^{ntK} = \frac{\hbar^2}{2hc} a_2^{(2)} \left(\hbar^4 \, {}^{\alpha\beta}_{(4)}Z^{aann} + \hbar^4 \, {}^{\alpha\beta}_{(4)}Z^{bbnn} + {}^{\alpha\beta}_{(4)}Z_{aann} + {}^{\alpha\beta}_{(4)}Z_{bbnn} + \hbar^2 \, {}^{\alpha\beta}_{(4)}Z^{aa}_{nn} + \hbar^2 \, {}^{\alpha\beta}_{(4)}Z^{nn}_{aa} + \hbar^2 \, {}^{\alpha\beta}_{(4)}Z^{bb}_{nn} + \hbar^2 \, {}^{\alpha\beta}_{(4)}Z^{nn}_{bb}\right)$$

$$\begin{aligned}\mu_1^{tt'J} = \frac{\hbar^2}{4hc} a_1^{(2)} \Big(&\hbar^4 \, {}^{\alpha\beta}_{(4)}Z^{aacc} + \hbar^4 \, {}^{\alpha\beta}_{(4)}Z^{aadd} + \hbar^4 \, {}^{\alpha\beta}_{(4)}Z^{bbcc} + \hbar^4 \, {}^{\alpha\beta}_{(4)}Z^{bbdd} + {}^{\alpha\beta}_{(4)}Z_{aacc} + {}^{\alpha\beta}_{(4)}Z_{aadd} \\ &+ {}^{\alpha\beta}_{(4)}Z_{bbcc} + {}^{\alpha\beta}_{(4)}Z_{bbdd} + \hbar^2 \, {}^{\alpha\beta}_{(4)}Z^{cc}_{aa} + \hbar^2 \, {}^{\alpha\beta}_{(4)}Z^{aa}_{cc} + \hbar^2 \, {}^{\alpha\beta}_{(4)}Z^{dd}_{aa} + \hbar^2 \, {}^{\alpha\beta}_{(4)}Z^{aa}_{dd} \\ &+ \hbar^2 \, {}^{\alpha\beta}_{(4)}Z^{cc}_{bb} + \hbar^2 \, {}^{\alpha\beta}_{(4)}Z^{bb}_{cc} + \hbar^2 \, {}^{\alpha\beta}_{(4)}Z^{dd}_{bb} + \hbar^2 \, {}^{\alpha\beta}_{(4)}Z^{bb}_{dd}\Big)\end{aligned}$$

TABLE XLV-3b . Continued

$$\mu_2^{tt'J} = \frac{\hbar^4}{4hc}\, a_1^{(2)} \left({}^{\alpha\beta}_{(4)}Z^{cd}_{ab} + {}^{\alpha\beta}_{(4)}Z^{ab}_{cd} - {}^{\alpha\beta}_{(4)}Z^{bc}_{ad} - {}^{\alpha\beta}_{(4)}Z^{ad}_{bc} \right)$$

$$\mu_1^{tt'K} = \frac{\hbar^2}{4hc}\, a_2^{(2)} \Big(\hbar^4\, {}^{\alpha\beta}_{(4)}Z^{aacc} + \hbar^4\, {}^{\alpha\beta}_{(4)}Z^{aadd} + \hbar^4\, {}^{\alpha\beta}_{(4)}Z^{bbcc} + \hbar^4\, {}^{\alpha\beta}_{(4)}Z^{bbdd} + {}^{\alpha\beta}_{(4)}Z_{aacc} + {}^{\alpha\beta}_{(4)}Z_{aadd}$$

$$+ {}^{\alpha\beta}_{(4)}Z_{bbcc} + {}^{\alpha\beta}_{(4)}Z_{bbdd} + \hbar^2\, {}^{\alpha\beta}_{(4)}Z^{cc}_{aa} + \hbar^2\, {}^{\alpha\beta}_{(4)}Z^{aa}_{cc} + \hbar^2\, {}^{\alpha\beta}_{(4)}Z^{dd}_{aa} + \hbar^2\, {}^{\alpha\beta}_{(4)}Z^{aa}_{dd}$$

$$+ \hbar^2\, {}^{\alpha\beta}_{(4)}Z^{cc}_{bb} + \hbar^2\, {}^{\alpha\beta}_{(4)}Z^{bb}_{cc} + \hbar^2\, {}^{\alpha\beta}_{(4)}Z^{dd}_{bb} + \hbar^2\, {}^{\alpha\beta}_{(4)}Z^{bb}_{dd} \Big)$$

$$\mu_2^{tt'K} = \frac{\hbar^4}{4hc}\, a_2^{(2)} \left({}^{\alpha\beta}_{(4)}Z^{cd}_{ab} + {}^{\alpha\beta}_{(4)}Z^{ab}_{cd} - {}^{\alpha\beta}_{(4)}Z^{bc}_{ad} - {}^{\alpha\beta}_{(4)}Z^{ad}_{bc} \right)$$

$$\mu_4^{J} = \frac{\hbar^2}{hc} \Big\{ \frac{3}{8} \sum_n a_1^{(2)} \left(\hbar^4\, {}^{\alpha\beta}_{(4)}Z^{nnnn} + {}^{\alpha\beta}_{(4)}Z_{nnnn} - \hbar^2\, {}^{\alpha\beta}_{(4)}Z^{nn}_{nn} \right)$$

$$+ \frac{1}{16} \sum_t a_1^{(2)} \Big(3\hbar^4\, {}^{\alpha\beta}_{(4)}Z^{aaaa} + 3\, {}^{\alpha\beta}_{(4)}Z_{aaaa} + 3\hbar^4\, {}^{\alpha\beta}_{(4)}Z^{bbbb} + 3\, {}^{\alpha\beta}_{(4)}Z_{bbbb} + \hbar^4\, {}^{\alpha\beta}_{(4)}Z^{aabb}$$

$$+ {}^{\alpha\beta}_{(4)}Z_{aabb} - 7\hbar^2\, {}^{\alpha\beta}_{(4)}Z^{aa}_{aa} - 7\hbar^2\, {}^{\alpha\beta}_{(4)}Z^{bb}_{bb} + 3\hbar^2\, {}^{\alpha\beta}_{(4)}Z^{bb}_{aa} + 3\hbar^2\, {}^{\alpha\beta}_{(4)}Z^{aa}_{bb} - 5\hbar^2\, {}^{\alpha\beta}_{(4)}Z^{ab}_{ab} \Big)$$

$$- \frac{1}{4} \Big[\sum_{\substack{t\sigma, t'\sigma' \\ t<t'}} a_1^{(2)} \left(\hbar^2\, {}^{\alpha\beta}_{(4)}Z^{t\sigma, t'\sigma'}_{t\sigma, t'\sigma'} \right) + \sum_{t\sigma, n} a_1^{(2)} \left({}^{\alpha\beta}_{(4)}Z^{t\sigma, n}_{t\sigma, n} \right) + \sum_{\substack{nn' \\ n<n'}} a_1^{(2)} \left(\hbar^2\, {}^{\alpha\beta}_{(4)}Z^{nn'}_{nn'} \right) \Big] \Big\}$$

TABLE XLV-3b . Continued

$$\mu_4^K = \frac{\hbar^2}{hc}\Big\{\frac{3}{8}\sum_n a_2^{(2)}\Big(\hbar^4\,{}^{\alpha\beta}_{(4)}Z^{nnnn} + {}^{\alpha\beta}_{(4)}Z_{nnnn} - \hbar^2\,{}^{\alpha\beta}_{(4)}Z^{nn}_{nn}\Big)$$

$$+\frac{1}{16}\sum_t a_2^{(2)}\Big(3\hbar^4\,{}^{\alpha\beta}_{(4)}Z^{aaaa} + 3\,{}^{\alpha\beta}_{(4)}Z_{aaaa} + 3\hbar^4\,{}^{\alpha\beta}_{(4)}Z^{bbbb} + 3\,{}^{\alpha\beta}_{(4)}Z_{bbbb} + \hbar^4\,{}^{\alpha\beta}_{(4)}Z^{aabb} + {}^{\alpha\beta}_{(4)}Z_{aabb}$$

$$-7\hbar^2\,{}^{\alpha\beta}_{(4)}Z^{aa}_{aa} - 7\hbar^2\,{}^{\alpha\beta}_{(4)}Z^{bb}_{bb} + 3\hbar^2\,{}^{\alpha\beta}_{(4)}Z^{bb}_{aa} + 3\hbar^2\,{}^{\alpha\beta}_{(4)}Z^{aa}_{bb} - 5\hbar^2\,{}^{\alpha\beta}_{(4)}Z^{ab}_{ab}\Big)$$

$$-\frac{1}{4}\Big[\sum_{\substack{t\sigma,t'\sigma'\\ t<t'}} a_2^{(2)}\Big(\hbar^2\,{}^{\alpha\beta}_{(4)}Z^{t\sigma,t'\sigma'}_{t\sigma,t'\sigma'}\Big) + \sum_{t\sigma,n} a_2^{(2)}\Big(\hbar^2\,{}^{\alpha\beta}_{(4)}Z^{t\sigma,n}_{t\sigma,n}\Big) + \sum_{\substack{nn'\\ n<n'}} a_2^{(2)}\Big(\hbar^2\,{}^{\alpha\beta}_{(4)}Z^{nn'}_{nn'}\Big)\Big]\Big\}$$

TABLE XLV-4 . The coefficients μ occurring in the diagonal matrix elements of $h_{44}^{\dagger}$

$$\mu_5^J = \frac{\hbar^2}{2hc}\Big({}^{xx}_{(4)}Z + {}^{yy}_{(4)}Z\Big)$$

$$\mu_5^K = \frac{\hbar^2}{2hc}\Big(2\,{}^{zz}_{(4)}Z - {}^{xx}_{(4)}Z - {}^{yy}_{(4)}Z\Big)$$

TABLE XLV-5 . The coefficients μ occurring in the diagonal matrix elements of $h^{\dagger}_{45}$

$$\mu^t_1 = \frac{1}{32hc}\Big\{25\big(u(aaaaaa)+u(bbbbbb)\big)+5\big(u(aaaabb)+u(aabbbb)+v(aa,aaaa)+v(bb,bbbb)\big)$$
$$+\big(v(aa,bbbb)+v(bb,aaaa)+v(aa,aabb)+v(bb,aabb)+v(ab,aaab)+v(ab,abbb)\big)\Big\}$$

$$\mu^t_2 = \frac{1}{32hc}\Big\{-15\big(u(aaaaaa)+u(bbbbbb)\big)-3\big(u(aaaabb)+u(aabbbb)+v(aa,aaaa)+v(bb,bbbb)\big)$$
$$+9\big(v(aa,bbbb)+v(bb,aaaa)\big)+\big(v(aa,aabb)+v(bb,aabb)\big)-3\big(v(ab,aaab)+v(ab,abbb)\big)\Big\}$$

$$\mu^t_3 = \mu^t_3(1)+\mu^t_3(2)+\mu^t_3(3)+\mu^t_3(4) \quad , \text{ where}$$

$$\mu^t_3(1) = \frac{1}{32hc}\Big[35\,u(aaaaaa)+35\,u(bbbbbb)+7u(aaaabb)+7u(aabbbb)-41v(aa,aaaa)-41v(bb,bbbb)$$
$$+11v(aa,bbbb)-5v(aa,aabb)-5v(bb,aabb)-13v(ab,aaab)-13v(ab,abbb)\Big]$$

$$\mu^t_3(2) = +\frac{1}{16hc}\sum_{n}\Big[u(aa\,nnnn)+u(bbnnnn)+v(aa,nnnn)+v(bb,nnnn)-v(nn,aann)-v(nn,bbnn)\Big]$$

$$\mu^t_3(3) = -\frac{1}{8hc}\Bigg\{\sum_{\substack{c\\ \left(c=t'_1,\,t'_2,\,n;\ t'\neq t\right)}}\Big[3v(ac,aaac)+v(bc,aabc)+3v(bc,bbbc)+v(ac,abbc)\Big]$$
$$-\sum_{\substack{cd\\ \left(c=t'_1,t'_2\,;\ d=t''_1,t''_2;\ t'\neq t''\neq t\,;\ t'<t''\right)}}\Big[v(cd,cdaa)+v(cd,cdbb)\Big]-\sum_{\substack{nn'\\ n<n'}}\Big[v(nn',nn'aa)+v(nn',nn'bb)\Big]\Bigg\}$$

TABLE XLV-5 . Continued

$$\mu_3^t(4) = \frac{1}{32hc} \sum_{\substack{abc \\ \left(\substack{a=t'_1,\ b=t'_2 \\ c=t_1,\ t_2 \\ t \text{ fixed, sum over } t'\neq t}\right)}} \Big[3u(aaaacc) + 3u(bbbbcc) + 3v(cc,aaaa) + 3v(cc,bbbb) + u(cc,aabb) + u(aabbcc)$$

$$- 7v(aa,aacc) - 7v(bb,bbcc) + 3v(aa,bbcc) + 3v(bb,aacc) - 5v(ab,abcc)\Big]$$

$$\mu_1^n = \frac{1}{2hc}\left[5u(nnnnnn) + v(nn,nnnn)\right]$$

$$\mu_2^n = \mu_2^n(1) + \mu_2^n(2) + \mu_2^n(3) + \mu_2^n(4) \quad , \text{ where}$$

$$\mu_2^n(1) = \frac{1}{8hc}\left[25u(nnnnnn) - 19v(nn,nnnn)\right]$$

$$\mu_2^n(2) = \frac{3}{8hc}\sum_{n'\neq n}\left[u(nnn'n'n'n') + v(nn,n'n'n'n') - v(n'n',nnn'n')\right]$$

$$\mu_2^n(3) = -\frac{1}{4hc}\Bigg[3\sum_{n'\neq n} v(nn',nnnn') + 3\sum_{\substack{a \\ \left(\substack{a=t_1,t_2 \\ \text{sum over } t}\right)}} v(na,nnna) + \sum_{\substack{n'n'' \\ \left(\substack{n'\neq n''\neq n \\ n'<n''}\right)}} v(n'n'',n'n''nn) + \sum_{\substack{ab \\ \left(\substack{a=t_1,t_2;\ b=t'_1,t'_2 \\ \text{sum over } tt' \\ t<t'}\right)}} v(ab,abnn)\Bigg]$$

$$\mu_2^n(4) = \frac{1}{16hc}\sum_t\Big\{3u(aaaann) + 3u(bbbbnn) + 3v(nn,aaaa) + 3v(nn,bbbb) + u(aabbnn) + v(nn,aabb)$$

$$- 7v(aa,aann) - 7v(bb,bbnn) + 3v(aa,bbnn) + 3v(bb,aann) - 5v(ab,abnn)\Big\}$$

TABLE XLV-5 . Continued

$$\mu_1^{tt'} = \frac{1}{32hc} \sum_{\substack{a \\ (a=t_1,t_2 \\ t \text{ fixed})}} \Big[9u(ccccaa) + 9u(ddddaa) + 9v(aa,cccc) + 9v(aa,dddd) + 3u(ccddaa) + 3v(aa,ccdd)$$
$$+ 3v(cc,ccaa) + 3v(dd,ddaa) + v(cc,ddaa) + v(dd,ccaa) + v(cd,cdaa)\Big]$$

$$\mu_2^{tt'} = \frac{1}{32hc} \sum_{\substack{a \\ (a=t_1,t_2 \\ t \text{ fixed})}} \Big[-3u(ccccaa) - 3u(ddddaa) - 3v(aa,cccc) - 3v(aa,dddd) - u(ccddaa) - v(aa,ccdd)$$
$$- v(cc,ccaa) - v(dd,ddaa) + 5v(cc,ddaa) + 5v(dd,ccaa) - 3v(cd,cdaa)\Big]$$

$$\mu_3^{tt'} = \frac{1}{16hc} \Big[3v(ca,dbbb) - 3v(da,cbbb) - 3v(cb,daaa) + 3v(db,caaa) + v(ca,daab) - v(da,caab)$$
$$- v(cb,dabb) + v(db,cabb)\Big]$$

$$\mu_1^{nt} = \frac{1}{16hc} \Big[9u(aaaann) + 9u(bbbbnn) + 9v(nn,aaaa) + 9v(nn,bbbb) + 3u(aabbnn) + 3v(nn,aabb)$$
$$+ 3v(aa,aann) + 3v(bb,bbnn) + v(aa,bbnn) + v(bb,aann) + v(ab,abnn)\Big]$$

$$\mu_2^{nt} = \frac{1}{16hc} \Big[-3u(aaaann) - 3u(bbbbnn) - 3v(nn,aaaa) - 3v(nn,bbbb) - u(aabbnn) - v(nn,aabb)$$
$$- v(aa,aann) - v(bb,bbnn) + 5v(aa,bbnn) + 5v(bb,aann) - 3v(ab,abnn)\Big]$$

TABLE XLV-5 . Continued

$$\mu_3^{nt} = \frac{1}{4hc}\left[+3u(nnnnaa) + 3u(nnnnbb) + 3v(aa,nnnn) + 3v(bb,nnnn) + v(nn,nnaa) + v(nn,nnbb)\right]$$

$$\mu^{nn'} = \frac{1}{2hc}\left[3u(nnnnn'n') + 3v(n'n',nnnn) + v(nn,nnn'n')\right]$$

$$\mu_1^{tt't''} = \frac{1}{8hc}\sum_{abc}\left[u(aabbcc) + v(aa,bbcc) + v(bb,aacc) + v(cc,aabb)\right]$$

$$\begin{pmatrix} a = t_1, t_2 \\ b = t_1', t_2' \\ c = t_1'', t_2'' \\ t, t', t'' \text{ fixed} \end{pmatrix}$$

$$\mu_2^{tt't''} = \frac{1}{8hc}\left[v(ac,bdee) + v(bd,acee) + v(ac,bdff) + v(bd,acff) - v(ad,bcee) - v(bc,adee) - v(ad,bcff) - v(bc,adff)\right]$$

$$\mu^{nn'n''} = \frac{1}{hc}\left[u(nnn'n'n''n'') + v(nn,n'n'n''n'') + v(n'n',nnn''n'') + v(n''n'',nnn'n')\right]$$

$$\mu_1^{tt'n} = \frac{1}{4hc}\sum_{ab}\left[u(aabbnn) + v(nn,aabb) + v(aa,nnbb) + v(bb,nnaa)\right]$$

$$\begin{pmatrix} a = t_1, t_2 \\ b = t_1', t_2' \\ t, t' \text{ fixed} \end{pmatrix}$$

TABLE XLV-5 . Continued

$$\mu_2^{tt'n} = \frac{1}{4hc}\left[v(ac, bdnn) + v(bd, acnn) - v(ad, bcnn) - v(bc, adnn)\right]$$

$$\mu^{nn't} = \frac{1}{2hc}\Big[u(nnn'n'aa) + u(nnn'n'bb) + v(nn, n'n'aa) + v(nn, n'n'bb) + v(n'n', nnaa) + v(n'n', nnbb)$$
$$+ v(aa, nnn'n') + v(bb, nnn'n')\Big]$$

TABLE XLV-6 . The coefficients μ occurring in the diagonal matrix elements of $h^{\dagger}_{46}$

$$\mu_4^t = \frac{1}{2}\left[\hbar^2\left({}_{(4)}Z^{aa} + {}_{(4)}Z^{bb}\right) + {}_{(4)}Z_{aa} + {}_{(4)}Z_{bb}\right]$$

$$\mu_3^n = \hbar^2\, {}_{(4)}Z^{nn} + {}_{(4)}Z_{nn}$$

TABLE XLVI-1 . The coefficients occurring in the off-diagonal matrix elements of $h^{\dagger}_{41}$

$$
\begin{aligned}
f_6^{\pm} = \frac{\hbar^6}{64hc}\Big[\Big\{&({}^{xx,xxxx}{}_{\varpi} - {}^{yy,yyyy}{}_{\varpi}) + ({}^{xx,yyyy}{}_{\varpi} - {}^{yy,xxxx}{}_{\varpi}) + 2({}^{yy,xxyy}{}_{\varpi} - {}^{xx,xxyy}{}_{\varpi}) + ({}^{yy,xyyx}{}_{\varpi} - {}^{xx,xyyx}{}_{\varpi}) \\
&+ ({}^{yy,yxxy}{}_{\varpi} - {}^{xx,yxxy}{}_{\varpi}) + 2({}^{yy,xyxy}{}_{\varpi} - {}^{xx,xyxy}{}_{\varpi}) + 4({}^{xy,xyyy}{}_{\varpi} - {}^{xy,yxxx}{}_{\varpi}) + 4({}^{xy,yyxy}{}_{\varpi} - {}^{xy,xxyx}{}_{\varpi}) \\
&+ \tfrac{1}{2}({}^{xxx,xxx}{}_{\gamma} - {}^{yyy,yyy}{}_{\gamma}) - 2({}^{\{xxx,xyy\}}{}_{\gamma} - {}^{\{yyy,xxy\}}{}_{\gamma}) - ({}^{\{xxx,yxy\}}{}_{\gamma} - {}^{\{yyy,xyx\}}{}_{\gamma}) \\
&\qquad + 2({}^{xyy,xyy}{}_{\gamma} - {}^{xxy,xxy}{}_{\gamma}) + 2({}^{\{xyy,yxy\}}{}_{\gamma} - {}^{\{xxy,xyx\}}{}_{\gamma}) + \tfrac{1}{2}({}^{xyx,xyx}{}_{\gamma} - {}^{yxy,yxy}{}_{\gamma})\Big\} \\
\mp i\Big\{&-2({}^{xx,xyyy}{}_{\varpi} - {}^{yy,xyyy}{}_{\varpi}) + 2({}^{xx,yxxx}{}_{\varpi} - {}^{yy,yxxx}{}_{\varpi}) + 2({}^{xx,xxyx}{}_{\varpi} - {}^{yy,xxyx}{}_{\varpi}) \\
&-2({}^{xx,yyxy}{}_{\varpi} - {}^{yy,yyxy}{}_{\varpi}) + 2({}^{xy,xxxx}{}_{\varpi} + {}^{xy,yyyy}{}_{\varpi}) - 2({}^{xy,xyyx}{}_{\varpi} + {}^{xy,yxxy}{}_{\varpi}) \\
&-4({}^{xy,xxyy}{}_{\varpi} + {}^{xy,xyxy}{}_{\varpi}) + 2({}^{\{xxx,xxy\}}{}_{\gamma} + {}^{\{yyy,xyy\}}{}_{\gamma}) - {}^{\{xxx,yyy\}}{}_{\gamma} \\
&\qquad + ({}^{\{xxx,xyx\}}{}_{\gamma} + {}^{\{yyy,yxy\}}{}_{\gamma}) - 4{}^{\{xyy,xxy\}}{}_{\gamma} - 2({}^{\{xyy,xyx\}}{}_{\gamma} + {}^{\{xxy,yxy\}}{}_{\gamma}) - {}^{\{yxy,xyx\}}{}_{\gamma}\Big\}\Big]
\end{aligned}
$$

$$
\begin{aligned}
f_4^{J\pm} = \frac{\hbar^6}{32hc}\Big[\Big\{&+3({}^{xx,xxxx}{}_{\varpi} + {}^{yy,yyyy}{}_{\varpi}) - ({}^{xx,yyyy}{}_{\varpi} + {}^{yy,xxxx}{}_{\varpi}) - 2({}^{xx,xxyy}{}_{\varpi} + {}^{yy,xxyy}{}_{\varpi}) \\
&-({}^{xx,xyyx}{}_{\varpi} + {}^{yy,yxxy}{}_{\varpi}) - ({}^{xx,yxxy}{}_{\varpi} + {}^{yy,xyyx}{}_{\varpi}) - 2({}^{xx,xyxy}{}_{\varpi} + {}^{yy,xyxy}{}_{\varpi}) \cdots\cdot
\end{aligned}
$$

TABLE XLVI-1 . Continued

$$\left(\mathfrak{f}_4^{J\pm}\right) \;\cdots\; -4\left({}^{xy,xyyy}{}_{\overline{\omega}} + {}^{xy,yxxx}{}_{\overline{\omega}}\right) - 4\left({}^{xy,xxyx}{}_{\overline{\omega}} + {}^{xy,yyxy}{}_{\omega}\right) + \frac{3}{2}\left({}^{xxx,xxx}{}_{\gamma} + {}^{yyy,yyy}{}_{\gamma}\right)$$

$$-2\left({}^{\{xxx,xyy\}}{}_{\gamma} + {}^{\{yyy,xxy\}}{}_{\gamma}\right) - \left({}^{\{xxx,yxy\}}{}_{\gamma} + {}^{\{yyy,xyx\}}{}_{\gamma}\right) - 2\left({}^{\{xyy,yxy\}}{}_{\gamma} + {}^{\{xxy,xyx\}}{}_{\gamma}\right)$$

$$-\left({}^{xyy,xyy}{}_{\gamma} + {}^{xxy,xxy}{}_{\gamma}\right) - \frac{1}{2}\left({}^{yxy,yxy}{}_{\gamma} + {}^{xyx,xyx}{}_{\gamma}\right)\Big\}$$

$$\mp i\Big\{\; 4\left({}^{xx,yxxx}{}_{\overline{\omega}} - {}^{yy,xyyy}{}_{\overline{\omega}}\right) + 4\left({}^{xx,xxyx}{}_{\overline{\omega}} - {}^{yy,yyxy}{}_{\overline{\omega}}\right) + 4\left({}^{xy,xxxx}{}_{\overline{\omega}} - {}^{xy,yyyy}{}_{\overline{\omega}}\right)$$

$$+4\left({}^{\{xxx,xxy\}}{}_{\gamma} - {}^{\{yyy,xyy\}}{}_{\gamma}\right) + 2\left({}^{\{xxx,xyx\}}{}_{\gamma} - {}^{\{yyy,yxy\}}{}_{\gamma}\right)\Big\}\Big]$$

$$\mathfrak{f}_4^{K\pm} = \frac{\hbar^6}{32hc}\Big[\Big\{ -\frac{3}{4}\left({}^{xx,xxxx}{}_{\overline{\omega}} + {}^{yy,yyyy}{}_{\overline{\omega}}\right) + \frac{1}{4}\left({}^{xx,yyyy}{}_{\overline{\omega}} + {}^{yy,xxxx}{}_{\overline{\omega}}\right) + \frac{1}{2}\left({}^{xx,xxyy}{}_{\overline{\omega}} + {}^{yy,xxyy}{}_{\overline{\omega}}\right)$$

$$+\frac{1}{4}\left({}^{xx,xyyx}{}_{\overline{\omega}} + {}^{yy,yxxy}{}_{\overline{\omega}}\right) + \frac{1}{4}\left({}^{xx,yxxy}{}_{\overline{\omega}} + {}^{yy,xyyx}{}_{\overline{\omega}}\right) + \frac{1}{2}\left({}^{xx,xyxy}{}_{\overline{\omega}} + {}^{yy,xyxy}{}_{\overline{\omega}}\right)$$

$$+\frac{1}{2}\left({}^{xx,xzzx}{}_{\overline{\omega}} + {}^{yy,yzzy}{}_{\overline{\omega}}\right) - \frac{1}{2}\left({}^{xx,yzzy}{}_{\overline{\omega}} + {}^{yy,xzzx}{}_{\overline{\omega}}\right) + \left({}^{xx,xxzz}{}_{\overline{\omega}} + {}^{yy,yyzz}{}_{\overline{\omega}}\right)$$

$$-\left({}^{xx,yyzz}{}_{\overline{\omega}} + {}^{yy,xxzz}{}_{\overline{\omega}}\right) + \frac{1}{2}\left({}^{xx,zxxz}{}_{\overline{\omega}} + {}^{yy,zyyz}{}_{\overline{\omega}}\right) - \frac{1}{2}\left({}^{xx,zyyz}{}_{\overline{\omega}} + {}^{yy,zxxz}{}_{\overline{\omega}}\right)$$

$$+\left({}^{xx,xzxz}{}_{\overline{\omega}} + {}^{yy,yzyz}{}_{\overline{\omega}}\right) - \left({}^{xx,yzyz}{}_{\overline{\omega}} + {}^{yy,xzxz}{}_{\overline{\omega}}\right) + \frac{1}{2}\left({}^{zz,xxxx}{}_{\overline{\omega}} + {}^{zz,yyyy}{}_{\overline{\omega}}\right) - {}^{zz,xxyy}{}_{\overline{\omega}} \;\cdots$$

TABLE XLVI-1 . Continued

$$\left(f_4^{K\pm}\right) \quad \cdots - \tfrac{1}{2}\left({}^{zz,xyyz}{}_{\overline{\omega}}\right) - {}^{zz,xyxy}{}_{\overline{\omega}} + 2\left({}^{xz,xxxz}{}_{\overline{\omega}} + {}^{yz,yyyz}{}_{\overline{\omega}}\right) + 2\left({}^{xz,xxzx}{}_{\overline{\omega}} + {}^{yz,yyzy}{}_{\overline{\omega}}\right)$$

$$-2\left({}^{xz,yxzy}{}_{\overline{\omega}} + {}^{yz,xyzx}{}_{\overline{\omega}}\right) - 2\left({}^{xz,xzyy}{}_{\overline{\omega}} + {}^{yz,yzxx}{}_{\overline{\omega}}\right) - 2\left({}^{xz,zxyy}{}_{\overline{\omega}} + {}^{yz,zyxx}{}_{\overline{\omega}}\right)$$

$$-2\left({}^{xz,xyyz}{}_{\overline{\omega}} + {}^{yz,yxxz}{}_{\overline{\omega}}\right) - 2\left({}^{xz,xyzy}{}_{\overline{\omega}} + {}^{yz,yxzx}{}_{\overline{\omega}}\right) - 2\left({}^{xz,zyxy}{}_{\overline{\omega}} + {}^{yz,zxyx}{}_{\overline{\omega}}\right)$$

$$+\left({}^{xy,xyyy}{}_{\overline{\omega}} + {}^{xy,yxxx}{}_{\overline{\omega}}\right) + \left({}^{xy,xxyx}{}_{\overline{\omega}} + {}^{xy,yyxy}{}_{\overline{\omega}}\right) - 2\ {}^{xy,zxyz}{}_{\overline{\omega}} - 2\left({}^{xy,yxzz}{}_{\overline{\omega}} + {}^{xy,xyzz}{}_{\overline{\omega}}\right)$$

$$-2\,{}^{xy,xzzy}{}_{\overline{\omega}} - 2\,{}^{xy,zyzx}{}_{\overline{\omega}} - 2\,{}^{xy,yzxz}{}_{\overline{\omega}} - \tfrac{3}{8}\left({}^{xxx,xxx}{}_{\gamma} + {}^{yyy,yyy}{}_{\gamma}\right) + \tfrac{1}{2}\left(\{xxx,xyy\}_{\gamma} + \{yyy,xxy\}_{\gamma}\right)$$

$$+\tfrac{1}{4}\left({}^{xxx,yxy}{}_{\gamma} + {}^{yyy,xyx}{}_{\gamma}\right) + \tfrac{1}{2}\left(\{xyy,yxy\}_{\gamma} + \{xxy,xyx\}_{\gamma}\right) + \tfrac{1}{2}\left({}^{xyy,xyy}{}_{\gamma} + {}^{xxy,xxy}{}_{\gamma}\right)$$

$$+\tfrac{1}{8}\left({}^{yxy,yxy}{}_{\gamma} + {}^{xyx,xyx}{}_{\gamma}\right) + \left(\{xxx,xzz\}_{\gamma} + \{yyy,yzz\}_{\gamma}\right) + \tfrac{1}{2}\left(\{xxx,xzx\}_{\gamma} + \{yyy,zyz\}_{\gamma}\right)$$

$$-2\left(\{xyy,xzz\}_{\gamma} + \{xxy,yzz\}_{\gamma}\right) - \left(\{xyy,zxz\}_{\gamma} + \{xxy,zyz\}_{\gamma}\right) - \left(\{yxy,xzz\}_{\gamma} + \{xyx,yzz\}_{\gamma}\right)$$

$$-\tfrac{1}{2}\left(\{yxy,zxz\}_{\gamma} + \{xyx,zyz\}_{\gamma}\right) + \left({}^{xxz,xxz}{}_{\gamma} + {}^{yyz,yyz}{}_{\gamma}\right) - 2\,\{xxz,yyz\}_{\gamma} + \left(\{xxz,xzx\}_{\gamma} + \{yyz,yzy\}_{\gamma}\right)$$

$$-\left(\{xxz,yzy\}_{\gamma} + \{yyz,xzx\}_{\gamma}\right) + \tfrac{1}{4}\left({}^{xzx,xzx}{}_{\gamma} + {}^{yzy,yzy}{}_{\gamma}\right) - \tfrac{1}{2}\,\{xzx,yzy\}_{\gamma}$$

$$-\left({}^{xyz,xyz}{}_{\gamma} + {}^{xzy,xzy}{}_{\gamma} + {}^{yxz,yxz}{}_{\gamma} + 2\,\{xyz,yxz\}_{\gamma} + 2\,\{xyz,xzy\}_{\gamma} + 2\,\{yxz,xzy\}_{\gamma}\right)\Big\} \cdots$$

TABLE XLVI-1 . Continued

$$\left(\mathfrak{f}_4^{K\pm}\right) \;\ldots\; \mp i\Big\{-\left(\{xxx,xxy\}_\gamma - \{yyy,xyy\}_\gamma\right) - \tfrac{1}{2}\left(\{xxx,xyx\}_\gamma - \{yyy,yxy\}_\gamma\right) + \left(\{xxx,yzz\}_\gamma - \{yyy,xzz\}_\gamma\right)$$

$$+\tfrac{1}{2}\left(\{xxx,zyz\}_\gamma - \{yyy,zxz\}_\gamma\right) + \tfrac{1}{2}\{xyy,xxy\}_\gamma - 2\left(\{xyy,yzz\}_\gamma - \{xxy,xzz\}_\gamma\right)$$

$$+2\left(\{xxz,xyz\}_\gamma + \{xxz,yxz\}_\gamma + \{xxz,xzy\}_\gamma - \{yyz,xyz\}_\gamma - \{yyz,yxz\}_\gamma - \{yyz,xzy\}_\gamma\right)$$

$$+\left(\{xzx,xyz\}_\gamma + \{xzx,yxz\}_\gamma + \{xzx,xzy\}_\gamma - \{yzy,xyz\}_\gamma - \{yzy,yxz\}_\gamma - \{yzy,xzy\}_\gamma\right)$$

$$-\left(xx,yxxx_{\bar{\omega}} - yy,xyyy_{\bar{\omega}}\right) - \left(xx,xxyx_{\bar{\omega}} - yy,yyxy_{\bar{\omega}}\right) + \left(xx,zxyz_{\bar{\omega}} - yy,zxyz_{\bar{\omega}}\right) + \left(xx,yxzz_{\bar{\omega}} - yy,yxzz_{\bar{\omega}}\right)$$

$$+\left(xx,xyzz_{\bar{\omega}} - yy,xyzz_{\bar{\omega}}\right) + \left(xx,xzzy_{\bar{\omega}} - yy,xzzy_{\bar{\omega}}\right) + \left(xx,zyxz_{\bar{\omega}} - yy,zyzx_{\bar{\omega}}\right) + \left(xx,yzxz_{\bar{\omega}} - yy,yzxz_{\bar{\omega}}\right)$$

$$-\left(zz,xyyy_{\bar{\omega}} - zz,yxxx_{\bar{\omega}}\right) + \left(zz,xxyx_{\bar{\omega}} - zz,yyxy_{\bar{\omega}}\right) - 2\left(xz,yyyz_{\bar{\omega}} - yz,xxxz_{\bar{\omega}}\right) - 2\left(xz,yyzy_{\bar{\omega}} - yz,xxzx_{\bar{\omega}}\right)$$

$$+2\left(xz,xyzx_{\bar{\omega}} - yz,yxzy_{\bar{\omega}}\right) + 2\left(xz,yzxx_{\bar{\omega}} - yz,xzyy_{\bar{\omega}}\right) + 2\left(xz,zyxx_{\bar{\omega}} - yz,zxyy_{\bar{\omega}}\right)$$

$$+2\left(xz,yxxz_{\bar{\omega}} - yz,xyyz_{\bar{\omega}}\right) + 2\left(xz,yxzx_{\bar{\omega}} - yz,xyzy_{\bar{\omega}}\right) + 2\left(xz,zxyx_{\bar{\omega}} - yz,zyxy_{\bar{\omega}}\right)$$

$$-\left(xy,xxxx_{\bar{\omega}} - xy,yyyy_{\bar{\omega}}\right) + \left(xy,xzzx_{\bar{\omega}} - xy,yzzy_{\bar{\omega}}\right) + 2\left(xy,xxzz_{\bar{\omega}} - xy,yyzz_{\bar{\omega}}\right)$$

$$+\left(xy,zxxz_{\bar{\omega}} - xy,zyyz_{\bar{\omega}}\right) + 2\left(xy,xzxz_{\bar{\omega}} - xy,yzyz_{\bar{\omega}}\right)\Big\}\Big]$$

TABLE XLVI-1 . Continued

$$\mathfrak{f}_4^{\pm} = \frac{\hbar^6}{2hc}\left\{\left(a_{44}^{(6)} + b_{44}^{(6)}\right) \mp i\left(a_{45}^{(6)} + b_{45}^{(6)}\right)\right\}$$

$$\mathfrak{f}_3^{J\pm} = \frac{\hbar^6}{2hc}\left\{\left(a_{36}^{(6)} + b_{36}^{(6)}\right) \mp i\left(a_{37}^{(6)} + b_{37}^{(6)}\right)\right\}$$

$$\mathfrak{f}_3^{K\pm} = \frac{\hbar^6}{2hc}\left\{\left(a_{38}^{(6)} + b_{38}^{(6)}\right) \mp i\left(a_{39}^{(6)} + b_{39}^{(6)}\right)\right\}$$

$$\mathfrak{f}_3^{\pm} = \frac{\hbar^6}{2hc}\left\{\left(a_{34}^{(6)} + b_{34}^{(6)}\right) \mp i\left(a_{35}^{(6)} + b_{35}^{(6)}\right)\right\}$$

TABLE XLVI-2 . The coefficients occurring in the off-diagonal matrix elements of $h_{42}^{\dagger}$

$$\mathfrak{f}_4^{n\pm} = \frac{\hbar^4}{32hc}\Big[\Big\{\big({}^{xxxx}v(nn) + {}^{yyyy}v(nn)\big) - 2\,{}^{xxyy}v(nn) - \big({}^{xyyx}v(nn) + {}^{yxxy}v(nn)\big) - 2\,{}^{xyxy}v(nn)\Big\}$$
$$\mp i\Big\{\big({}^{yxxx}v(nn) - {}^{xyyy}v(nn)\big) + \big({}^{xxyx}v(nn) - {}^{yyxy}v(nn)\big)\Big\}\Big]$$

$$\mathfrak{f}_3^{n\pm} = \frac{\hbar^4}{16hc}\Big[\Big\{{}^{xxxz}v(nn) + {}^{xxzx}v(nn) - {}^{yxzy}v(nn) - {}^{xzyy}v(nn) - {}^{zxyy}v(nn)$$
$$- {}^{xyyz}v(nn) - {}^{xyzy}v(nn) - {}^{zyxy}v(nn)\Big\}$$
$$\mp i\Big\{-{}^{yyyz}v(nn) - {}^{yyzy}v(nn) + {}^{xyzx}v(nn) + {}^{yzxx}v(nn) + {}^{zyxx}v(nn)$$
$$+ {}^{yxxz}v(nn) + {}^{yxzx}v(nn) + {}^{zxyx}v(nn)\Big\}\Big]$$

$$\mathfrak{f}_4^{t\pm} = \frac{\hbar^4}{16hc}\Big[\Big\{\big({}^{xxxx}u^{+} + {}^{yyyy}u^{+}\big) - 2\,{}^{xxyy}u^{+} - \big({}^{xyyx}u^{+} + {}^{yxxy}u^{+}\big) - 2\,{}^{xyxy}u^{+}\Big\}$$
$$\mp i\Big\{\big({}^{yxxx}u^{+} - {}^{xyyy}u^{+}\big) + \big({}^{xyyx}u^{+} - {}^{yyxy}u^{+}\big)\Big\}\Big]$$

$$\mathfrak{f}_3^{t\pm} = \frac{\hbar^4}{8hc}\Big[\Big\{{}^{xxxz}u^{+} + {}^{xxzx}u^{+} - {}^{yxzy}u^{+} - {}^{xzyy}u^{+} - {}^{zxyy}u^{+} - {}^{xyyz}u^{+} - {}^{xyzy}u^{+} - {}^{zyxy}u^{+}\Big\}$$
$$\mp i\Big\{-{}^{yyyz}u^{+} - {}^{yyzy}u^{+} + {}^{xyzx}u^{+} + {}^{yzxx}u^{+} + {}^{zyxx}u^{+} + {}^{yxxz}u^{+} + {}^{yxzx}u^{+} + {}^{zxyx}u^{+}\Big\}\Big]$$

TABLE XVLI-2 . Continued

$$
\begin{aligned}
f_{12}^{t,J\pm} = \frac{\hbar^4}{16hc}\Bigg[\Big\{ & 3({}^{xxxz}u^- + {}^{xxzx}u^-) + {}^{yxzy}u^- + {}^{xzyy}u^- + {}^{zxyy}u^- + {}^{xyyz}u^- + {}^{xyzy}u^- + {}^{zyxy}u^- \\
& -3({}^{yyyz}u + {}^{yyzy}u) - {}^{xyzx}u - {}^{yzxx}u - {}^{zyxx}u - {}^{yxxz}u - {}^{yxzx}u - {}^{zxyx}u \Big\} \\
\mp i\Big\{ & 3({}^{xxxz}u + {}^{xxzx}u) + {}^{yxzy}u + {}^{xzyy}u + {}^{zxyy}u + {}^{xyyz}u + {}^{xyzy}u + {}^{zyxy}u + 3({}^{yyyz}u^- + {}^{yyzy}u^-) \\
& + {}^{xyzx}u^- + {}^{yzxx}u^- + {}^{zyxx}u^- + {}^{yxxz}u^- + {}^{yxzx}u^- + {}^{zxyx}u^- \Big\}\Bigg]
\end{aligned}
$$

$$
\begin{aligned}
f_{12}^{t,K\pm} = \frac{\hbar^4}{64hc}\Bigg[\Big\{ & 4\,{}^{xzzz}u^- - 3\,{}^{xxxz}u^- - 3\,{}^{xxzx}u^- + 4\,{}^{zzxz}u^- - {}^{yxzy}u^- - {}^{xzyy}u^- - {}^{zxyy}u^- - {}^{xyyz}u^- - {}^{xyzy}u^- - {}^{zyxy}u^- \\
& -4\,{}^{yzzz}u + 3\,{}^{yyyz}u + 3\,{}^{yyzy}u - 4\,{}^{zzyz}u + {}^{xyzx}u + {}^{yzxx}u + {}^{zyxx}u + {}^{yxxz}u + {}^{yxzx}u + {}^{zxyx}u \Big\} \\
\mp i\Big\{ & 4\,{}^{xzzz}u - 3\,{}^{xxxz}u - 3\,{}^{xxzx}u + 4\,{}^{zzxz}u - {}^{yxzy}u - {}^{xzyy}u - {}^{zxyy}u - {}^{xyyz}u - {}^{xyzy}u - {}^{zyxy}u \\
& +4\,{}^{yzzz}u^- - 3\,{}^{yyyz}u^- - 3\,{}^{yyzy}u^- + 4\,{}^{zzyz}u^- - {}^{xyzx}u^- - {}^{yzxx}u^- - {}^{zyxx}u^- - {}^{yxxz}u^- - {}^{yxzx}u^- - {}^{zxyx}u^- \Big\}\Bigg]
\end{aligned}
$$

TABLE XVLI-2 . Continued

$$f^{t\pm}_{12} = \frac{\hbar^4}{64hc}\Big[\Big\{12\,{}^{xzzz}u^- - 5\,{}^{xxxz}u^- - 21\,{}^{xxzx}u^- - 4\,{}^{zzxz}u^- - 23\,{}^{yxzy}u^- + 9\,{}^{xzyy}u^- + 9\,{}^{zxyy}u^-$$

$$-7\,{}^{xyyz}u^- - 7\,{}^{xyzy}u^- - 7\,{}^{xyxy}u^- - 12\,{}^{yzzz}u + 5\,{}^{yyyz}u + 21\,{}^{yyzy} + 4\,{}^{zzyz}u + 23\,{}^{xyzx}u$$

$$-9\,{}^{yzxx}u - 9\,{}^{zyxx}u + 7\,{}^{yxxz}u + 7\,{}^{yxzx}u + 7\,{}^{zxyx}u\Big\}$$

$$\mp i\Big\{12\,{}^{xzzz}u - 5\,{}^{xxxz}u - 21\,{}^{xxzx}u - 4\,{}^{zzxz}u - 23\,{}^{yxzy}u + 9\,{}^{xzyy}u + 9\,{}^{zxyy}u - 7\,{}^{xyyz}u$$

$$-7\,{}^{xyzy}u - 7\,{}^{zyxy}u + 12\,{}^{yzzz}u^- - 5\,{}^{yyyz}u^- - 21\,{}^{yyzy}u^- - 4\,{}^{zzyz}u^- - 23\,{}^{xyzx}u^-$$

$$+9\,{}^{yzxx}u^- + 9\,{}^{zyxx}u^- - 7\,{}^{yxxz}u^- - 7\,{}^{yxzx}u^- - 7\,{}^{zxyx}u^-\Big\}\Big]$$

$$f^{t,J\pm}_{22^+} = \frac{\hbar^4}{32hc}\Big[\Big\{-{}^{xxxx}u^- + {}^{yyyy}u^- + {}^{xzzx}u^- - {}^{yzzy}u^- + 2\,{}^{xxzz}u^- - 2\,{}^{yyzz}u^- + {}^{zxxz}u^- - {}^{zyyz}u^- + 2\,{}^{xzxz}u^- - 2\,{}^{yzyz}u^-$$

$$-{}^{xyyy}u - {}^{yxxx}u - {}^{xxyx}u - {}^{yyxy}u + 2\,{}^{zxyz}u + 2\,{}^{yxzz}u + 2\,{}^{xyzz}u + 2\,{}^{xzzy}u + 2\,{}^{zyzx}u + 2\,{}^{yzxz}u\Big\}$$

$$\pm i\Big\{-{}^{xxxx}u + {}^{yyyy}u + {}^{xzzx}u - {}^{yzzy}u + 2\,{}^{xxzz}u - 2\,{}^{yyzz}u + {}^{zxxz}u - {}^{zyyz}u + 2\,{}^{xzxz}u$$

$$-2\,{}^{yzyz}u + {}^{xyyy}u^- + {}^{yxxx}u^- + {}^{xxyx}u^- + {}^{yyxy}u^- - 2\,{}^{zxyz}u^- - 2\,{}^{yxzz}u^-$$

$$-2\,{}^{xyzz}u^- - 2\,{}^{xzzy}u^- - 2\,{}^{zyzx}u^- - 2\,{}^{yzxz}u^-\Big\}\Big]$$

TABLE XLVI-2 . Continued

$$f^{t,K\pm}_{22+} = \frac{\hbar^4}{8hc}\Big[\Big\{-{}^{xxxx}u^{-} + {}^{yyyy}u^{-} + {}^{xyyx}u^{-} - {}^{yxxy}u^{-} + 2\,{}^{xxzz}u^{-} - 2\,{}^{yyzz}u^{-} - {}^{zxxz}u^{-} + {}^{zyyz}u^{-} - 2\,{}^{xxyx}u$$

$$- 2\,{}^{yyxy}u - 2\,{}^{zxyz}u + 2\,{}^{yxzz}u + 2\,{}^{xyzz}u\Big\}$$

$$\pm i\Big\{-{}^{xxxx}u + {}^{yyyy}u + {}^{xyyx}u - {}^{yxxy}u + 2\,{}^{xxzz}u - 2\,{}^{yyzz}u - {}^{zxxz}u + {}^{zyyz}u$$

$$+ 2\,{}^{xxyx}u^{-} + 2\,{}^{yyxy}u^{-} + 2\,{}^{zxyz}u^{-} - 2\,{}^{yxzz}u^{-} - 2\,{}^{xyzz}u^{-}\Big\}\Big]$$

$$f^{t\pm}_{22+} = \frac{\hbar^4}{8hc}\Big[\Big\{{}^{xxxx}u^{-} - {}^{yyyy}u^{-} + {}^{xyyy}u + {}^{yxxx}u + {}^{xxyx}u + {}^{yyxy}u\Big\}$$

$$\pm i\Big\{{}^{xxxx}u - {}^{yyyy}u - {}^{xyyy}u^{-} - {}^{yxxx}u^{-} - {}^{xxyx}u^{-} - {}^{yyxy}u^{-}\Big\}\Big]$$

$$f^{t,J\pm}_{22-} = \frac{\hbar^4}{32hc}\Big[\Big\{-{}^{xxxx}u^{-} + {}^{yyyy}u^{-} + {}^{xzzx}u^{-} - {}^{yzzy}u^{-} + 2\,{}^{xxzz}u^{-} - 2\,{}^{yyzz}u^{-} + {}^{zxxz}u^{-} - {}^{zyyz}u^{-} + 2\,{}^{xzxz}u^{-} - 2\,{}^{yzyz}u^{-}$$

$$+ {}^{xyyy}u + {}^{yxxx}u + {}^{xxyx}u + {}^{yyxy}u - 2\,{}^{zxyz}u - 2\,{}^{yxzz}u - 2\,{}^{xyzz}u - 2\,{}^{xzzy}u - 2\,{}^{zyzx}u - 2\,{}^{yzxz}u\Big\}$$

$$\mp i\Big\{-{}^{xxxx}u + {}^{yyyy}u + {}^{xzzx}u - {}^{yzzy}u + 2\,{}^{xxzz}u - 2\,{}^{yyzz}u + {}^{zxxz}u - {}^{zyyz}u + 2\,{}^{xzxz}u - 2\,{}^{yzyz}u$$

$$- {}^{xyyy}u^{-} - {}^{yxxx}u^{-} - {}^{xxyx}u^{-} - {}^{yyxy}u^{-} + 2\,{}^{zxyz}u^{-} + 2\,{}^{yxzz}u^{-}$$

$$+ 2\,{}^{xyzz}u^{-} + 2\,{}^{xzzy}u^{-} + 2\,{}^{zyzx}u^{-} + 2\,{}^{yzxz}u^{-}\Big\}\Big]$$

TABLE XLVI-2 . Continued

$$f^{t,\kappa\pm}_{22-} = \frac{\hbar^4}{8hc}\Big[\Big\{-{}^{xxxx}u^{-} + {}^{yyyy}u^{-} + {}^{xyyx}u^{-} - {}^{yxxy}u^{-} + 2\,{}^{xxzz}u^{-} - 2\,{}^{yyzz}u^{-} - {}^{zxxz}u^{-} + {}^{zyyz}u^{-} + 2\,{}^{xxyx}u$$

$$+ 2\,{}^{yyxy}u + 2\,{}^{zxyz}u - 2\,{}^{yxzz}u - 2\,{}^{xyzz}u\Big\}$$

$$\mp i\Big\{-{}^{xxxx}u + {}^{yyyy}u + {}^{xyyx}u - {}^{yxxy}u + 2\,{}^{xxzz}u - 2\,{}^{yyzz}u - {}^{zxxz}u + {}^{zyyz}u - 2\,{}^{xxyx}u^{-}$$

$$- 2\,{}^{yyxy}u^{-} - 2\,{}^{zxyz}u^{-} + 2\,{}^{yxzz}u^{-} + 2\,{}^{xyzz}u^{-}\Big\}\Big]$$

$$f^{t\pm}_{22-} = \frac{\hbar^4}{8hc}\Big[\Big\{{}^{xxxx}u^{-} - {}^{yyyy}u^{-} - {}^{xyyy}u - {}^{yxxx}u - {}^{xxyx}u - {}^{yyxy}u\Big\}$$

$$\mp i\Big\{{}^{xxxx}u - {}^{yyyy}u + {}^{xyyy}u^{-} + {}^{yxxx}u^{-} + {}^{xxyx}u^{-} + {}^{yyxy}u^{-}\Big\}\Big]$$

$$f^{t\pm}_{42} = \frac{\hbar^4}{32hc}\Big[\Big\{{}^{xxxx}u^{-} + {}^{yyyy}u^{-} - 2\,{}^{xxyy}u^{-} - {}^{xyyx}u^{-} - {}^{yxxy}u^{-} - 2\,{}^{xyxy}u^{-} + 2\,{}^{xyyy}u - 2\,{}^{yxxx}u - 2\,{}^{xxyx}u + 2\,{}^{yyxy}u\Big\}$$

$$\mp i\Big\{{}^{xxxx}u + {}^{yyyy}u - 2\,{}^{xxyy}u - {}^{xyyx}u - {}^{yxxy}u - 2\,{}^{xyxy}u - 2\,{}^{xyyy}u^{-} + 2\,{}^{yxxx}u^{-} + 2\,{}^{xxyx}u^{-} - 2\,{}^{yyxy}u^{-}\Big\}\Big]$$

TABLE XLVI-3 . The coefficients occurring in the off-diagonal matrix elements of $h^{\dagger}_{43}$

$$f^{t,t\pm}_{12} = \frac{\hbar^2}{2hc}\left[\left\{{}^{xz}w(aa,bb) - {}^{yz}w(ab)\right\} \mp i\left\{{}^{xz}w(ab) + {}^{yz}w(aa,bb)\right\}\right]$$

$$f^{t,t\pm}_{22+} = \frac{\hbar^2}{2hc}\left[\left\{\frac{1}{2}\left({}^{xx}w(aa,bb) - {}^{yy}w(aa,bb)\right) + {}^{xy}w(ab)\right\} \pm i\left\{\frac{1}{2}\left({}^{xx}w(ab) - {}^{yy}w(ab)\right) - {}^{xy}w(aa,bb)\right\}\right]$$

$$f^{t,t\pm}_{22-} = \frac{\hbar^2}{2h}\left[\left\{\frac{1}{2}\left({}^{xx}w(aa,bb) - {}^{yy}w(aa,bb)\right) - {}^{xy}w(ab)\right\} \mp i\left\{\frac{1}{2}\left({}^{xx}w(ab) - {}^{yy}w(ab)\right) + {}^{xy}w(aa,bb)\right\}\right]$$

$$f^{t,t'\pm}_{12} = \frac{\hbar^2}{4hc}\left[\left({}^{xz}w(abcd) - {}^{yz}v(abcd)\right) \mp i\left({}^{xz}v(abcd) + {}^{yz}w(abcd)\right)\right]$$

$$f^{t,t'\pm}_{22+} = \frac{\hbar^2}{4hc}\left[\left\{\frac{1}{2}\left({}^{xx}w(abcd) - {}^{yy}w(abcd)\right) + {}^{xy}v(abcd)\right\} \pm i\left\{\frac{1}{2}\left({}^{xx}v(abcd) - {}^{yy}v(abcd)\right) - {}^{xy}w(abcd)\right\}\right]$$

$$f^{t,t'\pm}_{22-} = \frac{\hbar^2}{4hc}\left[\left\{\frac{1}{2}\left({}^{xx}w(abcd) - {}^{yy}w(abcd)\right) - {}^{xy}v(abcd)\right\} \mp i\left\{\frac{1}{2}\left({}^{xx}v(abcd) - {}^{yy}v(abcd)\right) + {}^{xy}w(abcd)\right\}\right]$$

$$f^{t,n\pm}_{12} = \frac{\hbar^2}{8hc}\left[\left\{{}^{xz}w(ab,nn) - {}^{yz}v(ab,nn)\right\} \mp i\left\{{}^{xz}v(ab,nn) + {}^{yz}w(ab,nn)\right\}\right]$$

$$f^{t,n\pm}_{22+} = \frac{\hbar^2}{8hc}\left[\left\{\frac{1}{2}\left({}^{xx}w(ab,nn) - {}^{yy}w(ab,nn)\right) + {}^{xy}v(ab,nn)\right\} \pm i\left\{\frac{1}{2}\left({}^{xx}v(ab,nn) - {}^{yy}v(ab,nn)\right) - {}^{xy}w(ab,nn)\right\}\right]$$

$$f^{t,n\pm}_{22-} = \frac{\hbar^2}{8hc}\left[\left\{\frac{1}{2}\left({}^{xx}w(ab,nn) - {}^{yy}w(ab,nn)\right) - {}^{xy}v(ab,nn)\right\} \mp i\left\{\frac{1}{2}\left({}^{xx}v(ab,nn) - {}^{yy}v(ab,nn)\right) + {}^{xy}w(ab,nn)\right\}\right]$$

TABLE XLVI-3 . Continued

$$f_{14}^{t\pm} = \frac{\hbar^2}{4hc}\left[\left\{{}^{xz}r(aabb) + {}^{yz}r(ab)\right\} \pm i\left\{{}^{xz}r(ab) - {}^{yz}r(aabb)\right\}\right]$$

$$f_{24}^{t\pm} = \frac{\hbar^2}{4hc}\left[\left\{\frac{1}{2}\left({}^{xx}r(aabb) - {}^{yy}r(aabb)\right) - {}^{xy}r(ab)\right\} \pm i\left\{\frac{1}{2}\left({}^{xx}r(ab) - {}^{yy}r(ab)\right) + {}^{xy}r(aabb)\right\}\right]$$

$$f_{122}^{tt'\pm} = \frac{\hbar^2}{8hc}\left[\left\{{}^{xz}r_{3,3}(abcd) - {}^{xz}r_{4,4}(abcd) + {}^{yz}r_{4,3}(abcd) + {}^{yz}r_{3,4}(abcd)\right\}\right.$$
$$\left.\mp i\left\{{}^{yz}r_{3,3}(abcd) - {}^{yz}r_{4,4}(abcd) - {}^{xz}r_{4,3}(abcd) - {}^{xz}r_{3,4}(abcd)\right\}\right]$$

$$f_{222}^{tt'\pm} = \frac{\hbar^2}{8hc}\left[\left\{\frac{1}{2}\left({}^{xx}r_{3,3}(abcd) - {}^{yy}r_{3,3}(abcd)\right) - \frac{1}{2}\left({}^{xx}r_{4,4}(abcd) - {}^{yy}r_{4,4}(abcd)\right) - {}^{xy}r_{4,3}(abcd) - {}^{xy}r_{3,4}(abcd)\right\}\right.$$
$$\left.\mp i\left\{{}^{xy}r_{3,3}(abcd) - {}^{xy}r_{4,4}(abcd) + \frac{1}{2}\left({}^{xx}r_{4,3}(abcd) - {}^{yy}r_{4,3}(abcd)\right) + \frac{1}{2}\left({}^{xx}r_{3,4}(abcd) - {}^{yy}r_{3,4}(abcd)\right)\right\}\right]$$

$$g_{22+}^{tt',J\pm} = \frac{\hbar^4}{8hc}\left[\left\{{}^{xx}r_{3,3}(abcd) + {}^{yy}r_{3,3}(abcd) - {}^{xx}r_{4,4}(abcd) - {}^{yy}r_{4,4}(abcd)\right\}\right.$$
$$\left.\pm i\left\{{}^{xx}r_{3,4}(abcd) + {}^{yy}r_{3,4}(abcd) + {}^{xx}r_{4,3}(abcd) + {}^{yy}r_{4,3}(abcd)\right\}\right]$$

$$g_{22+}^{tt',K\pm} = \frac{\hbar^4}{8hc}\left[\left\{2\,{}^{zz}r_{3,3}(abcd) - {}^{xx}r_{3,3}(abcd) - {}^{yy}r_{3,3}(abcd) - 2\,{}^{zz}r_{4,4}(abcd) + {}^{xx}r_{4,4}(abcd) + {}^{yy}r_{4,4}(abcd)\right\}\right.$$
$$\left.\pm i\left\{2\,{}^{zz}r_{3,4}(abcd) - {}^{xx}r_{3,4}(abcd) - {}^{yy}r_{3,4}(abcd) + 2\,{}^{zz}r_{4,3}(abcd) - {}^{xx}r_{4,3}(abcd) - {}^{yy}r_{4,3}(abcd)\right\}\right]$$

TABLE XLVI-3 . Continued

$$g_{22-}^{tt',J\pm} = \frac{\hbar^4}{8hc}\Big[\Big\{{}^{xx}r_{3,3}(abcd) + {}^{yy}r_{3,3}(abcd) + {}^{xx}r_{4,4}(abcd) + {}^{yy}r_{4,4}(abcd)\Big\}$$

$$\mp i\Big\{{}^{xx}r_{3,4}(abcd) + {}^{yy}r_{3,4}(abcd) - {}^{xx}r_{4,3}(abcd) - {}^{yy}r_{4,3}(abcd)\Big\}\Big]$$

$$g_{22-}^{tt',K\pm} = \frac{\hbar^4}{8hc}\Big[\Big\{2\,{}^{zz}r_{3,3}(abcd) - {}^{xx}r_{3,3}(abcd) - {}^{yy}r_{3,3}(abcd) + 2\,{}^{zz}r_{4,4}(abcd) - {}^{xx}r_{4,4}(abcd) - {}^{yy}r_{4,4}(abcd)\Big\}$$

$$\mp i\Big\{2\,{}^{zz}r_{3,4}(abcd) - {}^{xx}r_{3,4}(abcd) - {}^{yy}r_{3,4}(abcd) - 2\,{}^{zz}r_{4,3}(abcd) + {}^{xx}r_{4,3}(abcd) + {}^{yy}r_{4,3}(abcd)\Big\}\Big]$$

$$g_{4}^{t,J\pm} = \frac{1}{64hc}\Big[3\hbar^6\,{}^{xx}_{(4)}Z^{aaaa} + 3\hbar^6\,{}^{yy}_{(4)}Z^{aaaa} + 3\hbar^2\,{}^{xx}_{(4)}Z_{aaaa} + 3\hbar^2\,{}^{yy}_{(4)}Z_{aaaa} + 3\hbar^6\,{}^{xx}_{(4)}Z^{bbbb}$$

$$+ 3\hbar^6\,{}^{yy}_{(4)}Z^{bbbb} + 3\hbar^2\,{}^{xx}_{(4)}Z_{bbbb} + 3\hbar^2\,{}^{yy}_{(4)}Z_{bbbb} - 3\hbar^6\,{}^{xx}_{(4)}Z^{aabb} - 3\hbar^6\,{}^{yy}_{(4)}Z^{aabb}$$

$$- 3\hbar^2\,{}^{xx}_{(4)}Z_{aabb} - 3\hbar^2\,{}^{yy}_{(4)}Z_{aabb} + \hbar^4\,{}^{xx}_{(4)}Z^{aa}_{aa} + \hbar^4\,{}^{yy}_{(4)}Z^{aa}_{aa} + \hbar^4\,{}^{xx}_{(4)}Z^{bb}_{bb} + \hbar^4\,{}^{yy}_{(4)}Z^{bb}_{bb}$$

$$- \hbar^4\,{}^{xx}_{(4)}Z^{aa}_{bb} - \hbar^4\,{}^{yy}_{(4)}Z^{aa}_{bb} - \hbar^4\,{}^{xx}_{(4)}Z^{bb}_{aa} - \hbar^4\,{}^{yy}_{(4)}Z^{bb}_{aa} - \hbar^4\,{}^{xx}_{(4)}Z^{ab}_{ab} - \hbar^4\,{}^{yy}_{(4)}Z^{ab}_{ab}$$

$$\pm i\Big\{3\hbar^6\,{}^{xx}_{(4)}Z^{aaab} + 3\hbar^6\,{}^{yy}_{(4)}Z^{aaab} - 3\hbar^6\,{}^{xx}_{(4)}Z^{abbb} - 3\hbar^6\,{}^{yy}_{(4)}Z^{abbb} + \hbar^4\,{}^{xx}_{(4)}Z^{ab}_{aa}$$

$$+ \hbar^4\,{}^{yy}_{(4)}Z^{ab}_{aa} - \hbar^4\,{}^{xx}_{(4)}Z^{ab}_{bb} - \hbar^4\,{}^{yy}_{(4)}Z^{ab}_{bb} + \hbar^4\,{}^{xx}_{(4)}Z^{aa}_{ab} + \hbar^4\,{}^{yy}_{(4)}Z^{aa}_{ab} - \hbar^4\,{}^{xx}_{(4)}Z^{bb}_{ab} - \hbar^4\,{}^{yy}_{(4)}Z^{bb}_{ab}\Big\}\Big]$$

TABLE XLVI-3 . Continued

$$
\begin{aligned}
g_4^{t,K\pm} = \frac{1}{64hc}\Big[\, & 6\hbar^6\, {}^{zz}_{(4)}Z^{aaaa} - 3\hbar^6\, {}^{xx}_{(4)}Z^{aaaa} - 3\hbar^6\, {}^{yy}_{(4)}Z^{aaaa} + 6\hbar^2\, {}^{zz}_{(4)}Z_{aaaa} - 3\hbar^2\, {}^{xx}_{(4)}Z_{aaaa} \\
& - 3\hbar^2\, {}^{yy}_{(4)}Z_{aaaa} + 6\hbar^6\, {}^{zz}_{(4)}Z^{bbbb} - 3\hbar^6\, {}^{xx}_{(4)}Z^{bbbb} - 3\hbar^6\, {}^{yy}_{(4)}Z^{bbbb} + 6\hbar^2\, {}^{zz}_{(4)}Z_{bbbb} \\
& - 3\hbar^2\, {}^{xx}_{(4)}Z_{bbbb} - 3\hbar^2\, {}^{yy}_{(4)}Z_{bbbb} - 6\hbar^6\, {}^{zz}_{(4)}Z^{aabb} + 3\hbar^6\, {}^{xx}_{(4)}Z^{aabb} + 3\hbar^6\, {}^{yy}_{(4)}Z^{aabb} \\
& - 6\hbar^2\, {}^{zz}_{(4)}Z_{aabb} + 3\hbar^2\, {}^{xx}_{(4)}Z_{aabb} + 3\hbar^2\, {}^{yy}_{(4)}Z_{aabb} + 2\hbar^4\, {}^{zz}_{(4)}Z^{aa}_{aa} - \hbar^4\, {}^{xx}_{(4)}Z^{aa}_{aa} - \hbar^4\, {}^{yy}_{(4)}Z^{aa}_{aa} \\
& + 2\hbar^4\, {}^{zz}_{(4)}Z^{bb}_{bb} - \hbar^4\, {}^{xx}_{(4)}Z^{bb}_{bb} - \hbar^4\, {}^{yy}_{(4)}Z^{bb}_{bb} - 2\hbar^4\, {}^{zz}_{(4)}Z^{aa}_{bb} + \hbar^4\, {}^{xx}_{(4)}Z^{aa}_{bb} + \hbar^4\, {}^{yy}_{(4)}Z^{aa}_{bb} \\
& - 2\hbar^4\, {}^{zz}_{(4)}Z^{bb}_{aa} + \hbar^4\, {}^{xx}_{(4)}Z^{bb}_{aa} + \hbar^4\, {}^{yy}_{(4)}Z^{bb}_{aa} - 2\hbar^4\, {}^{zz}_{(4)}Z^{ab}_{ab} + \hbar^4\, {}^{xx}_{(4)}Z^{ab}_{ab} + \hbar^4\, {}^{yy}_{(4)}Z^{ab}_{ab} \\
& \pm i\Big\{ 6\hbar^6\, {}^{zz}_{(4)}Z^{aaab} - 3\hbar^6\, {}^{xx}_{(4)}Z^{aaab} - 3\hbar^6\, {}^{yy}_{(4)}Z^{aaab} - 6\hbar^6\, {}^{zz}_{(4)}Z^{abbb} + 3\hbar^6\, {}^{xx}_{(4)}Z^{abbb} \\
& + 3\hbar^6\, {}^{yy}_{(4)}Z^{abbb} + 2\hbar^4\, {}^{zz}_{(4)}Z^{ab}_{aa} - \hbar^4\, {}^{xx}_{(4)}Z^{ab}_{aa} - \hbar^4\, {}^{yy}_{(4)}Z^{ab}_{aa} - 2\hbar^4\, {}^{zz}_{(4)}Z^{ab}_{bb} + \hbar^4\, {}^{xx}_{(4)}Z^{ab}_{bb} \\
& + \hbar^4\, {}^{yy}_{(4)}Z^{ab}_{bb} + 2\hbar^4\, {}^{zz}_{(4)}Z^{aa}_{ab} - \hbar^4\, {}^{xx}_{(4)}Z^{aa}_{ab} - \hbar^4\, {}^{yy}_{(4)}Z^{aa}_{ab} - 2\hbar^4\, {}^{zz}_{(4)}Z^{bb}_{ab} + \hbar^4\, {}^{xx}_{(4)}Z^{bb}_{ab} + \hbar^4\, {}^{yy}_{(4)}Z^{bb}_{ab} \Big\}\Big]
\end{aligned}
$$

TABLE XLVI-5 . The coefficients occurring in the off-diagonal matrix elements of $h^{\dagger}_{45}$

$$
\begin{aligned}
g^{tt',t\pm}_{22+} \quad & \frac{1}{64hc}\Big[\Big\{ 6u(bbbbdd) + 6u(aaaacc) + 6v(cc,aaaa) + 6v(dd,bbbb) - 6u(bbbbcc) - 6u(aaaadd) \\
& - 6v(dd,aaaa) - 6v(cc,bbbb) + 2v(dd,bbbb) + 2v(cc,aaaa) - 2v(cc,bbbb) - 2v(dd,aaaa) \\
& - 3u(aaabcd) - 3u(abbbcd) - 3v(cd,aaab) - 3v(cd,abbb) - v(aa,abcd) - v(ab,aacd) \\
& - v(bb,abcd) - v(ab,bbcd)\Big\} \pm i\Big\{6u(aaaacd) - 6u(bbbbcd) + 6v(cd,aaaa) - 6v(cd,bbbb) \\
& + 2v(aa,aacd) - 2v(bb,bbcd) + 3u(aaabcc) + 3u(abbbcc) - 3u(aaabdd) - 3u(abbbdd) \\
& + 3v(cc,aaab) + 3v(cc,abbb) - 3v(dd,aaab) - 3v(dd,abbb) + v(aa,abcc) + v(ab,aacc) \\
& + v(bb,abcc) + v(ab,bbcc) - v(aa,abdd) - v(ab,aadd) - v(bb,abdd) - v(ab,bbdd)\Big\}\Big] \\
g^{tt',t\pm}_{22-} = \; & \frac{1}{64hc}\Big[\Big\{ 6u(bbbbdd) + 6u(aaaacc) + 6v(cc,aaaa) + 6v(dd,bbbb) - 6u(bbbbcc) - 6u(aaaadd) \\
& - 6v(dd,aaaa) - 6v(cc,bbbb) + 2v(dd,bbbb) + 2v(cc,aaaa) - 2v(cc,bbbb) - 2v(dd,aaaa) \\
& + 3u(aaabcd) + 3u(abbbcd) + 3v(cd,aaab) + 3v(cd,abbb) + v(aa,abcd) + v(ab,aacd) \ldots.
\end{aligned}
$$

TABLE XLVI-5 . Continued

$\left(g_{22^-}^{tt',t\pm}\right)$ $\quad \cdots + v(bb,abcd) + v(ab,bbcd)\Big\} \mp i\Big\{-6u(bbbbcd) + 6u(aaaacd) + 6v(cd,aaaa)$

$-6v(cd,bbbb) + 2v(aa,aacd) - 2v(bb,bbcd) - 3u(aaabcc) - 3u(abbbcc) + 3u(aaabdd)$

$+3u(abbbdd) - 3v(cc,aaab) - 3v(cc,abbb) + 3v(dd,aaab) + 3v(dd,abbb) - v(aa,abcc)$

$-v(ab,aacc) - v(bb,abcc) - v(ab,bbcc) + v(aa,abdd) + v(ab,aadd) + v(bb,abdd) + v(ab,bbdd)\Big\}\Big]$

$g_{22^+}^{tt',t'\pm}$ is obtained from $g_{22^+}^{tt',t\pm}$ with the substitutions

$$a \to c\ ,\quad b \to d\ ,\quad c \to a\ ,\quad d \to b$$

$g_{22^-}^{tt',t'\pm}$ is obtained from $g_{22^-}^{tt',t\pm}$ with the substitutions

$$a \to c\ ,\quad b \to d\ ,\quad c \to a\ ,\quad d \to b\ ,\quad \mp i \to \pm i$$

TABLE XLVI-5 . Continued

$$
\begin{aligned}
g_{22+}^{tt',t''\pm} = \frac{1}{32hc}\Big[\Big\{ & u(bbddee)+u(bbddff)+u(aaccee)+u(aaccff)+v(ee,bbdd)+v(ff,bbdd)+v(ee,aacc) \\
& +v(ff,aacc)+v(bb,ddee)+v(dd,bbee)+v(bb,ddff)+v(dd,bbff)+v(aa,ccee)+v(cc,aaee) \\
& +v(aa,ccff)+v(cc,aaff)-u(bbccee)-u(bbccff)-u(aaddee)-u(aaddff)-v(ee,bbcc) \\
& -v(ff,bbcc)-v(ee,aadd)-v(ff,aadd)-v(bb,ccee)-v(cc,bbee)-v(bb,ccff)-v(cc,bbff) \\
& -v(aa,ddee)-v(dd,aaee)-v(aa,ddff)-v(dd,aaff)-v(abcdee)-v(abcdff)-v(ee,abcd) \\
& \qquad -v(ff,abcd)-v(ab,cdee)-v(ab,cdff)-v(cd,abee)-v(cd,abff)\Big\} \\
& \pm i\Big\{u(aacdff)+u(aacdee)+v(ff,aacd)+v(ee,aacd)+v(aa,cdff)+v(aa,cdee) \\
& +v(cd,aaee)+v(cd,aaff)-u(bbcdff)-u(bbcdee)-v(ff,bbcd)-v(ee,bbcd)-v(bb,cdff) \\
& -v(bb,cdee)-v(cd,bbee)-v(cd,bbff)+u(abccff)+u(abccee)+v(ff,abcc)+v(ee,abcc) \\
& +v(cc,abff)+v(cc,abee)+v(ab,ccee)+v(ab,ccff)-u(abddff)-u(abddee)-v(ff,abdd) \\
& \qquad -v(ee,abdd)-v(dd,abff)-v(dd,abee)-v(ab,ddee)-v(ab,ddff)\Big\}\Big]
\end{aligned}
$$

TABLE XLVI-5 . Continued

$$
\begin{aligned}
g_{22^-}^{tt',t''\pm} = \frac{1}{32hc}\Big[\Big\{ & u(bbddee)+u(bbddff)+u(aaccee)+u(aaccff)+v(ee,bbdd)+v(ff,bbdd)+v(ee,aacc) \\
& +v(ff,aacc)+v(bb,ddee)+v(dd,bbee)+v(bb,ddff)+v(dd,bbff)+v(aa,ccee)+v(cc,aaee) \\
& +v(aa,ccff)+v(cc,aaff)-u(bbccee)-u(bbccff)-u(aaddee)-u(aaddff)-v(ee,bbcc) \\
& -v(ff,bbcc)-v(ee,aadd)-v(ff,aadd)-v(bb,ccee)-v(cc,bbee)-v(bb,ccff)-v(cc,bbff) \\
& -v(aa,ddee)-v(dd,aaee)-v(aa,ddff)-v(dd,aaff)+u(abcdee)+u(abcdff)+v(ee,abcd) \\
& +v(ff,abcd)+v(ab,cdee)+v(ab,cdff)+v(cd,abee)+v(cd,abff)\Big\} \\
\mp i\Big\{ & u(aacdff)+u(aacdee)+v(ff,aacd)+v(ee,aacd)+v(aa,cdff)+v(aa,cdee)+v(cd,aaee) \\
& +v(cd,aaff)-u(bbcdff)-u(bbcdee)-v(ff,bbcd)-v(ee,bbcd)-v(bb,cdff)-v(bb,cdee) \\
& -v(cd,bbee)-v(cd,bbff)-u(abccff)-u(abccee)-v(ff,abcc)-v(ee,abcc)-v(cc,abff) \\
& -v(cc,abee)-v(ab,ccee)-v(ab,ccff)+u(abddff)+u(abddee)+v(ff,abdd)+v(ee,abdd) \\
& +v(dd,abff)+v(dd,abee)+v(ab,ddee)+v(ab,ddff)\Big\}\Big]
\end{aligned}
$$

TABLE XLVI-5 . Continued

$$g_{22^+}^{tt',n\pm} = \frac{1}{16hc}\Big[\Big\{u(bbddnn)+u(aaccnn)+v(nn,bbdd)+v(nn,aacc)+v(bb,ddnn)+v(dd,bbnn)+v(aa,ccnn)$$
$$+v(cc,aann)-u(bbccnn)-u(aaddnn)-v(nn,bbcc)-v(nn,aadd)-v(bb,ccnn)-v(cc,bbnn)$$
$$-v(aa,ddnn)-v(dd,aann)-u(abcdnn)-v(nn,abcd)-v(ab,cdnn)-v(cd,abnn)\Big\}$$
$$\pm i\Big\{u(aacdnn)+v(nn,aacd)+v(aa,cdnn)+v(cd,aann)-u(bbcdnn)-v(nn,bbcd)$$
$$-v(bb,cdnn)-v(cd,bbnn)+u(abccnn)+v(nn,abcc)+v(cc,abnn)+v(ab,ccnn)$$
$$-u(abddnn)-v(nn,abdd)-v(dd,abnn)-v(ab,ddnn)\Big\}\Big]$$

$$g_{22^-}^{tt',n\pm} = \frac{1}{16hc}\Big[\Big\{u(bbddnn)+u(aaccnn)+v(nn,bbdd)+v(nn,aacc)+v(bb,ddnn)+v(dd,bbnn)$$
$$+v(aa,ccnn)+v(cc,aann)-u(bbccnn)-u(aaddnn)-v(nn,bbcc)-v(nn,aadd)$$
$$-v(bb,ccnn)-v(cc,bbnn)-v(aa,ddnn)-v(dd,aann)+u(abcdnn)+v(nn,abcd)$$
$$+v(ab,cdnn)+v(cd,abnn)\Big\}\ldots.$$

TABLE XLVI-5 . Continued

$$\left(g_{22-}^{tt',n\pm}\right) \quad \ldots\ \mp i\Big\{u(aacdnn)+v(nn,aacd)+v(aa,cdnn)+v(cd,aann)-u(bbcdnn)-v(nn,bbcd)$$

$$-v(bb,cdnn)-v(cd,bbnn)-u(abccnn)-v(nn,abcc)-v(cc,abnn)-v(ab,ccnn)+u(abddnn)$$

$$+v(nn,abdd)+v(dd,abnn)+v(ab,ddnn)\Big\}\Big]\Big]$$

$$g_4^{t,t\pm} = \frac{1}{64hc}\Big[15u(aaaaaa)+15u(bbbbbb)-5u(aaaabb)-5u(aabbbb)+3v(aa,aaaa)+3v(bb,bbbb)$$

$$-v(aa,bbbb)-v(bb,aaaa)-v(aa,aabb)-v(bb,aabb)-v(ab,aaab)-v(ab,abbb)$$

$$\mp i\Big\{-10u(aaaaab)+10u(abbbbb)-2v(ab,aaaa)+2v(ab,bbbb)-2v(aa,aaab)+2v(bb,abbb)\Big\}\Big]$$

$$g_4^{t,t'\pm} = \frac{1}{64hc}\Big[3\hbar^6\ {}_{(4)}Z^{aaaacc}+3\hbar^6\ {}_{(4)}Z^{aaaadd}+3\hbar^4\ {}_{(4)}Z^{aaaa}_{cc}+3\hbar^4\ {}_{(4)}Z^{aaaa}_{dd}+3\hbar^2\ {}_{(4)}Z^{cc}_{aaaa}$$

$$+3\hbar^2\ {}_{(4)}Z^{dd}_{aaaa}+3\ {}_{(4)}Z_{aaaacc}+3\ {}_{(4)}Z_{aaaadd}+3\hbar^6\ {}_{(4)}Z^{bbbbcc}+3\hbar^6\ {}_{(4)}Z^{bbbbdd}$$

$$+3\hbar^4\ {}_{(4)}Z^{bbbb}_{cc}+3\hbar^4\ {}_{(4)}Z^{bbbb}_{dd}+3\hbar^2\ {}_{(4)}Z^{cc}_{bbbb}+3\hbar^2\ {}_{(4)}Z^{dd}_{bbbb}+3\ {}_{(4)}Z_{bbbbcc}+3\ {}_{(4)}Z_{bbbbdd}$$

$$-3\hbar^6\ {}_{(4)}Z^{aabbcc}-3\hbar^4\ {}_{(4)}Z^{aabb}_{cc}-3\hbar^6\ {}_{(4)}Z^{aabbdd}-3\hbar^4\ {}_{(4)}Z^{aabb}_{dd}-3\hbar^2\ {}_{(4)}Z^{cc}_{aabb}\ \ldots$$

TABLE XLVI-5 . Continued

$$\left(g_4^{t,t'\pm}\right) \quad \cdots\cdots -3\hbar^2\,{}_{(4)}Z^{dd}_{aabb} - 3\,{}_{(4)}Z_{aabbcc} - 3\,{}_{(4)}Z_{aabbdd} + \hbar^4\,{}_{(4)}Z^{aacc}_{aa} + \hbar^4\,{}_{(4)}Z^{aadd}_{aa}$$

$$+\hbar^2\,{}_{(4)}Z^{aa}_{aacc} + \hbar^2\,{}_{(4)}Z^{aa}_{aadd} + \hbar^4\,{}_{(4)}Z^{bbcc}_{bb} + \hbar^2\,{}_{(4)}Z^{bbdd}_{bb} + \hbar^2\,{}_{(4)}Z^{bb}_{bbcc} + \hbar^2\,{}_{(4)}Z^{bb}_{bbdd}$$

$$-\hbar^4\,{}_{(4)}Z^{aacc}_{bb} - \hbar^4\,{}_{(4)}Z^{aadd}_{bb} - \hbar^4\,{}_{(4)}Z^{bbcc}_{aa} - \hbar^4\,{}_{(4)}Z^{bbdd}_{aa} - \hbar^2\,{}_{(4)}Z^{bb}_{aacc} - \hbar^2\,{}_{(4)}Z^{bb}_{aadd}$$

$$-\hbar^4\,{}_{(4)}Z^{abcc}_{ab} - \hbar^4\,{}_{(4)}Z^{abdd}_{ab} - \hbar^2\,{}_{(4)}Z^{ab}_{abcc} - \hbar^2\,{}_{(4)}Z^{ab}_{abdd}$$

$$\pm i\Big\{3\hbar^6\,{}_{(4)}Z^{aaabcc} + 3\hbar^6\,{}_{(4)}Z^{aaabdd} + 3\hbar^4\,{}_{(4)}Z^{aaab}_{cc} + 3\hbar^4\,{}_{(4)}Z^{aaab}_{dd} - 3\hbar^6\,{}_{(4)}Z^{abbbcc}$$

$$-3\hbar^6\,{}_{(4)}Z^{abbbdd} - 3\hbar^4\,{}_{(4)}Z^{abbb}_{cc} - 3\hbar^4\,{}_{(4)}Z^{abbb}_{dd} + \hbar^4\,{}_{(4)}Z^{abcc}_{aa} + \hbar^4\,{}_{(4)}Z^{abdd}_{aa} + \hbar^2\,{}_{(4)}Z^{ab}_{aacc}$$

$$+\hbar^2\,{}_{(4)}Z^{ab}_{aadd} - \hbar^4\,{}_{(4)}Z^{abcc}_{bb} - \hbar^4\,{}_{(4)}Z^{abdd}_{bb} - \hbar^2\,{}_{(4)}Z^{ab}_{bbcc} - \hbar^2\,{}_{(4)}Z^{ab}_{bbdd} + \hbar^4\,{}_{(4)}Z^{aacc}_{ab}$$

$$+\hbar^4\,{}_{(4)}Z^{aadd}_{ab} + \hbar^2\,{}_{(4)}Z^{aa}_{abcc} + \hbar^2\,{}_{(4)}Z^{aa}_{abdd} - \hbar^4\,{}_{(4)}Z^{bbcc}_{ab} - \hbar^4\,{}_{(4)}Z^{bbdd}_{ab} - \hbar^2\,{}_{(4)}Z^{bb}_{abcc} - \hbar^2\,{}_{(4)}Z^{bb}_{abdd}\Big\}\Bigg]$$

$$g_4^{t,n\pm} = \frac{1}{32hc}\Big[3\hbar^6\,{}_{(4)}Z^{aaaann} + 3\hbar^4\,{}_{(4)}Z^{aaaa}_{nn} + 3\hbar^2\,{}_{(4)}Z^{nn}_{aaaa} + 3\,{}_{(4)}Z_{aaaann} + 3\hbar^6\,{}_{(4)}Z^{bbbbnn}$$

$$+3\hbar^4\,{}_{(4)}Z^{bbbb}_{nn} + 3\hbar^2\,{}_{(4)}Z^{nn}_{bbbb} + 3\,{}_{(4)}Z_{bbbbnn} - 3\hbar^6\,{}_{(4)}Z^{aabbnn} - 3\hbar^4\,{}_{(4)}Z^{aabb}_{nn}$$

$$-3\hbar^2\,{}_{(4)}Z^{nn}_{aabb} - 3\,{}_{(4)}Z_{aabbnn} + \hbar^4\,{}_{(4)}Z^{aann}_{aa} + \hbar^2\,{}_{(4)}Z^{aa}_{aann} + \hbar^4\,{}_{(4)}Z^{bbnn}_{bb} + \hbar^2\,{}_{(4)}Z^{bb}_{bbnn} \cdots\cdots$$

TABLE XLVI-5 . Continued

$$\left(g_4^{t,n\pm}\right) \quad \ldots\ldots -\hbar^4\,{}_{(4)}Z^{aann}_{bb} - \hbar^2\,{}_{(4)}Z^{aa}_{bbnn} - \hbar^4\,{}_{(4)}Z^{bbnn}_{aa} - \hbar^2\,{}_{(4)}Z^{bb}_{aann} - \hbar^4\,{}_{(4)}Z^{abnn}_{ab} - \hbar^2\,{}_{(4)}Z^{ab}_{abnn}$$

$$\pm i\Big\{3\hbar^6\,{}_{(4)}Z^{aaabnn} + 3\hbar^4\,{}_{(4)}Z^{aaab}_{nn} - 3\hbar^6\,{}_{(4)}Z^{abbbnn} - 3\hbar^4\,{}_{(4)}Z^{abbb}_{nn} + \hbar^4\,{}_{(4)}Z^{abnn}_{aa}$$

$$+\hbar^2\,{}_{(4)}Z^{ab}_{aann} - \hbar^4\,{}_{(4)}Z^{abnn}_{bb} - \hbar^2\,{}_{(4)}Z^{ab}_{bbnn} + \hbar^4\,{}_{(4)}Z^{aann}_{ab} + \hbar^2\,{}_{(4)}Z^{aa}_{abnn} - \hbar^4\,{}_{(4)}Z^{bbnn}_{ab} - \hbar^2\,{}_{(4)}Z^{bb}_{abnn}\Big\}\Big]$$

$$g_{24}^{tt'\pm} = \frac{1}{128hc}\Big[\Big\{3u(bbcccc) + 3u(bbdddd) + 3v(bb,cccc) + 3v(bb,dddd) - 3u(bbccdd) - 3v(bb,ccdd)$$

$$+v(cc,bbcc) + v(dd,bbdd) - v(cc,bbdd) - v(dd,bbcc) - v(cd,bbcd) - 3u(aacccc) - 3u(aadddd)$$

$$-3v(aa,cccc) - 3v(aa,dddd) + 3u(aaccdd) + 3v(aa,ccdd) - v(cc,aacc) - v(dd,aadd)$$

$$+v(cc,aadd) + v(dd,aacc) + v(cd,aacd) + 3u(abcccd) - 3u(abcddd) + 3v(ab,cccd)$$

$$-3v(ab,cddd) + v(cc,abcd) - v(dd,abcd) + v(cd,abcc) - v(cd,abdd)\Big\}$$

$$\pm i\Big\{-3u(abcccc) - 3u(abdddd) - 3v(ab,cccc) - 3v(ab,dddd) + 3u(abccdd) + 3v(ab,ccdd)$$

$$-v(cc,abcc) - v(dd,abdd) + v(cc,abdd) + v(dd,abcc) + v(cd,abcd) + 3u(bbcccd) - 3u(bbcddd)\ldots\ldots$$

TABLE XLVI-5 . Continued

$$\left(g_{24}^{tt'\pm}\right) \quad \cdots\cdots + 3v(bb,cccd) - 3v(bb,cddd) + v(cc,bbcd) + v(cd,bbcc) - v(dd,bbcd) - v(cd,bbdd) - 3u(aacccd)$$

$$+3u(aacddd) - v(aa,cccd) + v(aa,cddd) - v(cc,aacd) - v(cd,aacc) + v(dd,aacd) + v(cd,aadd)\Big]\Big]$$

$$g_{222}^{tt't''\pm} = \frac{1}{64hc}\Big[-\hbar^6\,{}_{(4)}Z^{aaccee} + \hbar^6\,{}_{(4)}Z^{aaccff} - \hbar^4\,{}_{(4)}Z^{aacc}_{ee} + \hbar^4\,{}_{(4)}Z^{aacc}_{ff} - \hbar^4\,{}_{(4)}Z^{aaee}_{cc} + \hbar^4\,{}_{(4)}Z^{aaff}_{cc}$$

$$-\hbar^2\,{}_{(4)}Z^{aa}_{ccee} + \hbar^2\,{}_{(4)}Z^{aa}_{ccff} - \hbar^6\,{}_{(4)}Z^{bbddee} + \hbar^6\,{}_{(4)}Z^{bbddff} - \hbar^4\,{}_{(4)}Z^{bbdd}_{ee} + \hbar^4\,{}_{(4)}Z^{bbdd}_{ff}$$

$$-\hbar^4\,{}_{(4)}Z^{bbee}_{dd} + \hbar^4\,{}_{(4)}Z^{bbff}_{dd} - \hbar^2\,{}_{(4)}Z^{bb}_{ddee} + \hbar^2\,{}_{(4)}Z^{bb}_{ddff} - \hbar^4\,{}_{(4)}Z^{ccee}_{aa} + \hbar^4\,{}_{(4)}Z^{ccff}_{aa}$$

$$-\hbar^2\,{}_{(4)}Z^{cc}_{aaee} + \hbar^2\,{}_{(4)}Z^{cc}_{aaff} - \hbar^2\,{}_{(4)}Z^{ee}_{aacc} + \hbar^2\,{}_{(4)}Z^{ff}_{aacc} - {}_{(4)}Z_{aaccee} + {}_{(4)}Z_{aaccff}$$

$$-\hbar^4\,{}_{(4)}Z^{ddee}_{bb} + \hbar^4\,{}_{(4)}Z^{ddff}_{bb} - \hbar^2\,{}_{(4)}Z^{dd}_{bbee} + \hbar^2\,{}_{(4)}Z^{dd}_{bbff} - \hbar^2\,{}_{(4)}Z^{ff}_{bbcc} + \hbar^2\,{}_{(4)}Z^{ee}_{bbcc}$$

$$+\hbar^6\,{}_{(4)}Z^{aaddee} - \hbar^6\,{}_{(4)}Z^{aaddff} + \hbar^4\,{}_{(4)}Z^{aadd}_{ee} - \hbar^4\,{}_{(4)}Z^{aadd}_{ff} + \hbar^4\,{}_{(4)}Z^{aaee}_{dd}$$

$$-\hbar^4\,{}_{(4)}Z^{aaff}_{dd} + \hbar^2\,{}_{(4)}Z^{aa}_{ddee} - \hbar^2\,{}_{(4)}Z^{aa}_{ddff} + \hbar^6\,{}_{(4)}Z^{bbccee} - \hbar^6\,{}_{(4)}Z^{bbccff} + \hbar^4\,{}_{(4)}Z^{bbcc}_{ee}$$

$$-\hbar^4\,{}_{(4)}Z^{bbcc}_{ff} + \hbar^4\,{}_{(4)}Z^{bbee}_{cc} - \hbar^4\,{}_{(4)}Z^{bbff}_{cc} + \hbar^2\,{}_{(4)}Z^{bb}_{ccee} - \hbar^2\,{}_{(4)}Z^{bb}_{ccff} + \hbar^4\,{}_{(4)}Z^{ddee}_{aa} \cdots\cdots$$

TABLE XLVI-5 . Continued

$\left(g_{222}^{tt't''\pm}\right)$

$$\ldots -\hbar^4\,{}_{(4)}Z^{ddff}_{aa} + \hbar^2\,{}_{(4)}Z^{dd}_{aaee} - \hbar^2\,{}_{(4)}Z^{dd}_{aaff} + \hbar^2\,{}_{(4)}Z^{ee}_{aadd} - \hbar^2\,{}_{(4)}Z^{ff}_{aadd} + {}_{(4)}Z_{aaddee}$$

$$- {}_{(4)}Z_{aaddff} + \hbar^4\,{}_{(4)}Z^{ccee}_{bb} - \hbar^4\,{}_{(4)}Z^{ccff}_{bb} + \hbar^2\,{}_{(4)}Z^{cc}_{bbee} - \hbar^2\,{}_{(4)}Z^{cc}_{bbff} + {}_{(4)}Z_{bbccee}$$

$$- {}_{(4)}Z_{bbccff} - \hbar^2\,{}_{(4)}Z^{ee}_{bbdd} + \hbar^2\,{}_{(4)}Z^{ff}_{bbdd} - {}_{(4)}Z_{bbddee} + {}_{(4)}Z_{bbddff} - \Big(\hbar^6\,{}_{(4)}Z^{aacdef}$$

$$+ \hbar^4\,{}_{(4)}Z^{aacd}_{ef} + \hbar^4\,{}_{(4)}Z^{aaef}_{cd} + \hbar^2\,{}_{(4)}Z^{aa}_{cdef} - \hbar^6\,{}_{(4)}Z^{bbcdef} - \hbar^4\,{}_{(4)}Z^{bbcd}_{ef} - \hbar^4\,{}_{(4)}Z^{bbef}_{cd}$$

$$- \hbar^2\,{}_{(4)}Z^{bb}_{cdef} + \hbar^4\,{}_{(4)}Z^{cdef}_{aa} + \hbar^2\,{}_{(4)}Z^{cd}_{aaef} + \hbar^2\,{}_{(4)}Z^{ef}_{aacd} + {}_{(4)}Z_{aacdef} - \hbar^4\,{}_{(4)}Z^{cdef}_{bb}$$

$$- \hbar^2\,{}_{(4)}Z^{cd}_{bbef} - \hbar^2\,{}_{(4)}Z^{ef}_{bbcd} - {}_{(4)}Z_{bbcdef}\Big) - \Big(\qquad\Big)_{\substack{ab\to ef\\ ef\to ab}} - \Big(\qquad\Big)_{\substack{ab\to cd\\ cd\to ab}}$$

$$\pm i\Big\{\hbar^6\,{}_{(4)}Z^{abcdef} + \hbar^4\,{}_{(4)}Z^{abcd}_{ef} + \hbar^4\,{}_{(4)}Z^{abef}_{cd} + \hbar^2\,{}_{(4)}Z^{ab}_{cdef} + \hbar^4\,{}_{(4)}Z^{cdef}_{ab} + \hbar^2\,{}_{(4)}Z^{cd}_{abef}$$

$$+ \hbar^2\,{}_{(4)}Z^{ef}_{abcd} + {}_{(4)}Z_{abcdef} - \Big(\Big(\hbar^6\,{}_{(4)}Z^{aaabcc} + \hbar^4\,{}_{(4)}Z^{aacc}_{ab} - \hbar^6\,{}_{(4)}Z^{aaabdd} - \hbar^4\,{}_{(4)}Z^{aadd}_{ab}$$

$$+ \hbar^4\,{}_{(4)}Z^{aaab}_{cc} + \hbar^2\,{}_{(4)}Z^{aa}_{abcc} - \hbar^4\,{}_{(4)}Z^{aaab}_{dd} - \hbar^4\,{}_{(4)}Z^{aa}_{abdd} + \hbar^4\,{}_{(4)}Z^{abcc}_{aa} + \hbar^2\,{}_{(4)}Z^{cc}_{aaab} \ldots$$

TABLE XLVI-5 . Continued

$$\left(g_{222}^{tt't''\pm}\right) \quad \ldots - \hbar^4 \, {}_{(4)}Z^{abdd}_{aa} - \hbar^2 \, {}_{(4)}Z^{dd}_{aaab} + \hbar^2 \, {}_{(4)}Z^{ab}_{aacc} + {}_{(4)}Z_{aaabcc} - \hbar^2 \, {}_{(4)}Z^{ab}_{aadd} - {}_{(4)}Z_{aaabdd}$$

$$- \hbar^6 \, {}_{(4)}Z^{abbbcc} - \hbar^4 \, {}_{(4)}Z^{bbcc}_{ab} + \hbar^6 \, {}_{(4)}Z^{abbbdd} + \hbar^4 \, {}_{(4)}Z^{bbdd}_{ab} - \hbar^4 \, {}_{(4)}Z^{abbb}_{cc} - \hbar^2 \, {}_{(4)}Z^{bb}_{abcc}$$

$$+ \hbar^4 \, {}_{(4)}Z^{abbb}_{dd} + \hbar^2 \, {}_{(4)}Z^{bb}_{abdd} - \hbar^4 \, {}_{(4)}Z^{abcc}_{bb} - \hbar^2 \, {}_{(4)}Z^{cc}_{abbb} + \hbar^4 \, {}_{(4)}Z^{abdd}_{bb} + \hbar^2 \, {}_{(4)}Z^{dd}_{abbb}$$

$$- \hbar^2 \, {}_{(4)}Z^{ab}_{bbcc} - {}_{(4)}Z_{abbbcc} + \hbar^2 \, {}_{(4)}Z^{ab}_{bbdd} + {}_{(4)}Z_{abbbdd} \Big)\Big) - \Big(\Big(\qquad \Big)\Big)_{\substack{ab \to ef \\ ef \to ab}}$$

$$- \Big(\Big(\qquad \Big)\Big)_{\substack{cd \to ef \\ ef \to cd}} \Big\} \Big]$$

$$g_6^{t\pm} = \frac{1}{128hc} \Big[-5u(aaaaaa) + 5u(bbbbbb) + 5u(aaaabb) - 5u(aabbbb) - v(aa,aaaa) + v(bb,bbbb)$$

$$- v(aa,bbbb) + v(bb,aaaa) + v(aa,aabb) - v(bb,aabb) + v(ab,aaab) - v(ab,abbb)$$

$$\pm i \Big\{ 5u(aaaaab) + 5u(abbbbb) - 5u(aaabbb) - v(ab,aabb) + v(ab,aaaa) + v(ab,bbbb)$$

$$+ v(aa,aaab) - v(bb,aaab) - v(aa,abbb) + v(bb,abbb) \Big\} \Big]$$

TABLE XLVII-a . Definition of the symbols appearing in the Tables XLV et XLVI

$$\{\alpha\beta\gamma,\delta\varepsilon\eta\}_\gamma = \frac{1}{2}\left({}^{\alpha\beta\gamma,\delta\varepsilon\eta}{}_\gamma + {}^{\delta\varepsilon\eta,\alpha\beta\gamma}{}_\gamma\right)$$

$${}^{\alpha\beta\gamma\delta}u^{\pm} = {}^{\alpha\beta\gamma\delta}u^{\pm}(aa,bb) = \frac{1}{2}\hbar^2\left({}^{\alpha\beta\gamma\delta}_{(4)}Z^{bb} \pm {}^{\alpha\beta\gamma\delta}_{(4)}Z^{aa}\right) + \frac{1}{2}\left({}^{\alpha\beta\gamma\delta}_{(4)}Z_{bb} \pm {}^{\alpha\beta\gamma\delta}_{(4)}Z_{aa}\right)$$

$${}^{\alpha\beta\gamma\delta}v(nn) = 2\left(\hbar^2\, {}^{\alpha\beta\gamma\delta}_{(4)}Z^{nn} + {}^{\alpha\beta\gamma\delta}_{(4)}Z_{nn}\right)$$

$$a_1^{(2)}({}^{\alpha\beta}\Gamma) = \frac{1}{2}\left({}^{xx}\Gamma + {}^{yy}\Gamma\right)$$

$$a_2^{(2)}({}^{\alpha\beta}\Gamma) = \frac{1}{2}\left(2\,{}^{zz}\Gamma - {}^{xx}\Gamma - {}^{yy}\Gamma\right)$$

$$u(aaaaaa) = \hbar^6\, {}_{(4)}Z^{aaaaaa} + {}_{(4)}Z_{aaaaaa}, \text{ etc....}$$

$$v(aa,bbbb) = \hbar^4\, {}_{(4)}Z^{bbbb}_{aa} + \hbar^2\, {}_{(4)}Z^{aa}_{bbbb}$$

$$v(aa,aabb) = \hbar^4\, {}_{(4)}Z^{aabb}_{aa} + \hbar^2\, {}_{(4)}Z^{aa}_{aabb}, \text{ etc....}$$

$${}^{\alpha\beta\gamma\delta}u = -\frac{1}{2}\left(\hbar^2\, {}^{\alpha\beta\gamma\delta}_{(4)}Z^{t_1t_2} + {}^{\alpha\beta\gamma\delta}_{(4)}Z_{t_1t_2}\right)$$

TABLE XLVII-a . Continued

$${}^{\alpha\beta}w(aa,bb) = \frac{3}{8}\hbar^4\left({}^{\alpha\beta}_{(4)}Z^{bbbb} - {}^{\alpha\beta}_{(4)}Z^{aaaa}\right) + \frac{3}{8}\left({}^{\alpha\beta}_{(4)}Z_{bbbb} - {}^{\alpha\beta}_{(4)}Z_{aaaa}\right) + \frac{\hbar^2}{8}\left({}^{\alpha\beta}_{(4)}Z^{bb}_{bb} - {}^{\alpha\beta}_{(4)}Z^{aa}_{aa}\right)$$

$${}^{\alpha\beta}w(ab) = -\frac{3}{16}\hbar^4\left({}^{\alpha\beta}_{(4)}Z^{aaab} + {}^{\alpha\beta}_{(4)}Z^{abbb}\right) - \frac{3}{16}\left({}^{\alpha\beta}_{(4)}Z_{aaab} + {}^{\alpha\beta}_{(4)}Z_{abbb}\right)$$
$$- \frac{\hbar^2}{16}\left({}^{\alpha\beta}_{(4)}Z^{ab}_{aa} + {}^{\alpha\beta}_{(4)}Z^{ab}_{bb} + {}^{\alpha\beta}_{(4)}Z^{aa}_{ab} + {}^{\alpha\beta}_{(4)}Z^{bb}_{ab}\right)$$

$${}^{\alpha\beta}w(abcd) = \frac{1}{4}\Big[\hbar^4\left({}^{\alpha\beta}_{(4)}Z^{bbcc} + {}^{\alpha\beta}_{(4)}Z^{bbdd} - {}^{\alpha\beta}_{(4)}Z^{aacc} - {}^{\alpha\beta}_{(4)}Z^{aadd}\right) + \left({}^{\alpha\beta}_{(4)}Z_{bbcc} + {}^{\alpha\beta}_{(4)}Z_{bbdd} - {}^{\alpha\beta}_{(4)}Z_{aacc} - {}^{\alpha\beta}_{(4)}Z_{aadd}\right)$$
$$+ \hbar^2\left({}^{\alpha\beta}_{(4)}Z^{cc}_{bb} + {}^{\alpha\beta}_{(4)}Z^{bb}_{cc} + {}^{\alpha\beta}_{(4)}Z^{dd}_{bb} + {}^{\alpha\beta}_{(4)}Z^{bb}_{dd} - {}^{\alpha\beta}_{(4)}Z^{cc}_{aa} - {}^{\alpha\beta}_{(4)}Z^{aa}_{cc} + {}^{\alpha\beta}_{(4)}Z^{dd}_{aa} - {}^{\alpha\beta}_{(4)}Z^{aa}_{dd}\right)\Big]$$

$${}^{\alpha\beta}v(abcd) = -\frac{1}{4}\Big[\hbar^4\left({}^{\alpha\beta}_{(4)}Z^{abcc} + {}^{\alpha\beta}_{(4)}Z^{abdd}\right) + {}^{\alpha\beta}_{(4)}Z_{abcc} + {}^{\alpha\beta}_{(4)}Z_{abdd} + \hbar^2\left({}^{\alpha\beta}_{(4)}Z^{cc}_{ab} + {}^{\alpha\beta}_{(4)}Z^{ab}_{cc} + {}^{\alpha\beta}_{(4)}Z^{dd}_{ab} + {}^{\alpha\beta}_{(4)}Z^{ab}_{dd}\right)\Big]$$

$${}^{\alpha\beta}w(ab,nn) = 2\Big[\hbar^4\left({}^{\alpha\beta}_{(4)}Z^{bbnn} - {}^{\alpha\beta}_{(4)}Z^{aann}\right) + \left({}^{\alpha\beta}_{(4)}Z_{bbnn} - {}^{\alpha\beta}_{(4)}Z_{aann}\right) + \hbar^2\left({}^{\alpha\beta}_{(4)}Z^{bb}_{nn} + {}^{\alpha\beta}_{(4)}Z^{nn}_{bb} - {}^{\alpha\beta}_{(4)}Z^{nn}_{aa} - {}^{\alpha\beta}_{(4)}Z^{aa}_{nn}\right)\Big]$$

$${}^{\alpha\beta}v(ab,nn) = -2\Big[\hbar^4\,{}^{\alpha\beta}_{(4)}Z^{abnn} + {}^{\alpha\beta}_{(4)}Z_{abnn} + \hbar^2\left({}^{\alpha\beta}_{(4)}Z^{nn}_{ab} + {}^{\alpha\beta}_{(4)}Z^{ab}_{nn}\right)\Big]$$

$${}^{\alpha\beta}r(aabb) = \frac{1}{16}\Big[3\hbar^4\left({}^{\alpha\beta}_{(4)}Z^{aaaa} + {}^{\alpha\beta}_{(4)}Z^{bbbb}\right) + 3\left({}^{\alpha\beta}_{(4)}Z_{aaaa} + {}^{\alpha\beta}_{(4)}Z_{bbbb}\right) - 3\left(\hbar^4\,{}^{\alpha\beta}_{(4)}Z^{aabb} + {}^{\alpha\beta}_{(4)}Z_{aabb}\right)$$
$$+ \hbar^2\left({}^{\alpha\beta}_{(4)}Z^{aa}_{aa} + {}^{\alpha\beta}_{(4)}Z^{bb}_{bb} - {}^{\alpha\beta}_{(4)}Z^{bb}_{aa} - {}^{\alpha\beta}_{(4)}Z^{aa}_{bb}\right)\Big]$$

TABLE XLVII-a . Continued

$$ {}^{\alpha\beta}r(ab) = \frac{1}{16}\left[3\hbar^4\left({}^{\alpha\beta}_{(4)}Z^{aaab} - {}^{\alpha\beta}_{(4)}Z^{abbb}\right) + 3\left({}^{\alpha\beta}_{(4)}Z_{aaab} - {}^{\alpha\beta}_{(4)}Z_{abbb}\right) + \hbar^2\left({}^{\alpha\beta}_{(4)}Z^{bb}_{aa} + {}^{\alpha\beta}_{(4)}Z^{aa}_{bb} - {}^{\alpha\beta}_{(4)}Z^{ab}_{bb} - {}^{\alpha\beta}_{(4)}Z^{bb}_{ab}\right)\right] $$

$$ {}^{\alpha\beta}r_{3,3}(abcd) = \frac{1}{4}\left[\hbar^4\left({}^{\alpha\beta}_{(4)}Z^{bbdd} + {}^{\alpha\beta}_{(4)}Z^{aacc} - {}^{\alpha\beta}_{(4)}Z^{aadd} - {}^{\alpha\beta}_{(4)}Z^{bbcc}\right) + \left({}^{\alpha\beta}_{(4)}Z_{bbdd} + {}^{\alpha\beta}_{(4)}Z_{aacc} - {}^{\alpha\beta}_{(4)}Z_{aadd} - {}^{\alpha\beta}_{(4)}Z_{bbcc}\right)\right. $$
$$ \left. + \hbar^2\left({}^{\alpha\beta}_{(4)}Z^{dd}_{bb} + {}^{\alpha\beta}_{(4)}Z^{bb}_{dd} + {}^{\alpha\beta}_{(4)}Z^{cc}_{aa} + {}^{\alpha\beta}_{(4)}Z^{aa}_{cc} - {}^{\alpha\beta}_{(4)}Z^{cc}_{bb} - {}^{\alpha\beta}_{(4)}Z^{bb}_{cc} - {}^{\alpha\beta}_{(4)}Z^{dd}_{aa} - {}^{\alpha\beta}_{(4)}Z^{aa}_{dd}\right)\right] $$

$$ {}^{\alpha\beta}r_{4,4}(abcd) = \frac{1}{4}\left[\left(\hbar^4\,{}^{\alpha\beta}_{(4)}Z^{abcd} + {}^{\alpha\beta}_{(4)}Z_{abcd}\right) + \hbar^2\left({}^{\alpha\beta}_{(4)}Z^{cd}_{ab} + {}^{\alpha\beta}_{(4)}Z^{ab}_{cd}\right)\right] $$

$$ {}^{\alpha\beta}r_{4,3}(abcd) = \frac{1}{4}\left[\hbar^4\left({}^{\alpha\beta}_{(4)}Z^{abcc} - {}^{\alpha\beta}_{(4)}Z^{abdd}\right) + \left({}^{\alpha\beta}_{(4)}Z_{abcc} - {}^{\alpha\beta}_{(4)}Z_{abdd}\right) + \hbar^2\left({}^{\alpha\beta}_{(4)}Z^{cc}_{ab} + {}^{\alpha\beta}_{(4)}Z^{ab}_{cc} - {}^{\alpha\beta}_{(4)}Z^{dd}_{ab} - {}^{\alpha\beta}_{(4)}Z^{ab}_{dd}\right)\right] $$

$$ {}^{\alpha\beta}r_{3,4}(abcd) = \frac{1}{4}\left[\hbar^4\left({}^{\alpha\beta}_{(4)}Z^{aacd} - {}^{\alpha\beta}_{(4)}Z^{bbcd}\right) + \left({}^{\alpha\beta}_{(4)}Z_{aacd} - {}^{\alpha\beta}_{(4)}Z_{bbcd}\right) + \hbar^2\left({}^{\alpha\beta}_{(4)}Z^{aa}_{cd} + {}^{\alpha\beta}_{(4)}Z^{cd}_{aa} - {}^{\alpha\beta}_{(4)}Z^{cd}_{bb} - {}^{\alpha\beta}_{(4)}Z^{bb}_{cd}\right)\right] $$

$$ a^{(6)}_{44} = -\frac{3}{64}\left({}^{xx,xxxx}_{\overline{\omega}} + {}^{yy,yyyy}_{\overline{\omega}}\right) + \frac{1}{64}\left({}^{xx,yyyy}_{\overline{\omega}} + {}^{yy,xxxx}_{\overline{\omega}}\right) + \frac{1}{32}\left({}^{xx,xxyy}_{\overline{\omega}} + {}^{yy,xxyy}_{\overline{\omega}}\right) $$
$$ + \frac{1}{64}\left({}^{xx,xyyx}_{\overline{\omega}} + {}^{yy,yxxy}_{\overline{\omega}}\right) + \frac{1}{64}\left({}^{xx,yxxy}_{\overline{\omega}} + {}^{yy,xyyx}_{\overline{\omega}}\right) + \frac{1}{32}\left({}^{xx,xyxy}_{\overline{\omega}} + {}^{yy,xyxy}_{\overline{\omega}}\right) $$
$$ + \frac{1}{32}\left({}^{xx,xzzx}_{\overline{\omega}} + {}^{yy,yzzy}_{\overline{\omega}}\right) - \frac{1}{32}\left({}^{xx,yzzy}_{\overline{\omega}} + {}^{yy,xzzx}_{\overline{\omega}}\right) + \frac{1}{16}\left({}^{xx,xxzz}_{\overline{\omega}} + {}^{yy,yyzz}_{\overline{\omega}}\right) $$
$$ - \frac{1}{16}\left({}^{xx,yyzz}_{\overline{\omega}} + {}^{yy,xxzz}_{\overline{\omega}}\right) + \frac{1}{32}\left({}^{xx,zxxz}_{\overline{\omega}} + {}^{yy,zyyz}_{\overline{\omega}}\right) - \frac{1}{32}\left({}^{xx,zyyz}_{\overline{\omega}} + {}^{yy,zxxz}_{\overline{\omega}}\right) \ldots . $$

TABLE XLVII-a . Continued

$$\left(a^{(6)}_{44}\right)\ldots\ + \frac{1}{16}\left({}^{xx,xzxz}\overline{\omega} + {}^{yy,yzyz}\overline{\omega}\right) - \frac{1}{16}\left({}^{xx,yzyz}\overline{\omega} + {}^{yy,xzxz}\overline{\omega}\right) + \frac{1}{32}\left({}^{zz,xxxx}\overline{\omega} + {}^{zz,yyyy}\overline{\omega}\right)$$

$$- \frac{1}{16}\,{}^{zz,xxyy}\overline{\omega} - \frac{1}{32}\left({}^{zz,xyyx}\overline{\omega} + {}^{zz,yxxy}\overline{\omega}\right) - \frac{1}{16}\,{}^{zz,xyxy}\overline{\omega} + \frac{1}{8}\left({}^{xz,xxxz}\overline{\omega} + {}^{yz,yyyz}\overline{\omega}\right)$$

$$+ \frac{1}{8}\left({}^{xz,xxzx}\overline{\omega} + {}^{yz,yyzy}\overline{\omega}\right) - \frac{1}{8}\left({}^{xz,yxzy}\overline{\omega} + {}^{yz,xyzx}\overline{\omega}\right) - \frac{1}{8}\left({}^{xz,xzyy}\overline{\omega} + {}^{yz,yzxx}\overline{\omega}\right)$$

$$- \frac{1}{8}\left({}^{xz,zxyy}\overline{\omega} + {}^{yz,zyxx}\overline{\omega}\right) - \frac{1}{8}\left({}^{xz,xyzy}\overline{\omega} + {}^{yz,yxzx}\overline{\omega}\right) - \frac{1}{8}\left({}^{xz,zyxy}\overline{\omega} + {}^{yz,zxyx}\overline{\omega}\right)$$

$$+ \frac{1}{16}\left({}^{xy,xyyy}\overline{\omega} + {}^{xy,yxxx}\overline{\omega}\right) + \frac{1}{16}\left({}^{xy,xxyx}\overline{\omega} + {}^{xy,yyxy}\overline{\omega}\right) - \frac{1}{8}\,{}^{xy,zxyz}\overline{\omega}$$

$$- \frac{1}{8}\left({}^{xy,yxzz}\overline{\omega} + {}^{xy,xyzz}\overline{\omega}\right) - \frac{1}{8}\,{}^{xy,xzzy}\overline{\omega} - \frac{1}{8}\,{}^{xy,zyzx}\overline{\omega} - \frac{1}{8}\,{}^{xy,yzxz}\overline{\omega} - \frac{1}{8}\left({}^{xz,xyyz}\overline{\omega} + {}^{yz,yxxz}\overline{\omega}\right)$$

$$a^{(6)}_{45} = \frac{1}{16}\left({}^{xy,xzzx}\overline{\omega} - {}^{xy,yzzy}\overline{\omega}\right) - \frac{1}{16}\left({}^{xx,yxxx}\overline{\omega} - {}^{yy,xyyy}\overline{\omega}\right) - \frac{1}{16}\left({}^{xx,xxyx}\overline{\omega} - {}^{yy,yyxy}\overline{\omega}\right)$$

$$+ \frac{1}{8}\left({}^{xy,xzxz}\overline{\omega} - {}^{xy,yzyz}\overline{\omega}\right) + \frac{1}{16}\left({}^{xx,zxyz}\overline{\omega} - {}^{yy,zxyz}\overline{\omega}\right) + \frac{1}{16}\left({}^{xx,yxzz}\overline{\omega} - {}^{yy,yxzz}\overline{\omega}\right)$$

$$+ \frac{1}{16}\left({}^{xx,xyzz}\overline{\omega} - {}^{yy,xyzz}\overline{\omega}\right) + \frac{1}{16}\left({}^{xx,xzzy}\overline{\omega} - {}^{yy,xzzy}\overline{\omega}\right) + \frac{1}{16}\left({}^{xx,zyzx}\overline{\omega} - {}^{yy,zyzx}\overline{\omega}\right)$$

$$+ \frac{1}{16}\left({}^{xx,yzxz}\overline{\omega} - {}^{yy,yzxz}\overline{\omega}\right) - \frac{1}{16}\left({}^{zz,xyyy}\overline{\omega} - {}^{zz,yxxx}\overline{\omega}\right) + \frac{1}{16}\left({}^{zz,xxyx}\overline{\omega} - {}^{zz,yyxy}\overline{\omega}\right)\ldots$$

TABLE XLVII-a . Continued

$$\left(a^{(6)}_{45}\right)\ldots - \frac{1}{8}\left({}^{xz,yyyz}\overline{\omega} - {}^{yz,xxxz}\overline{\omega}\right) - \frac{1}{8}\left({}^{xz,yyzy}\overline{\omega} - {}^{yz,xxzx}\overline{\omega}\right) + \frac{1}{8}\left({}^{xz,xyzx}\overline{\omega} - {}^{yz,yxzy}\overline{\omega}\right)$$

$$+ \frac{1}{8}\left({}^{xz,yzxx}\overline{\omega} - {}^{yz,xzyy}\overline{\omega}\right) + \frac{1}{8}\left({}^{xz,zyxx}\overline{\omega} - {}^{yz,zxyy}\overline{\omega}\right) + \frac{1}{8}\left({}^{xz,yxxz}\overline{\omega} - {}^{yz,xyyz}\overline{\omega}\right)$$

$$+ \frac{1}{8}\left({}^{xz,yxzx}\overline{\omega} - {}^{yz,xyzy}\overline{\omega}\right) + \frac{1}{8}\left({}^{xz,zxyx}\overline{\omega} - {}^{yz,zyxy}\overline{\omega}\right) - \frac{1}{16}\left({}^{xy,xxxx}\overline{\omega} - {}^{xy,yyyy}\overline{\omega}\right)$$

$$+ \frac{1}{8}\left({}^{xy,xxzz}\overline{\omega} - {}^{xy,yyzz}\overline{\omega}\right) + \frac{1}{16}\left({}^{xy,zxxz}\overline{\omega} - {}^{xy,zyyz}\overline{\omega}\right)$$

$$b^{(6)}_{44} = -\frac{3}{128}\left({}^{xxx;xxx}\gamma + {}^{yyy,yyy}\gamma\right) + \frac{1}{32}\left({}^{xxx,xyy}\gamma + {}^{yyy,xxy}\gamma\right) + \frac{1}{64}\left({}^{xxx,yxy}\gamma + {}^{yyy,xyx}\gamma\right)$$

$$+ \frac{1}{32}\left({}^{xyy,yxy}\gamma + {}^{xxy,xyx}\gamma\right) + \frac{1}{32}\left({}^{xyy,xyy}\gamma + {}^{xxy,xxy}\gamma\right) + \frac{1}{128}\left({}^{yxy,yxy}\gamma + {}^{xyx,xyx}\gamma\right)$$

$$+ \frac{1}{16}\left({}^{xxx,xzz}\gamma + {}^{yyy,yzz}\gamma\right) + \frac{1}{32}\left({}^{xxx,zxz}\gamma + {}^{yyy,zyz}\gamma\right) - \frac{1}{8}\left({}^{xyy,xzz}\gamma + {}^{xxy,yzz}\gamma\right)$$

$$- \frac{1}{16}\left({}^{xyy,zxz}\gamma + {}^{xxy,zyz}\gamma\right) - \frac{1}{16}\left({}^{yxy,xzz}\gamma + {}^{xyx,yzz}\gamma\right) - \frac{1}{16}\,{}^{xzy,xzy}\gamma - \frac{1}{16}\,{}^{yxz,yxz}\gamma$$

$$- \frac{1}{32}\left({}^{yxy,zxz}\gamma + {}^{xyx,zyz}\gamma\right) + \frac{1}{8}\,{}^{xxz,xxz}\gamma - \frac{1}{8}\,{}^{xxz,yyz}\gamma + \frac{1}{16}\,{}^{xxz,xzx}\gamma - \frac{1}{16}\,{}^{xxz,yzy}\gamma$$

$$+ \frac{1}{16}\,{}^{yyz,yyz}\gamma - \frac{1}{16}\,{}^{yyz,xzx}\gamma + \frac{1}{16}\,{}^{yyz,yzy}\gamma + \frac{1}{64}\,{}^{xzx,xzx}\gamma - \frac{1}{32}\,{}^{xzx,yzy}\gamma$$

$$+ \frac{1}{64}\,{}^{yzy,yzy}\gamma - \frac{1}{16}\,{}^{xyz,xyz}\gamma - \frac{1}{8}\,{}^{xyz,yxz}\gamma - \frac{1}{8}\,{}^{xyz,xzy}\gamma - \frac{1}{8}\,{}^{yxz,xzy}\gamma$$

TABLE XLVII-a . Continued

$$
\begin{aligned}
b_{45}^{(6)} = & -\frac{1}{64}\left({}^{xxx,xxy}\gamma - {}^{yyy,xyy}\gamma\right) - \frac{1}{128}\left({}^{xxx,xyx}\gamma - {}^{yyy,yxy}\gamma\right) + \frac{1}{16}\left({}^{xxx,yzz}\gamma - {}^{yyy,xzz}\gamma\right) \\
& + \frac{1}{32}\left({}^{xxx,zyz}\gamma - {}^{yyy,zxz}\gamma\right) + \frac{3}{64}\left({}^{xyy,yyy}\gamma - {}^{xxy,xxx}\gamma\right) + \frac{1}{32}\,{}^{xyy,xxy}\gamma \\
& + \frac{1}{64}\left({}^{xyy,xyx}\gamma - {}^{xxy,yxy}\gamma\right) - \frac{1}{8}\left({}^{xyy,yzz}\gamma - {}^{xxy,xzz}\gamma\right) - \frac{1}{16}\left({}^{xyy,zyz}\gamma - {}^{xxy,zxz}\gamma\right) \\
& + \frac{3}{128}\left({}^{yxy,yyy}\gamma - {}^{xyx,xxx}\gamma\right) + \frac{1}{64}\left({}^{yxy,xxy}\gamma - {}^{xyx,xyy}\gamma\right) - \frac{1}{16}\left({}^{yxy,yzz}\gamma - {}^{xyx,xzz}\gamma\right) \\
& - \frac{1}{32}\left({}^{yxy,zyz}\gamma - {}^{xyx,zxz}\gamma\right) + \frac{1}{8}\,{}^{xxz,xyz}\gamma + \frac{1}{8}\,{}^{xxz,yxz}\gamma + \frac{1}{8}\,{}^{xxz,xzy}\gamma \\
& - \frac{1}{8}\,{}^{yyz,xyz}\gamma - \frac{1}{8}\,{}^{yyz,yxz}\gamma - \frac{1}{8}\,{}^{yyz,xzy}\gamma + \frac{1}{16}\,{}^{xzx,xyz}\gamma + \frac{1}{16}\,{}^{xzx,yxz}\gamma \\
& + \frac{1}{16}\,{}^{xzx,xzy}\gamma - \frac{1}{16}\,{}^{yzy,xyz}\gamma - \frac{1}{16}\,{}^{yzy,yxz}\gamma - \frac{1}{16}\,{}^{yzy,xzy}\gamma
\end{aligned}
$$

TABLE XLVII-b . Definition of the symbols appearing in the Tables XLV and XLVI

	$\mathfrak{B}_{\mathfrak{zzzx}'\mathfrak{xx}}$	$\mathfrak{B}_{\mathfrak{zzzx}'\mathfrak{hh}}$	$\mathfrak{B}_{\mathfrak{zxxx}'\mathfrak{xx}}$	$\mathfrak{B}_{\mathfrak{zxxx}'\mathfrak{hh}}$	$\mathfrak{B}_{\mathfrak{xzxx}'\mathfrak{xx}}$	$\mathfrak{B}_{\mathfrak{xzxx}'\mathfrak{hh}}$	$\mathfrak{B}_{\mathfrak{zxzz}'\mathfrak{xx}}$	$\mathfrak{B}_{\mathfrak{zxzz}'\mathfrak{hh}}$	$\mathfrak{B}_{\mathfrak{hzxh}'\mathfrak{xx}}$	$\mathfrak{B}_{\mathfrak{hzxh}'\mathfrak{hh}}$	$\mathfrak{B}_{\mathfrak{hhzx}'\mathfrak{xx}}$	$\mathfrak{B}_{\mathfrak{hhzx}'\mathfrak{hh}}$	$\mathfrak{B}_{\mathfrak{hhxz}'\mathfrak{xx}}$	
$a_{34}^{(6)}$	$+\frac{15}{16}$	$-\frac{15}{16}$	$-\frac{59}{64}$	$+\frac{23}{64}$	$-\frac{75}{64}$	$+\frac{39}{64}$	$+\frac{11}{16}$	$-\frac{11}{16}$	$-\frac{17}{64}$	$+\frac{53}{64}$	$+\frac{15}{64}$	$+\frac{21}{64}$	$+\frac{15}{64}$	.....
$a_{36}^{(6)}$	0	0	$+\frac{5}{16}$	$-\frac{1}{16}$	$+\frac{5}{16}$	$-\frac{1}{16}$	0	0	$-\frac{1}{16}$	$-\frac{3}{16}$	$-\frac{1}{16}$	$-\frac{3}{16}$	$-\frac{1}{16}$	.....
$a_{38}^{(6)}$	$+\frac{1}{16}$	$-\frac{1}{16}$	$-\frac{5}{64}$	$+\frac{1}{64}$	$-\frac{5}{64}$	$+\frac{1}{64}$	$+\frac{1}{16}$	$-\frac{1}{16}$	$+\frac{1}{64}$	$+\frac{3}{64}$	$+\frac{1}{64}$	$+\frac{3}{64}$	$+\frac{1}{64}$	.....

TABLE XLVII-b . Continued

		$\omega_{\mathfrak{hhxz}'\mathfrak{hh}}$	$\omega_{\mathfrak{zhhx}'\mathfrak{xx}}$	$\omega_{\mathfrak{zhhx}'\mathfrak{hh}}$	$\omega_{\mathfrak{hzhx}'\mathfrak{xx}}$	$\omega_{\mathfrak{hzhx}'\mathfrak{hh}}$	$\omega_{\mathfrak{hxhz}'\mathfrak{xx}}$	$\omega_{\mathfrak{hxhz}'\mathfrak{hh}}$	$\omega_{\mathfrak{zzzx}'\mathfrak{zz}}$	$\omega_{\mathfrak{zxxx}'\mathfrak{zz}}$	$\omega_{\mathfrak{xzxx}'\mathfrak{zz}}$	$\omega_{\mathfrak{zxzz}'\mathfrak{zz}}$	$\omega_{\mathfrak{hzxh}'\mathfrak{zz}}$	$\omega_{\mathfrak{hhzx}'\mathfrak{zz}}$	
$\left(a_{34}^{(6)}\right)$	....	$+\frac{21}{64}$	$-\frac{1}{64}$	$+\frac{37}{64}$	$-\frac{1}{64}$	$+\frac{37}{64}$	$-\frac{1}{64}$	$+\frac{37}{64}$	0	$+\frac{9}{16}$	$+\frac{9}{16}$	0	$-\frac{9}{16}$	$-\frac{9}{16}$	....
$\left(a_{36}^{(6)}\right)$	....	$-\frac{3}{16}$	$-\frac{1}{16}$	$-\frac{3}{16}$	$-\frac{1}{16}$	$-\frac{3}{16}$	$-\frac{1}{16}$	$-\frac{3}{16}$	0	0	0	0	0	0	....
$\left(a_{38}^{(6)}\right)$	....	$+\frac{3}{64}$	$+\frac{1}{64}$	$+\frac{3}{64}$	$+\frac{1}{64}$	$+\frac{3}{64}$	$+\frac{1}{64}$	$+\frac{3}{64}$	0	$+\frac{1}{16}$	$+\frac{1}{16}$	0	$-\frac{1}{16}$	$-\frac{1}{16}$	....

TABLE XLVII-b . Continued

		$\underline{\omega}_{\hbar\hbar x\xi,\xi\xi}$	$\underline{\omega}_{\xi\hbar\hbar x,\xi\xi}$	$\underline{\omega}_{\hbar\xi\hbar x,\xi\xi}$	$\underline{\omega}_{\hbar x\hbar\xi,\xi\xi}$	$\underline{\omega}_{\xi\xi\xi\hbar,\hbar x}$	$\underline{\omega}_{\xi\hbar\hbar\hbar,\hbar x}$	$\underline{\omega}_{\hbar\xi\hbar\hbar,\hbar x}$	$\underline{\omega}_{\xi\hbar\xi\xi,\hbar x}$	$\underline{\omega}_{x\xi\hbar x,\hbar x}$	$\underline{\omega}_{xx\xi\hbar,\hbar x}$	$\underline{\omega}_{xx\hbar\xi,\hbar x}$	$\underline{\omega}_{\xi xx\hbar,\hbar x}$	$\underline{\omega}_{x\xi x\hbar,\hbar x}$	
$\left(a^{(6)}_{34}\right)$	....	$-\frac{9}{16}$	$-\frac{9}{16}$	$-\frac{9}{16}$	$-\frac{9}{16}$	$-\frac{15}{8}$	$+\frac{41}{32}$	$+\frac{57}{32}$	$-\frac{11}{8}$	$+\frac{35}{32}$	$+\frac{3}{32}$	$+\frac{3}{32}$	$+\frac{19}{32}$	$+\frac{19}{32}$	....
$\left(a^{(6)}_{36}\right)$	....	0	0	0	0	0	$-\frac{3}{8}$	$-\frac{3}{8}$	0	$-\frac{1}{8}$	$-\frac{1}{8}$	$-\frac{1}{8}$	$-\frac{1}{8}$	$-\frac{1}{8}$	....
$\left(a^{(6)}_{38}\right)$	....	$-\frac{1}{16}$	$-\frac{1}{16}$	$-\frac{1}{16}$	$-\frac{1}{16}$	$-\frac{1}{8}$	$+\frac{3}{32}$	$+\frac{3}{32}$	$-\frac{1}{8}$	$+\frac{1}{32}$	$+\frac{1}{32}$	$+\frac{1}{32}$	$+\frac{1}{32}$	$+\frac{1}{32}$	....

TABLE XLVII-b . Continued

		$\mathcal{B}_{x\eta x\xi,\eta x}$	$\mathcal{B}_{xxxx,\xi x}$	$\mathcal{B}_{\eta\eta\eta\eta,\xi x}$	$\mathcal{B}_{\eta\eta xx,\xi x}$	$\mathcal{B}_{x\eta\eta x,\xi x}$	$\mathcal{B}_{\eta xx\eta,\xi x}$	$\mathcal{B}_{\eta x\eta x,\xi x}$	$\mathcal{B}_{x\xi\xi x,\xi x}$	$\mathcal{B}_{\eta\xi\xi\eta,\xi x}$	$\mathcal{B}_{\xi\xi xx,\xi x}$	$\mathcal{B}_{\xi\xi\eta\eta,\xi x}$	$\mathcal{B}_{\xi xx\xi,\xi x}$	$\mathcal{B}_{\xi\eta\eta\xi,\xi x}$	
$\left(a^{(6)}_{34}\right)$	....	$+\frac{19}{32}$	$-\frac{51}{64}$	$-\frac{43}{64}$	$+\frac{47}{32}$	$+\frac{63}{64}$	$+\frac{31}{64}$	$+\frac{47}{32}$	$-\frac{3}{16}$	$+\frac{3}{16}$	$+\frac{1}{8}$	$-\frac{1}{8}$	$-\frac{7}{16}$	$+\frac{7}{16}$	....
$\left(a^{(6)}_{36}\right)$	....	$-\frac{1}{8}$	$+\frac{5}{16}$	$-\frac{3}{16}$	$-\frac{1}{8}$	$-\frac{1}{16}$	$-\frac{1}{16}$	$-\frac{1}{8}$	0	0	0	0	0	0	....
$\left(a^{(6)}_{38}\right)$	....	$+\frac{1}{32}$	$-\frac{5}{64}$	$+\frac{3}{64}$	$+\frac{1}{32}$	$+\frac{1}{64}$	$+\frac{1}{64}$	$+\frac{1}{32}$	$+\frac{1}{16}$	$-\frac{1}{16}$	$+\frac{1}{8}$	$-\frac{1}{8}$	$+\frac{1}{16}$	$-\frac{1}{16}$	....

TABLE XLVII-b . Continued

		$\bar{\omega}_{\xi x\xi x',\xi x}$	$\bar{\omega}_{\xi\eta\xi\eta',\xi x}$	$\bar{\omega}_{\xi\xi\xi\xi',\xi x}$	$\bar{\omega}_{\eta\eta\eta x',\xi\eta}$	$\bar{\omega}_{xxx\eta',\xi\eta}$	$\bar{\omega}_{x\eta xx',\xi\eta}$	$\bar{\omega}_{\eta x\eta\eta',\xi\eta}$	$\bar{\omega}_{\xi\eta x\xi',\xi\eta}$	$\bar{\omega}_{\xi\xi x\eta',\xi\eta}$	$\bar{\omega}_{\xi\xi\eta x',\xi\eta}$	$\bar{\omega}_{\eta\xi\xi x',\xi\eta}$	$\bar{\omega}_{x\xi\eta\xi',\xi\eta}$	$\bar{\omega}_{\xi x\xi\eta',\xi\eta}$
$\left(a^{(6)}_{34}\right)$	....	$-\frac{3}{8}$	$+\frac{3}{8}$	0	$+\frac{41}{32}$	$-\frac{53}{32}$	$-\frac{37}{32}$	$+\frac{57}{32}$	$+\frac{7}{8}$	$-\frac{1}{8}$	$-\frac{1}{8}$	$+\frac{3}{8}$	$+\frac{3}{8}$	$+\frac{3}{8}$
$\left(a^{(6)}_{36}\right)$	....	0	0	0	$-\frac{3}{8}$	$-\frac{1}{8}$	$-\frac{1}{8}$	$-\frac{3}{8}$	0	0	0	0	0	0
$\left(a^{(6)}_{38}\right)$	....	$+\frac{1}{8}$	$-\frac{1}{8}$	0	$+\frac{3}{32}$	$+\frac{1}{32}$	$+\frac{1}{32}$	$+\frac{3}{32}$	$-\frac{1}{8}$	$-\frac{1}{8}$	$-\frac{1}{8}$	$-\frac{1}{8}$	$-\frac{1}{8}$	$-\frac{1}{8}$

TABLE XLVII-c . Definition of the symbols appearing in the Tables XLV and XLVI

	$\mathfrak{B}_{\mathfrak{z}\mathfrak{z}\mathfrak{z}\mathfrak{h},\mathfrak{x}\mathfrak{x}}$	$\mathfrak{B}_{\mathfrak{z}\mathfrak{z}\mathfrak{z}\mathfrak{h},\mathfrak{h}\mathfrak{h}}$	$\mathfrak{B}_{\mathfrak{z}\mathfrak{h}\mathfrak{h}\mathfrak{h},\mathfrak{x}\mathfrak{x}}$	$\mathfrak{B}_{\mathfrak{z}\mathfrak{h}\mathfrak{h}\mathfrak{h},\mathfrak{h}\mathfrak{h}}$	$\mathfrak{B}_{\mathfrak{h}\mathfrak{z}\mathfrak{h}\mathfrak{h},\mathfrak{x}\mathfrak{x}}$	$\mathfrak{B}_{\mathfrak{h}\mathfrak{z}\mathfrak{h}\mathfrak{h},\mathfrak{h}\mathfrak{h}}$	$\mathfrak{B}_{\mathfrak{z}\mathfrak{h}\mathfrak{z}\mathfrak{z},\mathfrak{x}\mathfrak{x}}$	$\mathfrak{B}_{\mathfrak{z}\mathfrak{h}\mathfrak{z}\mathfrak{z},\mathfrak{h}\mathfrak{h}}$	$\mathfrak{B}_{\mathfrak{x}\mathfrak{z}\mathfrak{h}\mathfrak{x},\mathfrak{x}\mathfrak{x}}$	$\mathfrak{B}_{\mathfrak{x}\mathfrak{z}\mathfrak{h}\mathfrak{x},\mathfrak{h}\mathfrak{h}}$	$\mathfrak{B}_{\mathfrak{x}\mathfrak{x}\mathfrak{z}\mathfrak{h},\mathfrak{x}\mathfrak{x}}$	$\mathfrak{B}_{\mathfrak{x}\mathfrak{x}\mathfrak{z}\mathfrak{h},\mathfrak{h}\mathfrak{h}}$	$\mathfrak{B}_{\mathfrak{x}\mathfrak{x}\mathfrak{h}\mathfrak{z},\mathfrak{x}\mathfrak{x}}$	
$a^{(6)}_{35}$	$+\frac{15}{16}$	$-\frac{15}{16}$	$-\frac{23}{64}$	$+\frac{59}{64}$	$-\frac{39}{64}$	$+\frac{75}{64}$	$+\frac{11}{16}$	$-\frac{11}{16}$	$-\frac{53}{64}$	$+\frac{17}{64}$	$-\frac{21}{64}$	$-\frac{15}{64}$	$-\frac{21}{64}$	.....
$a^{(6)}_{37}$	0	0	$+\frac{1}{16}$	$-\frac{5}{16}$	$+\frac{1}{16}$	$-\frac{5}{16}$	0	0	$+\frac{3}{16}$	$+\frac{1}{16}$	$+\frac{3}{16}$	$+\frac{1}{16}$	$+\frac{3}{16}$	.....
$a^{(6)}_{39}$	$+\frac{1}{16}$	$-\frac{1}{16}$	$-\frac{1}{64}$	$+\frac{5}{64}$	$-\frac{1}{64}$	$+\frac{5}{64}$	$+\frac{1}{16}$	$-\frac{1}{16}$	$-\frac{3}{64}$	$-\frac{1}{64}$	$-\frac{3}{64}$	$-\frac{1}{64}$	$-\frac{3}{64}$	.....

TABLE XLVII-c . Continued

		$\mathfrak{W}_{xx\mathfrak{h}\mathfrak{z}',\mathfrak{h}\mathfrak{h}}$	$\mathfrak{W}_{\mathfrak{z}xx\mathfrak{h}',xx}$	$\mathfrak{W}_{\mathfrak{z}xx\mathfrak{h}',\mathfrak{h}\mathfrak{h}}$	$\mathfrak{W}_{x\mathfrak{z}x\mathfrak{h}',xx}$	$\mathfrak{W}_{x\mathfrak{z}x\mathfrak{h}',\mathfrak{h}\mathfrak{h}}$	$\mathfrak{W}_{x\mathfrak{h}x\mathfrak{z}',xx}$	$\mathfrak{W}_{x\mathfrak{h}x\mathfrak{z}',\mathfrak{h}\mathfrak{h}}$	$\mathfrak{W}_{\mathfrak{z}\mathfrak{z}\mathfrak{z}\mathfrak{h}',\mathfrak{z}\mathfrak{z}}$	$\mathfrak{W}_{\mathfrak{z}\mathfrak{h}\mathfrak{h}\mathfrak{h}',\mathfrak{z}\mathfrak{z}}$	$\mathfrak{W}_{\mathfrak{h}\mathfrak{z}\mathfrak{h}\mathfrak{h}',\mathfrak{z}\mathfrak{z}}$	$\mathfrak{W}_{\mathfrak{z}\mathfrak{h}\mathfrak{z}\mathfrak{z}',\mathfrak{z}\mathfrak{z}}$	$\mathfrak{W}_{x\mathfrak{z}\mathfrak{h}x',\mathfrak{z}\mathfrak{z}}$	$\mathfrak{W}_{xx\mathfrak{z}\mathfrak{h}',\mathfrak{z}\mathfrak{z}}$	
$\left(a_{35}^{(6)}\right)$	....	$-\frac{15}{64}$	$-\frac{37}{64}$	$+\frac{1}{64}$	$-\frac{37}{64}$	$+\frac{1}{64}$	$-\frac{37}{64}$	$+\frac{1}{64}$	0	$-\frac{9}{16}$	$-\frac{9}{16}$	0	$+\frac{9}{16}$	$+\frac{9}{16}$	.....
$\left(a_{37}^{(6)}\right)$	....	$+\frac{1}{16}$	$+\frac{3}{16}$	$+\frac{1}{16}$	$+\frac{3}{16}$	$+\frac{1}{16}$	$+\frac{3}{16}$	$+\frac{1}{16}$	0	0	0	0	0	0	.....
$\left(a_{39}^{(6)}\right)$	....	$-\frac{1}{64}$	$+\frac{3}{64}$	$-\frac{1}{64}$	$-\frac{3}{64}$	$-\frac{1}{64}$	$-\frac{3}{64}$	$-\frac{1}{64}$	0	$-\frac{1}{16}$	$-\frac{1}{16}$	0	$+\frac{1}{16}$	$+\frac{1}{16}$	.....

TABLE XLVII-c . Continued

		$\underline{\omega}_{xx\hbar\xi,\xi\xi}$	$\underline{\omega}_{\xi xx\hbar,\xi\xi}$	$\underline{\omega}_{x\xi x\hbar,\xi\xi}$	$\underline{\omega}_{x\hbar x\xi,\xi\xi}$	$\underline{\omega}_{\xi\xi\xi x,\hbar x}$	$\underline{\omega}_{\xi xxx,\hbar x}$	$\underline{\omega}_{x\xi xx,\hbar x}$	$\underline{\omega}_{\xi x\xi\xi,\hbar x}$	$\underline{\omega}_{\hbar\xi x\hbar,\hbar x}$	$\underline{\omega}_{\hbar\hbar\xi x,\hbar x}$	$\underline{\omega}_{\hbar\hbar x\xi,\hbar x}$	$\underline{\omega}_{\xi\hbar\hbar x,\hbar x}$	$\underline{\omega}_{\hbar\xi\hbar x,\hbar x}$	
$\left(a^{(6)}_{35}\right)$	....	$+\frac{9}{16}$	$+\frac{9}{16}$	$+\frac{9}{16}$	$+\frac{9}{16}$	$+\frac{15}{8}$	$-\frac{41}{32}$	$-\frac{57}{32}$	$+\frac{11}{8}$	$-\frac{35}{32}$	$-\frac{3}{32}$	$-\frac{3}{32}$	$-\frac{19}{32}$	$-\frac{19}{32}$	.....
$\left(a^{(6)}_{37}\right)$	....	0	0	0	0	0	$+\frac{3}{8}$	$+\frac{3}{8}$	0	$+\frac{1}{8}$	$+\frac{1}{8}$	$+\frac{1}{8}$	$+\frac{1}{8}$	$+\frac{1}{8}$	.....
$\left(a^{(6)}_{39}\right)$	....	$+\frac{1}{16}$	$+\frac{1}{16}$	$+\frac{1}{16}$	$+\frac{1}{16}$	$+\frac{1}{8}$	$-\frac{3}{32}$	$-\frac{3}{32}$	$+\frac{1}{8}$	$-\frac{1}{32}$	$-\frac{1}{32}$	$-\frac{1}{32}$	$-\frac{1}{32}$	$-\frac{1}{32}$	.....

TABLE XLVII-c . Continued

		$\mathfrak{B}_{\mathfrak{h}\mathfrak{x}\mathfrak{h}\mathfrak{z},\mathfrak{h}\mathfrak{x}}$	$\mathfrak{B}_{\mathfrak{x}\mathfrak{x}\mathfrak{x}\mathfrak{x},\mathfrak{z}\mathfrak{h}}$	$\mathfrak{B}_{\mathfrak{h}\mathfrak{h}\mathfrak{h}\mathfrak{h},\mathfrak{z}\mathfrak{h}}$	$\mathfrak{B}_{\mathfrak{h}\mathfrak{h}\mathfrak{x}\mathfrak{x},\mathfrak{z}\mathfrak{h}}$	$\mathfrak{B}_{\mathfrak{x}\mathfrak{h}\mathfrak{h}\mathfrak{x},\mathfrak{z}\mathfrak{h}}$	$\mathfrak{B}_{\mathfrak{h}\mathfrak{x}\mathfrak{x}\mathfrak{h},\mathfrak{z}\mathfrak{h}}$	$\mathfrak{B}_{\mathfrak{h}\mathfrak{x}\mathfrak{h}\mathfrak{x},\mathfrak{z}\mathfrak{h}}$	$\mathfrak{B}_{\mathfrak{x}\mathfrak{z}\mathfrak{z}\mathfrak{x},\mathfrak{z}\mathfrak{h}}$	$\mathfrak{B}_{\mathfrak{h}\mathfrak{z}\mathfrak{z}\mathfrak{h},\mathfrak{z}\mathfrak{h}}$	$\mathfrak{B}_{\mathfrak{z}\mathfrak{z}\mathfrak{x}\mathfrak{x},\mathfrak{z}\mathfrak{h}}$	$\mathfrak{B}_{\mathfrak{z}\mathfrak{z}\mathfrak{h}\mathfrak{h},\mathfrak{z}\mathfrak{h}}$	$\mathfrak{B}_{\mathfrak{z}\mathfrak{x}\mathfrak{x}\mathfrak{z},\mathfrak{z}\mathfrak{h}}$	$\mathfrak{B}_{\mathfrak{z}\mathfrak{h}\mathfrak{h}\mathfrak{z},\mathfrak{z}\mathfrak{h}}$	
$\left(a_{35}^{(6)}\right)$	....	$-\frac{19}{32}$	$+\frac{43}{64}$	$+\frac{51}{64}$	$-\frac{47}{32}$	$-\frac{31}{64}$	$-\frac{63}{64}$	$-\frac{47}{32}$	$-\frac{3}{16}$	$+\frac{3}{16}$	$+\frac{1}{8}$	$-\frac{1}{8}$	$-\frac{7}{16}$	$+\frac{7}{16}$	.....
$\left(a_{37}^{(6)}\right)$	....	$+\frac{1}{8}$	$+\frac{3}{16}$	$-\frac{5}{16}$	$+\frac{1}{8}$	$+\frac{1}{16}$	$+\frac{1}{16}$	$+\frac{1}{8}$	0	0	0	0	0	0	.....
$\left(a_{39}^{(6)}\right)$	....	$-\frac{1}{32}$	$-\frac{3}{64}$	$+\frac{5}{64}$	$-\frac{1}{32}$	$-\frac{1}{64}$	$-\frac{1}{64}$	$-\frac{1}{32}$	$+\frac{1}{16}$	$-\frac{1}{16}$	$+\frac{1}{8}$	$-\frac{1}{8}$	$+\frac{1}{16}$	$-\frac{1}{16}$	.....

TABLE XLVII-c . Continued

		$\bar{\omega}_{\xi x\xi x,\xi\eta}$	$\bar{\omega}_{\xi\eta\xi\eta,\xi\eta}$	$\bar{\omega}_{\xi\xi\xi\xi,\xi\eta}$	$\bar{\omega}_{\eta\eta\eta x,\xi x}$	$\bar{\omega}_{xx\xi\eta,\xi x}$	$\bar{\omega}_{x\eta\eta x,\xi x}$	$\bar{\omega}_{\eta x\eta\eta,\xi x}$	$\bar{\omega}_{\xi\eta x\xi,\xi x}$	$\bar{\omega}_{\xi\xi x\eta,\xi x}$	$\bar{\omega}_{\xi\xi\eta x,\xi x}$	$\bar{\omega}_{\eta\xi\xi x,\xi x}$	$\bar{\omega}_{x\xi\eta\xi,\xi x}$	$\bar{\omega}_{\xi x\xi\eta,\xi x}$
$(a^{(6)}_{35})$	....	$-\frac{3}{8}$	$+\frac{3}{8}$	0	$+\frac{53}{32}$	$-\frac{41}{32}$	$-\frac{57}{32}$	$+\frac{37}{32}$	$-\frac{7}{8}$	$+\frac{1}{8}$	$+\frac{1}{8}$	$-\frac{3}{8}$	$-\frac{3}{8}$	$-\frac{3}{8}$
$(a^{(6)}_{37})$	....	0	0	0	$+\frac{1}{8}$	$+\frac{3}{8}$	$+\frac{3}{8}$	$+\frac{1}{8}$	0	0	0	0	0	0
$(a^{(6)}_{39})$	....	$+\frac{1}{8}$	$-\frac{1}{8}$	0	$-\frac{1}{32}$	$-\frac{3}{32}$	$-\frac{3}{32}$	$-\frac{1}{32}$	$+\frac{1}{8}$	$+\frac{1}{8}$	$+\frac{1}{8}$	$+\frac{1}{8}$	$+\frac{1}{8}$	$+\frac{1}{8}$

TABLE XLVII-d . Definition of the symbols appearing in the Tables XLV and XLVI

	$\lambda_{xzx'xxx}$	$\lambda_{hzh'xxx}$	$\lambda_{zxx'xxx}$	$\lambda_{zhh'xxx}$	$\lambda_{xzx'hhx}$	$\lambda_{hzh'hhx}$	$\lambda_{zxx'hhx}$	$\lambda_{zhh'hhx}$	$\lambda_{xzx'hxh}$	$\lambda_{hxh'hxh}$	$\lambda_{zxx'hxh}$	$\lambda_{zhh'hxh}$	$\lambda_{zzz'xxx}$	
$b^{(6)}_{34}$	$-\frac{67}{128}$	$-\frac{57}{128}$	$-\frac{59}{64}$	$-\frac{49}{64}$	$+\frac{63}{64}$	$+\frac{61}{64}$	$+\frac{55}{32}$	$+\frac{53}{32}$	$+\frac{55}{128}$	$+\frac{69}{128}$	$+\frac{47}{64}$	$+\frac{61}{64}$	$+\frac{27}{32}$	.....
$b^{(6)}_{36}$	$+\frac{5}{32}$	$-\frac{1}{32}$	$+\frac{5}{16}$	$-\frac{1}{16}$	$-\frac{1}{16}$	$-\frac{3}{16}$	$-\frac{1}{8}$	$-\frac{3}{8}$	$-\frac{1}{32}$	$-\frac{3}{32}$	$-\frac{1}{16}$	$-\frac{3}{16}$	0	.....
$b^{(6)}_{38}$	$-\frac{5}{128}$	$+\frac{1}{128}$	$-\frac{5}{64}$	$+\frac{1}{64}$	$+\frac{1}{64}$	$+\frac{3}{64}$	$+\frac{1}{32}$	$+\frac{3}{32}$	$+\frac{1}{128}$	$+\frac{3}{128}$	$+\frac{1}{64}$	$+\frac{3}{64}$	$+\frac{1}{32}$	.....

TABLE XLVII-d . Continued

		$\ell_{\xi\xi\xi'\eta\eta x}$	$\ell_{\xi\xi\xi'\eta x\eta}$	$\ell_{\xi xx'\xi\xi x}$	$\ell_{\xi\eta\eta'\xi\xi x}$	$\ell_{\xi\eta\eta'\xi\xi x}$	$\ell_{\eta\xi\eta'\xi\xi x}$	$\ell_{\xi xx'\xi x\xi}$	$\ell_{\xi\eta\eta'\xi x\xi}$	$\ell_{x\xi x'\xi x\xi}$	$\ell_{\eta\xi\eta'\xi x\xi}$	$\ell_{\xi\eta x'\eta\eta\eta}$	$\ell_{\xi x\eta'\eta\eta\eta}$	$\ell_{\eta\xi x'\eta\eta\eta}$	
$\left(b^{(6)}_{34}\right)$	....	$-\frac{54}{32}$	$-\frac{27}{32}$	$+\frac{1}{8}$	$-\frac{1}{8}$	$+\frac{1}{16}$	$-\frac{1}{16}$	$-\frac{1}{16}$	$+\frac{1}{16}$	$-\frac{1}{32}$	$+\frac{1}{32}$	$+\frac{5}{64}$	$+\frac{5}{64}$	$+\frac{5}{64}$	.....
$\left(b^{(6)}_{36}\right)$	....	0	0	0	0	0	0	0	0	0	0	$-\frac{3}{16}$	$-\frac{3}{16}$	$-\frac{3}{16}$	.....
$\left(b^{(6)}_{38}\right)$	....	$-\frac{1}{16}$	$-\frac{1}{32}$	$+\frac{1}{8}$	$-\frac{1}{8}$	$+\frac{1}{16}$	$-\frac{1}{16}$	$+\frac{1}{16}$	$-\frac{1}{16}$	$+\frac{1}{32}$	$-\frac{1}{32}$	$+\frac{3}{64}$	$+\frac{3}{64}$	$+\frac{3}{64}$	.....

TABLE XLVII-d . Continued

	$\lambda_{\mathfrak{zyx},\mathfrak{yxx}}$	$\lambda_{\mathfrak{zxy},\mathfrak{yxx}}$	$\lambda_{\mathfrak{yzx},\mathfrak{yxx}}$	$\lambda_{\mathfrak{zyx},\mathfrak{xyx}}$	$\lambda_{\mathfrak{zxy},\mathfrak{xyx}}$	$\lambda_{\mathfrak{yzx},\mathfrak{xyx}}$	$\lambda_{\mathfrak{zyx},\mathfrak{zzy}}$	$\lambda_{\mathfrak{zxy},\mathfrak{zzy}}$	$\lambda_{\mathfrak{yzx},\mathfrak{zzy}}$	$\lambda_{\mathfrak{zyx},\mathfrak{zyz}}$	$\lambda_{\mathfrak{zxy},\mathfrak{zyz}}$	$\lambda_{\mathfrak{yzx},\mathfrak{zyz}}$
$(b_{34}^{(6)})$	$-\frac{1}{32}$	$-\frac{1}{32}$	$-\frac{1}{32}$	$+\frac{7}{64}$	$+\frac{7}{64}$	$+\frac{7}{64}$	$-\frac{1}{8}$	$-\frac{1}{8}$	$-\frac{1}{8}$	$+\frac{1}{16}$	$+\frac{1}{16}$	$+\frac{1}{16}$
$(b_{36}^{(6)})$	$-\frac{1}{8}$	$-\frac{1}{8}$	$-\frac{1}{8}$	$-\frac{1}{16}$	$-\frac{1}{16}$	$-\frac{1}{16}$	0	0	0	0	0	0
$(b_{38}^{(6)})$	$+\frac{1}{32}$	$+\frac{1}{32}$	$+\frac{1}{32}$	$+\frac{1}{64}$	$+\frac{1}{64}$	$+\frac{1}{64}$	$-\frac{1}{8}$	$-\frac{1}{8}$	$-\frac{1}{8}$	$-\frac{1}{16}$	$-\frac{1}{16}$	$-\frac{1}{16}$

TABLE XLVII-e . Definition of the symbols appearing in the Tables XLV and XLVI

	$\lambda_{x\ell x',\hbar\hbar\hbar}$	$\lambda_{\hbar\ell\hbar',\hbar\hbar\hbar}$	$\lambda_{\ell xx',\hbar\hbar\hbar}$	$\lambda_{\ell\hbar\hbar',\hbar\hbar\hbar}$	$\lambda_{x\ell x',\hbar xx}$	$\lambda_{\hbar\ell\hbar',\hbar xx}$	$\lambda_{\ell xx',\hbar xx}$
$b^{(6)}_{35}$	$+\frac{57}{128}$	$+\frac{67}{128}$	$+\frac{49}{64}$	$+\frac{59}{64}$	$-\frac{61}{64}$	$-\frac{63}{64}$	$-\frac{53}{32}$
$b^{(6)}_{37}$	$+\frac{1}{32}$	$-\frac{5}{32}$	$+\frac{1}{16}$	$-\frac{5}{16}$	$+\frac{3}{16}$	$+\frac{1}{16}$	$+\frac{3}{8}$
$b^{(6)}_{39}$	$-\frac{1}{128}$	$+\frac{5}{128}$	$-\frac{1}{64}$	$+\frac{5}{64}$	$-\frac{3}{64}$	$-\frac{1}{64}$	$-\frac{3}{32}$

	$\lambda_{\ell\hbar\hbar',\hbar xx}$	$\lambda_{x\ell x',x\hbar x}$	$\lambda_{\hbar\ell\hbar',x\hbar x}$	$\lambda_{\ell xx',x\hbar x}$	$\lambda_{\ell\hbar\hbar',x\hbar x}$	$\lambda_{\ell\ell\ell',\hbar\hbar\hbar}$	
$b^{(6)}_{35}$	$-\frac{55}{32}$	$-\frac{69}{128}$	$-\frac{55}{128}$	$-\frac{61}{64}$	$-\frac{47}{64}$	$-\frac{27}{32}$	.....
$b^{(6)}_{37}$	$+\frac{1}{8}$	$+\frac{3}{32}$	$+\frac{1}{32}$	$+\frac{3}{16}$	$+\frac{1}{16}$	0	.....
$b^{(6)}_{39}$	$-\frac{1}{32}$	$-\frac{3}{128}$	$-\frac{1}{128}$	$-\frac{3}{64}$	$-\frac{1}{64}$	$-\frac{1}{32}$	.....

TABLE XLVII-e . Continued

		$\lambda_{\xi\xi\xi,\eta x x}$	$\lambda_{\xi\xi\xi,x\eta x}$	$\lambda_{\xi x x,\xi\xi\eta}$	$\lambda_{\xi\eta\eta,\xi\xi\eta}$	$\lambda_{\xi\eta\eta,\xi\xi\eta}$	$\lambda_{\eta\xi\eta,\xi\xi\eta}$	$\lambda_{\xi x x,\xi\eta\xi}$	$\lambda_{\xi\eta\eta,\xi\eta\xi}$	$\lambda_{x\xi x,\xi\eta\xi}$	$\lambda_{\eta\xi\eta,\xi\eta\xi}$	$\lambda_{\xi\eta x,x x x}$	$\lambda_{\xi x\eta,x x x}$	$\lambda_{\eta\xi x,x x x}$	
$\left(b^{(6)}_{35}\right)$	....	$+\frac{54}{32}$	$+\frac{27}{32}$	$+\frac{1}{8}$	$-\frac{1}{8}$	$+\frac{1}{16}$	$-\frac{1}{16}$	$-\frac{1}{16}$	$+\frac{1}{16}$	$-\frac{1}{32}$	$+\frac{1}{32}$	$-\frac{5}{64}$	$-\frac{5}{64}$	$-\frac{5}{64}$	.....
$\left(b^{(6)}_{37}\right)$	....	0	0	0	0	0	0	0	0	0	0	$+\frac{3}{16}$	$+\frac{3}{16}$	$+\frac{3}{16}$	.....
$\left(b^{(6)}_{39}\right)$	....	$+\frac{1}{16}$	$+\frac{1}{32}$	$+\frac{1}{8}$	$-\frac{1}{8}$	$+\frac{1}{16}$	$-\frac{1}{16}$	$+\frac{1}{16}$	$-\frac{1}{16}$	$+\frac{1}{32}$	$-\frac{1}{32}$	$-\frac{3}{64}$	$-\frac{3}{64}$	$-\frac{3}{64}$	.....

TABLE XLVII-e . Continued

	$\ell_{\mathfrak{z}\mathfrak{h}\mathfrak{x}',\mathfrak{h}\mathfrak{h}\mathfrak{x}}$	$\ell_{\mathfrak{z}\mathfrak{x}\mathfrak{h}',\mathfrak{h}\mathfrak{h}\mathfrak{x}}$	$\ell_{\mathfrak{h}\mathfrak{z}\mathfrak{x}',\mathfrak{h}\mathfrak{h}\mathfrak{x}}$	$\ell_{\mathfrak{z}\mathfrak{h}\mathfrak{x}',\mathfrak{h}\mathfrak{x}\mathfrak{h}}$	$\ell_{\mathfrak{z}\mathfrak{x}\mathfrak{h}',\mathfrak{h}\mathfrak{x}\mathfrak{h}}$	$\ell_{\mathfrak{h}\mathfrak{z}\mathfrak{x}',\mathfrak{h}\mathfrak{x}\mathfrak{h}}$	$\ell_{\mathfrak{z}\mathfrak{h}\mathfrak{x}',\mathfrak{z}\mathfrak{z}\mathfrak{x}}$	$\ell_{\mathfrak{z}\mathfrak{x}\mathfrak{h}',\mathfrak{z}\mathfrak{z}\mathfrak{x}}$	$\ell_{\mathfrak{h}\mathfrak{z}\mathfrak{x}',\mathfrak{z}\mathfrak{z}\mathfrak{x}}$	$\ell_{\mathfrak{z}\mathfrak{h}\mathfrak{x}',\mathfrak{z}\mathfrak{x}\mathfrak{z}}$	$\ell_{\mathfrak{z}\mathfrak{x}\mathfrak{h}',\mathfrak{z}\mathfrak{x}\mathfrak{z}}$	$\ell_{\mathfrak{h}\mathfrak{z}\mathfrak{x}',\mathfrak{z}\mathfrak{x}\mathfrak{z}}$
$\left(b_{35}^{(6)}\right)$	$+\frac{1}{32}$	$+\frac{1}{32}$	$+\frac{1}{32}$	$-\frac{7}{64}$	$-\frac{7}{64}$	$-\frac{7}{64}$	$+\frac{1}{8}$	$+\frac{1}{8}$	$+\frac{1}{8}$	$-\frac{1}{16}$	$-\frac{1}{16}$	$-\frac{1}{16}$
$\left(b_{37}^{(6)}\right)$	$+\frac{1}{8}$	$+\frac{1}{8}$	$+\frac{1}{8}$	$+\frac{1}{16}$	$+\frac{1}{16}$	$+\frac{1}{16}$	0	0	0	0	0	0
$\left(b_{39}^{(6)}\right)$	$-\frac{1}{32}$	$-\frac{1}{32}$	$-\frac{1}{32}$	$-\frac{1}{64}$	$-\frac{1}{64}$	$-\frac{1}{64}$	$+\frac{1}{8}$	$+\frac{1}{8}$	$+\frac{1}{8}$	$+\frac{1}{16}$	$+\frac{1}{16}$	$+\frac{1}{16}$

NOTES TO CHAPTER IX

1. Matrix elements of $h_4^\dagger$ have been computed for asymmetric molecules by Parker and his coworkers ([50] and references contained therein).
2. The expressions of coefficients appearing in Tables XLIII and XLIV in terms of molecular constants are very complicated. However, in the case of linear molecules, the coefficients μ_1^J and μ_1^{tJ}, μ_1^{nJ} (related to $h_{41}^\dagger$ and $h_{42}^\dagger$ respectively in Table XLIII) have been computed in references [53] and [22, 54].

PART 3

ENERGY LEVELS, FREQUENCIES OF ROTATION AND ROTATION-VIBRATION TRANSITIONS

CHAPTER X

MOLECULES WITH A THREEFOLD SYMMETRY

INTRODUCTION

We shall consider in this chapter axially symmetric molecules with a threefold symmetry and compute the wave numbers of transitions belonging to rotation and rotation-vibration spectra, assuming that no accidental resonances occur. Calculations will be carried out to fourth order in the latter case and to sixth order in the former. Results will be given for both cases (rotation and rotation-vibration transitions). As a first step, rotation-vibration energies will be given. For this purpose we shall first write down the Hamiltonian matrix, then factor it into submatrices and finally diagonalize the submatrices.

In writing down the Hamiltonian matrix, we shall use, as basic wave functions, the eigenfunctions ψ_0 of zeroth order Hamiltonian defined by

$$H_0 \psi_0 = E_0 \psi_0 \qquad (X,1)$$

and characterized by quantum numbers $J K M v_s \ldots l_t \ldots$. In this representation, H is diagonal with respect to J and M. It has furthermore been shown in Chapters II, III and IV that a practical diagonalization with respect to quantum numbers v_s (i.e. v_n and v_t) can be achieved by submiting the Hamiltonian to two contact transformations, i.e. by replacing H by

$$H^{\dagger} = H_0 + h_1' + h_2^{\dagger} + h_3^{\dagger} + h_4^{\dagger} + \ldots \qquad (X,2)$$

The first three terms on the right hand side of equation (X,2) are diagonal with respect to all quantum numbers v_s. The matrix elements of $h_3^\dagger$, $h_4^\dagger$ nondiagonal with respect to v_s will contribute to the energy in orders 6, 7 ... ; they can be ignored in a calculation performed to fourth order. On the other hand, since we want to compute rotation transitions up to sixth order, the influence of matrix elements of $h_3^\dagger$ nondiagonal in v_s should in principle be taken into account. However it can be shown that, for low values of J, only matrix elements originating from vibrational operators contribute to the energy in the sixth order and their contributions cancel out, for pure rotational transitions. Since the sixth order approximation is needed essentially for transitions observed in the microwave spectrum, where indeed the value of J is small, we shall be concerned in all instances with matrix elements of $H^\dagger$ which are diagonal in the quantum numbers J, M, v_s ... but possibly nondiagonal with respect to K, ℓ_t ... We shall write them as

$$(K\, \ell_t\, \ell_{t'} \dots |H^\dagger| K + \Delta K, \ell_t + \Delta\ell_t, \ell_{t'} + \Delta\ell_{t'} \dots) \qquad (X,3)$$

It can easily be shown from the arguments developed in Chapter VI that, among these matrix elements, only those for which

$$-\Delta K + \sum_t \Delta\ell_t = 0, \pm 3, \pm 6 \qquad (X,4)$$

are non-vanishing for a molecule with a threefold axis. The expression of these matrix elements in terms of quantum numbers can be found in Chapters VIII and IX.

However, the discussion on orders of magnitude presented in Chapter VII shows that certain matrix elements, although they originate from operators belonging to $h_2^\dagger$, $h_3^\dagger$ or $h_4^\dagger$ actually contribute to the energy with an index of magnitude larger than 4 and consequently can be neglected in a computation performed to fourth order. Let us recall that, according to Chapter VII, the true index of magnitude n of a matrix element depends upon the values of rotational quantum numbers ; this situation has lead to considering three cases :

$$J \simeq K \simeq 30$$

$$J \simeq 30 \; ; \; K \simeq 1$$

$$J \simeq K \simeq 1$$

We have studied the rotation-vibration energy in these three cases. In the first two cases (J large) , we have included contributions up to fourth order. In the third case (J and K small) we have included, up to sixth order, contributions originating from rotational operators in order to be effective in analysing the very accurate measurements of rotation lines in the microwave absorption spectrum.

A detailed discussion of the index of magnitude h of the contributions of all matrix elements to the energy can be found in Reference [8] where the various situations described in Chapter VII (weak resonances, doubling, second order strong resonances, accidentally strong essential resonances) are considered (see also Table XL in Chapter IX).

Due to symmetry properties of wave functions ψ_0 , the matrix $H^\dagger$ breaks up into one submatrix of species A and two submatrices of species E , these last two submatrices having identical eigenvalues. If the molecule belongs to one of the groups C_{3v} , D_3 , D_{3h} , D_{3d} (which is the case for most known molecules with a threefold axis), the matrix $H^\dagger$ can be obtained in a completely reduced form[1] by using new basic wave functions, which are symmetric and antisymmetric linear combinations of the functions ψ_0 :

$$\psi^{\pm}(|K|\, \ell_t \ell_{t'} \ldots) = \frac{1}{\sqrt{2}} \Big[\psi_0(K=|K|, \ell_t, \ell_{t'} \ldots) \pm \psi_0(K=-|K|, -\ell_t, -\ell_{t'} \ldots) \Big] \qquad (X,5)$$

This will have the effect to split the matrix of species A into two submatrices of species A_1 and A_2 .

The last step to obtain rotation-vibration energies is to diagonalize the submatrices A_1 A_2 E , which will be done by a perturbation method whenever possible, or otherwise by solving secular determinants.

ROTATION-VIBRATION ENERGY OF MOLECULES C_{3v}, D_3, D_{3h}, D_{3d}

The rotation-vibration energy has been computed for the following vibrational states :

- states in which no degenerate vibration is excited
- states in which one degenerate vibration t is excited by 1, 2 or 3 quanta (v_t = 1,2,3)
- states in which two degenerate vibrations t and t' are simultaneously excited by 1 quantum ($v_t = v_{t'} = 1$).

The number of excited non degenerate vibrations is, in each case, arbitrary.

For each vibrational state, three ranges have been considered for rotational quantum numbers :

A $J \simeq K \simeq 1$

B $J \simeq K \simeq 30$

C $J \simeq 30$, $K \simeq 1$

Rotation-vibration energies are given in Tables XLVIII for all instances where an explicit formula has been obtained. These results hold for any molecule whose equilibrium configuration belongs to one of the point groups C_{3v}, D_3, D_{3h}, D_{3d}.

In the labeling of Tables XLVIII the numbers 1 to 8 refer to the various vibrational states as follows :

1 $v_t = v_{t'} = \dots = 0$ — $v_n, v_{n'} \dots$ arbitrary

2 $v_t = 1$; $\ell_t = \pm 1$

3 $v_t = 2$; $\ell_t = 0$

4 $v_t = 2$; $\ell_t = \pm 2$

5 $v_t = 3$; $\ell_t = \pm 1$

6 $v_t = 3$; $\ell_t = \pm 3$

(cases 2–6: $v_{t'} = \dots = 0$)

7 $v_t = v_{t'} = 1$; $\ell_t = -\ell_{t'} = \pm 1$

8 $v_t = v_{t'} = 1$; $\ell_t = \ell_{t'} = \pm 1$

(cases 7–8: $v_{t''} = \dots = 0$; cases 1–8: $v_n, v_{n'} \dots$ arbitrary)

The letters A B C refer to the three ranges defined above for rotational quantum numbers.

The energies given in Tables XLVIII , 1A , 1B , 1C , 2A , 2B , 3A , 3B , 4A , 4B , 5A , 5B , 6A , 6B , 7B , and 8B have been obtained by using a perturbation method to diagonalize the Hamiltonian submatrices. The energies given in Tables XLVIII , 2C , 7A and 8A have been obtained by solving secular equations of degree two, which leads to expressions much more complicated than in the preceding case[2]. There are no Tables XLVIII , 3C to 8C , because when $v_t = 2, 3$ or $v_t = v_{t'} = 1$, with $J \simeq 30$, $K \simeq 1$, the computation of the energy requires the solution of secular equations with a degree larger than two and no general explicit formula can be derived.

In Tables XLVIII, the rotation-vibration energies are given in terms of rotational quantum numbers and coefficients which depend upon molecular parameters and vibrational quantum numbers. These coefficients can be expanded in power series with respect to vibrational quantum numbers as shown[3] in Table XLIX. In the approximation we use, we need 1, 2 or 3 terms (respectively designated as a, b and c) in the expansions. Using the expansions of the coefficients given in Table XLIX, one can expand the various contributions to the energy given in Tables XLVIII. The last three columns on the right side of these tables give for each contribution the indices of magnitude of the first, second and third terms obtained in such expansions. When the indices of magnitude of the third or second terms are too small, they are not given. Contributions to the energy are listed in Tables XLVIII up to fourth order. In Tables A however, where both J and K are small, contributions originating from rotational operators are given up to sixth order.

In each table, the rotation-vibration energy will be obtained by adding up the various contributions listed under each other. This procedure is straightforward in Tables XLVIII, 1 A, 1 B, 2 B, 3 A, 3 B, 4 B, 5 B, 6 B, 7 B and 8 B. In the other tables however, it has not been possible to give, for every contribution, a formula holding for any level. For the contributions to the energy listed about the end of these tables, different expressions must be used according to the values of quantum numbers.

In Tables XLVIII, 1 C, 2 A, 5 A and 6 A, three possibilities are considered for the last contributions to the energy as follows :

TABLE XLIII, 1 C

- - - - - - - - - -	
$\|K\| \neq 3$	
$\|K\| = 3$	ψ_3^+ state
	ψ_3^- state

TABLES XLVIII, 2 A AND 5 A

- - - - - - - - - -	
$K\ell_t \neq 1$	
$K\ell_t = 1$	$\psi_{1,1}^+$ state
	$\psi_{1,1}^-$ state

TABLE XLVIII, 6 A

- - - - - - - - - -	
$\|K\| \neq 0$	
$K = 0$	$\psi_{0,3}^+$ state
	$\psi_{0,3}^-$ state

In Table XLVIII, 1 C, the expression to be used for the last contribution to the energy is different for $|K| \neq 3$ and for $|K| = 3$. For the latter level where splitting A_1 A_2 occurs, the contribution is given first for the symmetric state ψ_3^+, then for the antisymmetric state ψ_3^-. The same situation is found in Tables XLVIII, 2 A and 5 A where the splitting A_1 A_2 occurs for the level $K\ell_t = 1$, and in Table 6 A where splitting A_1 A_2 occurs for $K = 0$.

In Tables XLVIII, 4 A and 8 A, the situation is more complex and the possibilities to be considered for the last two contributions can be summarized as follows :

TABLE XLVIII, 4 A

- - - - - - - - - - - - - - -		
$K\ell_t \neq -2$		(I)
$K\ell_t \neq 4$		(II)
$K\ell_t = 4$	$\psi_{2,2}^+$ state	(III)
	$\psi_{2,2}^-$ state	(IV)
$K\ell_t = -2$	$\psi_{1,-2}^+$ state	(V)
	$\psi_{1,-2}^-$ state	(VI)

TABLE XLVIII, 8 A

- - - - - - - - - - - - - - -	
$K\ell_t \neq -1$	
$K\ell_t \neq 2$	
$K\ell_t = 2$	$\psi_{2,1,1}^+$ state
	$\psi_{2,1,1}^-$ state
$K\ell_t = -1$	$\psi_{1,-1,-1}^+$ state
	$\psi_{1,-1,-1}^-$ state

In these tables, the last two contributions to the energy are obtained by selecting two out of the six expressions given at the end of the table. In Table XLVIII, 4 A for instance, the expressions (I) to (VI) appear as shown hereafter in the energy of the different states :

$K\ell_t \neq -2$ and 4 : (I) + (II)

$K\ell_t = 4$ $\begin{cases} \psi^+_{2,2} \text{ state} & : \text{(I) + (III)} \\ \psi^-_{2,2} \text{ state} & : \text{(I) + (IV)} \end{cases}$

$K\ell_t = -2$ $\begin{cases} \psi^+_{1,-2} \text{ state} & : \text{(II) + (V)} \\ \psi^-_{1,-2} \text{ state} & : \text{(II) + (VI)} \end{cases}$

In Table XLVIII, 2 C, the energies are expressed[2] in terms of the quantum number $L = K - \ell_t$, instead of K and ℓ_t. The various possibilities to be considered for the last contributions to the energy can be summarized as follows[4] according to the values of $|L|$.

TABLE XLVIII, 2 C

- -		
$\lvert L\rvert \neq 0$		(I)
$\lvert L\rvert \neq 0, 3$		(II)
$\lvert L\rvert = 3$	φ^+_a and φ^+_b states	(III)
	φ^-_a and φ^-_b states	(IV)
$L = 0$	$\psi^+_{1,1}$ state	(V)
	$\psi^-_{1,1}$ state	(VI)

One or two out of the six expressions (I) to (VI) appear in the energy of the different states, as shown hereafter :

$|L| \neq 0$ and 3 $\begin{cases} \text{level } E^+ & : \text{(I) + (II), with the upper signs} \\ \text{level } E^- & : \text{(I) + (II), with the lower signs} \end{cases}$

$|L| = 3$
- φ_a^+ state : (I) + (III), with the upper signs
- φ_b^+ state : (I) + (III), with the lower signs
- φ_a^- state : (I) + (IV), with the upper signs
- φ_b^- state : (I) + (IV), with the lower signs

$L = 0$
- $\psi_{1,1}^+$ state : (V)
- $\psi_{1,1}^-$ state : (VI)

In Table XLVIII, 7 A, the energies are not expressed[2] in terms of K, ℓ_t and $\ell_{t'}$, but in terms of $|K|$, with two values of the energy for each value of $|K|$. The cases to be considered for the last contributions to the energy are the following :

TABLE XLVIII, 7 A

- - - - - - - - - - - -		
$\|K\| \neq 0$		(I)
$K = 0$	$\psi_{0,1,-1}^+$ state	(II)
	$\psi_{0,1,-1}^-$ state	(III)

One out of the three expressions (I) to (III) appears in the energy as shown hereafter :

$|K| \neq 0$
- level E^+ : (I), with the upper signs
- level E^- : (I), with the lower signs

$K = 0$
- $\psi_{0,1,-1}^+$ state : (II)
- $\psi_{0,1,-1}^-$ state : (III)

At the present point, we found it convenient to use for certain coefficients appearing in the diagonal and nondiagonal matrix elements a notation different from the one used in Chapters VIII and IX.

The correspondance between these two notations is given in Table L, where the symbols written in the first column are the ones used in the present chapter, while those written in the second column are the corresponding symbols used in the preceding chapters.

FREQUENCIES OF ROTATION AND ROTATION-VIBRATION LINES FOR C_{3v} MOLECULES

Wave numbers have been computed for various rotation and rotation-vibration lines of C_{3v} molecules. These results hold for a number of simple molecules for which high resolution microwave or infrared absorption spectra can conveniently be obtained : X_3YZ molecules as methyl halides, X_3YZT molecules as methyl cyanide, X_3YZTU molecules as methyl acetylene and also pyramidal X_3Y molecules as arsine when inversion is neglected.

A . Selection Rules

Selection rules for dipole transitions have been derived by various authors [55,56,57,25][5] by considering a transition moment

$$\mu = \int_\tau \bar{\psi}_0'' \, M \, \psi_0' \, d\tau \tag{X,6}$$

where ψ_0' and ψ_0'' are zeroth order eigenfunctions defined by a set of quantum numbers $J K M v_s \ldots l_t \ldots$. When corrections to the energy of order larger than or equal to 2 are taken into account, the basic wave functions used in the practical computation of transition moments are often symmetric and antisymmetric linear combinations $\psi^{\pm}$ of the functions ψ_0 . The selection rules established from the relation (X,6) need then to be completed. Functions ψ^+ and ψ^- defined by equation (X,5) are basis of irreductible representations of the group C_{3v} , so that they can be labeled[1] as A_1 A_2 or E . By using symmetry properties of these functions, it has been possible to establish, for dipole transitions of C_{3v} molecules, the complete selection rules given in Table LI. These selection rules hold when inversion is neglected and when no accidentally strong resonance occurs (see Chapter VII).

B . Frequencies of Rotation Lines

Line frequencies have been computed for rotation transitions taking place in any one of the vibrational levels considered above. Results are given in Tables LII where the values of J and K have been assumed to be small (case A defined above) and where the numbering 1 ... 8 refers to the same different vibrational levels as indicated above for Tables XLVIII. Frequencies are expressed in terms of rotational quantum numbers J and K associated with the lower level of the transition. A symbol $\nu_{J,|K|}$ is used for the frequency in levels (1 and 3) where all ℓ_t are equal to zero ; a symbol $\nu_{J,K\ell_t}$ is used in the levels 2, 4, 5, 6 and 8 (where one ℓ_t is different from zero or where $\ell_t = \ell_{t'} = \pm 1$) ; symbols $\nu^+_{J,|K|}$ (or $\nu'^+_{J,|K|}$) and $\nu^-_{J,|K|}$ (or $\nu'^-_{J,|K|}$) are used for the frequency in the level 7 (where $\sum_t \ell_t = 0$ with $\ell_t = -\ell_{t'} = \pm 1$). The symbols $\nu^+_{J,|K|}$ (or $\nu'^+_{J,|K|}$) and $\nu^-_{J,|K|}$ (or $\nu'^-_{J|K|}$) correspond respectively to the cases where the upper level of the transition is $E^+_{(J+1,|K|)}$ or $E^-_{(J+1,|K|)}$, as defined in note 2 and given in Table XLVIII, 7 A .

The frequencies are obtained from the Tables LII by adding up the various contributions in the same manner as described for the energies given in Tables XLVIII ; here again it may happen that several cases need to be considered for the last contributions to the frequency according to the values of $|K|$ or $K\ell_t$. In Tables LII , 2 and 5 for instance, the last contribution to the frequency is different for the lines with $K\ell_t \neq 1$ and $K\ell_t = 1$. In the latter case, where a splitting of the line occurs, the contribution is given first for the $\psi^{+} \rightarrow \psi^{+}$ transition, then for the $\psi^{-} \rightarrow \psi^{-}$ transition. In Tables LII , 4 and 8, the situation for the last two contributions to the frequency is the same as in Tables XLVIII , 4 A and 8 A for the last two contributions to the energy.

In the vibrational state 7 $(v_t = v_{t'} = 1 \, , \, \ell_t = -\ell_{t'} = \pm 1)$, as a consequence of the strong vibrational resonance which characterizes this state, one is lead to consider two different kinds of rotational transitions as shown on figure (1) :

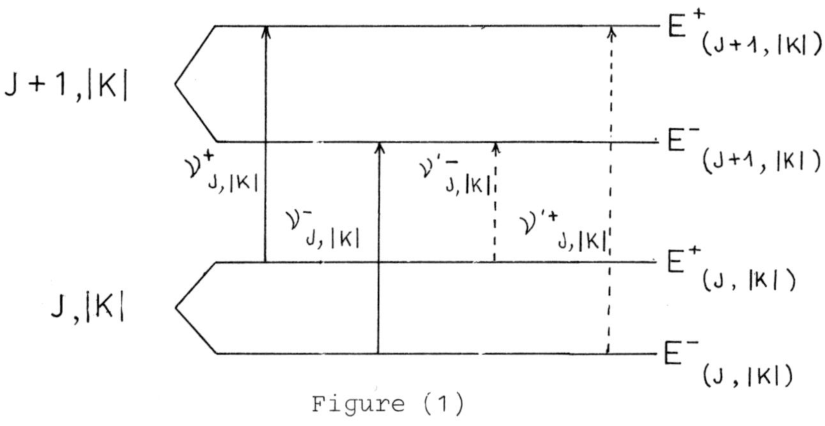

Figure (1)

(a) the two transitions drawn with unbroken lines in the figure obey the normal selection rules ;

(b) two new transitions (generally less intense) are shown with dashed lines. They occur as a consequence of the mixing of states due to the resonance.

Frequencies of these two kinds of transitions are given in Tables LII, 7a and LII, 7b respectively ; in these tables, the frequencies of the lines $\nu^{+}_{J,|K|}$ (or $\nu'^{+}_{J,|K|}$) and $\nu^{-}_{J,|K|}$ (or $\nu'^{-}_{J|K|}$) for $|K| \neq 0$ are obtained by using in the last contributions respectively the upper and lower signs.

Among others, Tables LII give the frequencies of rotation lines in the ground vibrational state of the molecule and in its low excited vibrational states : $v_s = 1, 2, 3$ and $v_s = v_{s'} = 1$, s being either n (non degenerate) or t (twofold degenerate).

C . Frequencies of Rotation-Vibration Lines

Line frequencies are given in Tables LIII for various rotation-vibration transitions. In the labeling of these tables, letters A B C refer to the range of rotational quantum numbers and numbers 1 to 8 characterize the upper vibrational level of the transition, with the same conventions as above for Tables XLVIII and LII; the lower level of the transition is always assumed to be a non degenerate level (of type 1 in the notation defined above). Furthermore, α and β refer to lines belonging to Q branches and R or P branches respectively. In Tables LIII, symbols v'' and v' represent all vibrational quantum numbers which characterize respectively the lower and the upper levels of the transitions. It has been assumed that every normal vibration belongs[6] to species A_1 or E . Quantum numbers J and K are those associated with the lower levels.

In parallel bands for which the upper level corresponds to zero values of all quantum numbers ℓ_t (such as ν_n , $2\nu_n$ or $2\nu_t^0$), symbols ${}^QQ_{J,|K|}$ ${}^QR_{J,|K|}$ ${}^QP_{J,|K|}$ are used for the various lines. In parallel bands where at least one quantum number ℓ_t is different from zero in the upper level (such as $3\nu_t^{\pm 3}$ or $\nu_t^{\pm 1} + \nu_{t'}^{\mp 1}$), symbols ${}^QQ_{J,K\ell_t}$ ${}^QR_{J,K\ell_t}$ ${}^QP_{J,K\ell_t}$ are used for the lines except where J and K are small in $\nu_t^{\pm 1} + \nu_{t'}^{\mp 1}$: in this case, ${}^QQ^+_{J,|K|}$ ${}^QR^+_{J,|K|}$ ${}^QP^+_{J,|K|}$ or ${}^QQ^-_{J,|K|}$ ${}^QR^-_{J,|K|}$ ${}^QP^-_{J,|K|}$ are used respectively when the upper level of the transition is $E^+_{(|K|)}$ or $E^-_{(|K|)}$.

In perpendicular bands, symbols ${}^RQ_{J,|K|}$ ${}^RR_{J,|K|}$ ${}^RP_{J,|K|}$ are used for lines with $\Delta |K| = +1$ and $|K| = 0,1,2,3 \ldots$; symbols ${}^PQ_{J,|K|}$ ${}^PR_{J,|K|}$ ${}^PP_{J,|K|}$ are used for lines with $\Delta |K| = -1$ and $|K| = 1,2,3 \ldots$

Furthermore, the following conventions have been used :

- in Tables LIII , 1β , 3β , 7β :

  $m = J + 1$ for ${}^{Q}R_{J,|K|}$ or ${}^{Q}R_{J,K\ell_t}$

  $m = -J$ for ${}^{Q}P_{J,|K|}$ or ${}^{Q}P_{J,K\ell_t}$

- in Tables LIII , 2α , 4α , 5α , 6α , 8α :

  the upper sign for ${}^{R}Q_{J,|K|}$

  the lower sign for ${}^{P}Q_{J,|K|}$

- in Tables LIII , 2β , 4β , 5β , 6β , 8β :

  $m = J + 1$ and the upper sign for ${}^{R}R_{J,|K|}$

  $m = J + 1$ and the lower sign for ${}^{P}R_{J,|K|}$

  $m = -J$ and the upper sign for ${}^{R}P_{J,|K|}$

  $m = -J$ and the lower sign for ${}^{P}P_{J,|K|}$

The frequencies are obtained from the Tables LIII by adding up the various contributions in the same manner as described for the energies given in Tables XLVIII and for the frequencies of rotational lines given in Tables LII ; here again it may happen that several cases need to be considered for the last contributions to the frequency according to the values of $|K|$. For instance, in the Tables LIII , 7 Aα and 7 Aβ , two cases need to be considered for the last contribution to the frequencies :

TABLES LIII, 7 Aα and 7 Aβ

- - - - - - - - - - - - - -		
$\|K\| \neq 0$		(I)
$K = 0$		(II)

The expressions to be used for this last contribution are the following :

${}^{Q}Q^{+}_{J,|K|}$ ${}^{Q}R^{+}_{J,|K|}$ ${}^{Q}P^{+}_{J,|K|}$ lines ($|K| \neq 0$) : (I), with the upper signs

${}^{Q}Q^{-}_{J,|K|}$ ${}^{Q}R^{-}_{J,|K|}$ ${}^{Q}P^{-}_{J,|K|}$ lines ($|K| \neq 0$) : (I), with the lower signs

${}^{Q}Q_{J,0}$ ${}^{Q}R_{J,0}$ ${}^{Q}P_{J,0}$ lines : (II)

Among others, Tables LIII give the line frequencies for the fundamental bands ν_n and ν_t, for the overtones $2\nu_n$, $2\nu_t$ ($//$ and $\perp$ components), $3\nu_n$, $3\nu_t$ ($\perp$ and $//$ components) and for the combination bands $\nu_n + \nu_{n'}$, $\nu_n + \nu_{t'}$, $\nu_t + \nu_{t'}$ ($//$ and $\perp$ components).

In Tables LII and LIII, orders of magnitude of the various contributions are given, as for the energies, in the columns on the right side. The contributions are given to fourth order in Tables LIII and to sixth order in Tables LII. Let us recall finally that coefficients B^{zz}_{v}, B^{xx}_{v}, $(B^{zz}\zeta_t)_v$ ... which appear in the expressions for the frequencies are given in Table XLIX in terms of vibrational quantum numbers and molecular parameters.

NOTES TO CHAPTER X

1 . The definition of symmetry species of rotation-vibration states, using the full symmetry group of the molecule is treated in detail in the Chapter III of reference [8].

2 . In the case of Table XLVIII, 2 C ($v_t = 1$; $J \simeq 30$, $K \simeq 1$), the ℓ-type rotational resonance is strong and makes resonance submatrices of degree two appear in the energy matrix. This resonance mixes the levels with the same value of $L = K - \ell_t$; then the energy is given by the sum or the difference of two expressions F_1 and F_2, as follows :

$$E^{\pm}_{(v_n \ldots, v_t, J, |L|)} = F_{1(v_n \ldots, v_t, J, |L|)} \pm F_{2(v_n \ldots, v_t, J, |L|)}$$

where only F_2 depends on L.

(Notes continued p. 322)

TABLES XLVIII . ROTATION-VIBRATION ENERGIES FOR MOLECULES BELONGING TO SYMMETRY GROUPS C_{3v}, D_3, D_{3h}, D_{3d}

TABLE XVLIII, 1A . $v_t = v_{t'} = \ldots = 0$; v_n, $v_{n'} \ldots$ arbitrary ; $J \simeq K \simeq 1$.

Contributions	h: a	b	c
$\frac{E_v}{hc}$	0	2	4
$+ B_v^{xx} J(J+1)$	2	4	6
$+ (B_v^{zz} - B_v^{xx}) K^2$	2	4	6
$- D_e^J J^2 (J+1)^2$	6		
$- D_e^{JK} J(J+1) K^2$	6		
$- D_e^K K^4$	6		

TABLE XLVIII, 1 B . $v_t = v_{t'} = \ldots = 0$; $v_n, v_{n'} \ldots$ arbitrary ; $J \simeq K \simeq 30$

Contributions	h: a	b	c
$\frac{E_v}{hc}$	0	2	4
$+ B_v^{xx} J(J+1)$	0	2	4
$+ (B_v^{zz} - B_v^{xx}) K^2$	0	2	4
$- D_v^J J^2 (J+1)^2$	2	4	
$- D_v^{JK} J(J+1) K^2$	2	4	
$- D_v^K K^4$	2	4	
$+ H^J J^3 (J+1)^3$	4		
$+ H^{JJK} J^2 (J+1)^2 K^2$	4		
$+ H^{JKK} J(J+1) K^4$	4		
$+ H^K K^6$	4		

TABLE XLVIII, 1 C. $v_t = v_{t'} = \ldots = 0$; $v_n, v_{n'}, \ldots$ arbitrary ; $J \simeq 30$, $K \simeq 1$

Contributions		h: a	b	c
$\frac{E_v}{hc}$		0	2	4
$+B_v^{xx} J(J+1)$		0	2	4
$+(B_v^{zz} - B_v^{xx})K^2$		2	4	
$-D_v^J J^2(J+1)^2$		2	4	
$-D_e^{JK} J(J+1)K^2$		4		
$+H^J J^3(J+1)^3$		4		
$\lvert K\rvert \neq 3$	$+\frac{(q_3)^2}{3(B_e^{zz} - B_e^{xx})}\Big\{(2K-3)\big[J(J+1)-K(K-1)\big]\big[J(J+1)-(K-1)(K-2)\big]\big[J(J+1)-(K-2)(K-3)\big]$ $-(2K+3)\big[J(J+1)-K(K+1)\big]\big[J(J+1)-(K+1)(K+2)\big]\big[J(J+1)-(K+2)(K+3)\big]\Big\}$	4		

$\|K\| = 3$	ψ_3^+ state	$+ f_6 J(J+1)[J(J+1)-2][J(J+1)-6]$ $- \frac{3(q_3)^2}{B_e^{zz} - B_e^{xx}}[J(J+1)-12][J(J+1)-20][J(J+1)-30]$	4		
	ψ_3^- state	$-\left(f_6 - \frac{2(q_3)^2}{B_e^{zz} - B_e^{xx}}\right) J(J+1)[J(J+1)-2][J(J+1)-6]$ $- \frac{3(q_3)^2}{B_e^{zz} - B_e^{xx}}[J(J+1)-12][J(J+1)-20][J(J+1)-30]$	4		

TABLE XLVIII, 2 A. $v_t = 1$, $|\ell_t| = 1$; $v_{t'} = \ldots = 0$; $v_n, v_{n'}, \ldots$ arbitrary; $J \simeq K \simeq 1$

Contributions	k		
	a	b	c
$\frac{E_v}{hc}$	0	2	4
$+B_v^{xx} J(J+1)$	2	4	6
$+(B_v^{zz} - B_v^{xx})K^2$	2	4	6
$-2(\zeta_t B^{zz})_v K\ell_t$	2	4	6
$-D_e^J J^2(J+1)^2$	6		
$-D_e^{JK} J(J+1)K^2$	6		
$-D_e^K K^4$	6		
$+\eta_t^J J(J+1)K\ell_t$	6		
$+\eta_t^K K^3\ell_t$	6		

		$-\dfrac{4(q_{12}^{t})^{2}(2K\ell_t+1)\left[J(J+1)-K\ell_t(K\ell_t+1)\right]}{B_e^{zz}-B_e^{xx}+2B_e^{z}\zeta_t^{z}}$	6		
$K\ell_t \neq 1$		$+\dfrac{(q_0^{t})^{2}\left[J(J+1)-K\ell_t(K\ell_t-1)\right]\left[J(J+1)-(K\ell_t-1)(K\ell_t-2)\right]}{(K\ell_t-1)(B_e^{zz}-B_e^{xx}-B_e^{zz}\zeta_t^{z})}$	6		
$K\ell_t = 1$	$\Psi_{1,1}^{+}$ state	$+2q_{eff}^{t}J(J+1)$	4	6	
	$\Psi_{1,1}^{-}$ state	$-2q_{eff}^{t}J(J+1)$	4	6	

TABLE XLVIII, 2 B. $v_t = 1$, $|\ell_t| = 1$; $v_{t'} = \ldots = 0$; $v_n, v_{n'} \ldots$ arbitrary ; $J \simeq K \simeq 30$

Contributions	h: a	b	c
$\frac{E_v}{hc}$	0	2	4
$+ B_v^{xx} J(J+1)$	0	2	4
$+ (B_v^{zz} - B_v^{xx}) K^2$	0	2	4
$- 2(\zeta_t B^{zz})_v K\ell_t$	1	3	
$- D_v^J J^2 (J+1)^2$	2	4	
$- D_v^{JK} J(J+1) K^2$	2	4	
$- D_v^K K^4$	2	4	
$+ \eta_t^J J(J+1) K\ell_t$	3		
$+ \eta_t^K K^3 \ell_t$	3		

$+H^{J}J^{3}(J+1)^{3}$	4		
$+H^{JJK}J^{2}(J+1)^{2}K^{2}$	4		
$+H^{JKK}J(J+1)K^{4}$	4		
$+H^{K}K^{6}$	4		
$-\dfrac{4(q_{12}^{t})^{2}(2K\ell_{t}+1)\left[J(J+1)-K\ell_{t}(K\ell_{t}+1)\right]}{B_{e}^{zz}-B_{e}^{xx}+2B_{e}^{zz}\zeta_{t}^{z}}$	4		

TABLE XLVIII, 2 C . $v_t = 1, |\ell_t| = 1 ; v_{t'} = \dots = 0 ; v_n , v_{n'} \dots$ arbitrary ; $J \simeq 30, K \simeq 1$

Contributions	h: a	h: b	h: c
$\frac{E_v}{hc}$	0	2	4
$+ B_v^{zz} - B_v^{xx} - 2(\zeta_t B^{zz})_v$	2	4	
$+ J(J+1)\left[B_v^{xx}\right.$	0	2	4
$\left.+ \eta_t^J - D_e^{JK}\right]$	4		
$- D_v^J J^2 (J+1)^2$	2	4	
$+ H^J J^3 (J+1)^3$	4		
$+(B_v^{zz} - B_v^{xx}) L^2$	2	4	
$- D_e^{JK} J(J+1) L^2$	4		
$\pm 2\sqrt{(B_e^{zz} - B_e^{xx} - B_e^{zz} \zeta_t^z)^2 L^2 + (q_0^t)^2 (a_{2,L-1}^+)^2}$	2		

$\lvert L\rvert \neq 0$ $(L = K - \ell_t)$	$\pm\Big\{2q_0^t(a^+_{2,L-1})^2\Big[\sum_s f^{t,s}_{22}(v_s+\frac{d_s}{2})+f^{t,J}_{22}J(J+1)\Big]-2(B_e^{zz}-B_e^{xx}$ $-B_e^{zz}\zeta_t^z)L^2\Big[\sum_s(\alpha_s^z-\alpha_s^x-\frac{1}{2}\eta_{ts})(v_s+\frac{d_s}{2})+(D_e^{JK}-\frac{1}{2}\eta_t^J)J(J+1)\Big]\Big\}$ $\times\Big[(B_e^{zz}-B_e^{xx}-B_e^{zz}\zeta_t^z)^2L^2+(q_0^t)^2(a^+_{2,L-1})^2\Big]^{-\frac{1}{2}}$	4			
	$+\sum_{i=+,-}(a^+_{3,L+1})^2\Big\{4(q_0^t)^2(a^+_{2,L-1})^2\big[q_3(2L+1)-d_3^t\big]-4q_{12}^t q_0^t(2L+3)A_L^{\pm}$ $+A_L^{\pm}A^i_{L+3}\big[q_3(2L+5)+d_3^t\big]\Big\}^2\Big[4(q_0^t)^2(a^+_{2,L-1})^2+(A_L^{\pm})^2\Big]^{-1}$ $\times\Big[4(q_0^t)^2(a^+_{2,L+2})^2+(A^i_{L+3})^2\Big]^{-1}\Big[T_L^{\pm}-T^i_{L+3}\Big]^{-1}$	4			
$\lvert L\rvert \neq 0,3$	$+\sum_{i=+,-}(a^+_{3,L-2})^2\Big\{4(q_0^t)^2(a^+_{2,L-4})^2\big[q_3(2L-5)-d_3^t\big]-4q_{12}^t q_0^t(2L-3)A^i_{L-3}$ $+A^i_{L-3}A_L^{\pm}\big[q_3(2L-1)+d_3^t\big]\Big\}^2\Big[4(q_0^t)^2(a^+_{2,L-4})^2+(A^i_{L-3})^2\Big]^{-1}$ $\times\Big[4(q_0^t)^2(a^+_{2,L-1})^2+(A_L^{\pm})^2\Big]^{-1}\Big[T_L^{\pm}-T^i_{L-3}\Big]^{-1}$	4			

$\lvert L \rvert = 3$	φ_a^+ and φ_b^+ states	$+ f_{42}^t \left[1 \mp \frac{3(B_e^{zz} - B_e^{xx} - B_e^{zz}\zeta_t^z)}{\sqrt{9(B_e^{zz} - B_e^{xx} - B_e^{zz}\zeta_t^z)^2 + (q_0^t)^2 (a_{2,2}^+)^2}} \right] J(J+1)\left[J(J+1) - 2\right]$ $+ \frac{(a_{3,1}^+)^2 \left[12 q_{12}^t q_0^t - 2 q_0^t (q_3 - d_3^t) J(J+1) - A_3^{\pm} (5 q_3 + d_3^t) \right]^2}{\left[4 (q_0^t)^2 (a_{2,2}^+)^2 + (A_3^{\pm})^2 \right] \left[T_3^{\pm} - T_0^- \right]}$	4		
	φ_a^- and φ_b^- states	$- f_{42}^t \left[1 \mp \frac{3(B_e^{zz} - B_e^{xx} - B_e^{zz}\zeta_t^z)}{\sqrt{9(B_e^{zz} - B_e^{xx} - B_e^{zz}\zeta_t^z)^2 + (q_0^t)^2 (a_{2,2}^+)^2}} \right] J(J+1)\left[J(J+1) - 2\right]$ $+ \frac{(a_{3,1}^+)^2 \left[12 q_{12}^t q_0^t + 2 q_0^t (q_3 - d_3^t) J(J+1) - A_3^{\pm} (5 q_3 + d_3^t) \right]^2}{\left[4 (q_0^t)^2 (a_{2,2}^+)^2 + (A_3^{\pm})^2 \right] \left[T_3^{\pm} - T_0^+ \right]}$	4		
	$\psi_{1,1}^+$ state	$+ 2 q_{eff}^t J(J+1)$	2	4	
		$+ 2 f_{22}^{t,J} J^2 (J+1)^2$	4		
		$+ \frac{(a_{3,1}^+)^2 \left[12 q_{12}^t q_0^t + 2 q_0^t (q_3 - d_3^t) J(J+1) - A_3^+ (5 q_3 + d_3^t) \right]^2}{\left[4 (q_0^t)^2 (a_{2,2}^+)^2 + (A_3^+)^2 \right] \left[T_0^+ - T_3^+ \right]}$	4		

$L = 0$		$+\dfrac{(a^{+}_{3,1})^2\left[12q^t_{12}q^t_0+2q^t_0(q_3-d^t_3)J(J+1)-A^{-}_3(5q_3+d^t_3)\right]^2}{\left[4(q^t_0)^2(a^{+}_{2,2})^2+(A^{-}_3)^2\right]\left[T^{+}_0-T^{-}_3\right]}$	4		
	$\psi^{-}_{1,1}$ state	$-2q^t_{eff}\,J(J+1)$	2	4	
		$-2f^{t,J}_{22}\,J^2(J+1)^2$	4		
		$+\dfrac{(a^{+}_{3,1})^2\left[12q^t_{12}q^t_0-2q^t_0(q_3-d^t_3)J(J+1)-A^{+}_3(5q_3+d^t_3)\right]^2}{\left[4(q^t_0)^2(a^{+}_{2,2})^2+(A^{+}_3)^2\right]\left[T^{-}_0-T^{+}_3\right]}$	4		
		$+\dfrac{(a^{+}_{3,1})^2\left[12q^t_{12}q^t_0-2q^t_0(q_3-d^t_3)J(J+1)-A^{-}_3(5q_3+d^t_3)\right]^2}{\left[4(q^t_0)^2(a^{+}_{2,2})^2+(A^{-}_3)^2\right]\left[T^{-}_0-T^{-}_3\right]}$	4		

$$A^{\pm}_L=-\left[2L(B^{zz}_e-B^{xx}_e-B^{zz}_e\zeta^z_t)\pm 2\sqrt{L^2(B^{zz}_e-B^{xx}_e-B^{zz}_e\zeta^z_t)^2+(q^t_0)^2(a^{+}_{2,L-1})^2}\right]$$

$$T^{\pm}_{L\neq 0}=(L^2+1)(B^{zz}_e-B^{xx}_e)-2B^{zz}_e\zeta^z_t\pm 2\sqrt{L^2(B^{zz}_e-B^{xx}_e-B^{zz}_e\zeta^z_t)^2+(q^t_0)^2(a^{+}_{2,L-1})^2}$$

$$T^{\pm}_{L=0}=B^{zz}_e-B^{xx}_e-2B^{zz}_e\zeta^z_t\pm 2q^t_0\,J(J+1)$$

TABLE XLVIII, 3 A . $v_t = 2$, $\ell_t = 0$; $v_{t'} = \ldots = 0$; $v_n, v_{n'}, \ldots$ arbitrary; $J \simeq K \simeq 1$

Contributions	k: a	b	c
$\dfrac{E_v}{hc}$	0	2	4
$+B_v^{xx} J(J+1)$	2	4	6
$+(B_v^{zz} - B_v^{xx})K^2$	2	4	6
$-D_e^J J^2 (J+1)^2$	6		
$-D_e^{JK} J(J+1) K^2$	6		
$-D_e^K K^4$	6		
$+\sum\limits_{\ell_t = +2,-2} \dfrac{2\left[q_{12}^t (K\ell_t - 1) + d_{12}^{t,t}\right]^2 \left[4J(J+1) - K\ell_t(K\ell_t - 2)\right]}{(K\ell_t - 1)(B_e^{zz} - B_e^{xx} + 2B_e^{zz}\zeta_t^z) - 2B_e^{zz}\zeta_t^z - 4x_{\ell_t\ell_t}}$	6		
$-\sum\limits_{\ell_t = +2,-2} \dfrac{(q_0^t)^2 \left[4J(J+1) - K\ell_t(K\ell_t + 2)\right]\left[4J(J+1) - (K\ell_t + 2)(K\ell_t + 4)\right]}{4(K\ell_t + 2)(B_e^{zz} - B_e^{xx} - B_e^{zz}\zeta_t^z) - 8B_e^{zz}\zeta_t^z + 8x_{\ell_t\ell_t}}$	6		

TABLE XLVIII, 3 B . $v_t = 2$, $\ell_t = 0$; $v_{t'} = \ldots = 0$; $v_n, v_{n'} \ldots$ arbitrary ; $J \simeq K \simeq 30$

Contributions	h: a	h: b	h: c
$\frac{E_v}{hc}$	0	2	4
$+B_v^{xx} J(J+1)$	0	2	4
$+(B_v^{zz} - B_v^{xx})K^2$	0	2	4
$-D_v^J J^2 (J+1)^2$	2	4	
$-D_v^{JK} J(J+1)K^2$	2	4	
$-D_v^K K^4$	2	4	
$+H^J J^3(J+1)^3 + H^{JJK} J^2 (J+1)^2 K^2 + H^{JKK} J(J+1)K^4 + H^K K^6$	4		
$+ \sum_{\ell_t = 2,-2} \frac{2(q_{12}^t)^2 (K\ell_t - 1)^2 \left[4J(J+1) - K\ell_t(K\ell_t - 2)\right]}{(K\ell_t - 1)(B_e^{zz} - B_e^{xx} + 2B_e^{zz}\zeta_t^z) - 2B_e^{zz}\zeta_t^z - 4x_{\ell_t \ell_t}}$	4		

TABLE XLVIII, 4 A . $v_t = 2$, $|\ell_t| = 2$; $v_{t'} = \dots = 0$; $v_n, v_{n'} \dots$ arbitrary ; $J \simeq K \simeq 1$

Contributions	h: a	b	c
$\frac{E_v}{hc}$	0	2	4
$+ B_v^{xx} J(J+1)$	2	4	6
$+ (B_v^{zz} - B_v^{xx}) K^2$	2	4	6
$- 2(\zeta_t B^{zz})_v K\ell_t$	2	4	6
$- D_e^J J^2 (J+1)^2$	6		
$- D_e^{JK} J(J+1) K^2$	6		
$- D_e^K K^4$	6		
$+ \eta_t^J J(J+1) K\ell_t$	6		
$+ \eta_t^K K^3 \ell_t$	6		

$K\ell_t \neq -2$	$-\dfrac{2\left[q_{12}^{t}(K\ell_t+1)+d_{12}^{t,t}\right]^2\left[4J(J+1)-K\ell_t(K\ell_t+2)\right]}{(K\ell_t+1)(B_e^{zz}-B_e^{xx}+2B_e^{zz}\zeta_t^{z})-2B_e^{zz}\zeta_t^{z}-4x_{\ell_t\ell_t}}$		6		
$K\ell_t \neq 4$	$+\dfrac{(q_0^{t})^2\left[4J(J+1)-K\ell_t(K\ell_t-2)\right]\left[4J(J+1)-(K\ell_t-2)(K\ell_t-4)\right]}{4(K\ell_t-2)(B_e^{zz}-B_e^{xx}-B_e^{zz}\zeta_t^{z})-8B_e^{zz}\zeta_t^{z}+8x_{\ell_t\ell_t}}$		6		
$K\ell_t = 4$	$\Psi_{2,2}^{+}$ state	$+\dfrac{4(q_0^{t})^2 J(J+1)\left[J(J+1)-2\right]}{B_e^{zz}-B_e^{xx}-2B_e^{zz}\zeta_t^{z}+x_{\ell_t\ell_t}}$	6		
	$\Psi_{2,2}^{-}$ state	0			
$K\ell_t = -2$	$\Psi_{1,-2}^{+}$ state	$+\ 8f_{24}^{t}J(J+1)$	6		
	$\Psi_{1,-2}^{-}$ state	$-\ 8\left[f_{24}^{t}-\dfrac{2(q_{12}^{t}-d_{12}^{t,t})^2}{B_e^{zz}-B_e^{xx}+4B_e^{zz}\zeta_t^{z}+4x_{\ell_t\ell_t}}\right]J(J+1)$	6		

TABLE XLVIII, 4 B . $v_t = 2$, $|\ell_t| = 2$; $v_{t'} = \ldots = 0$; v_n , $v_{n'}$, ... arbitrary ; $J \simeq K \simeq 30$

Contributions	h: a	b	c
$\frac{E_v}{hc}$	0	2	4
$+ B_v^{xx} J(J+1)$	0	2	4
$+ (B_v^{zz} - B_v^{xx})K^2$	0	2	4
$- 2(\zeta_t B^{zz})_v K\ell_t$	1	3	
$- D_v^J J^2 (J+1)^2$	2	4	
$- D_v^{JK} J(J+1)K^2$	2	4	
$- D_v^K K^4$	2	4	
$+ \eta_t^J J(J+1)K\ell_t$	3		
$+ \eta_t^K K^3 \ell_t$	3		

$+H^{J}J^{3}(J+1)^{3}$	4		
$+H^{JJK}J^{2}(J+1)^{2}K^{2}$	4		
$+H^{JKK}J(J+1)K^{4}$	4		
$+H^{K}K^{6}$	4		
$-\dfrac{2(q_{12}^{t})^{2}(K\ell_t+1)^{2}\left[4J(J+1)-K\ell_t(K\ell_t+2)\right]}{(B_e^{zz}-B_e^{xx}+2B_e^{zz}\zeta_t^{z})(K\ell_t+1)-2B_e^{zz}\zeta_t^{z}-4x_{\ell_t\ell_t}}$	4		

TABLE XLVIII, 5 A . $v_t = 3$, $|\ell_t| = 1$; $v_{t'} = \ldots = 0$; v_n , $v_{n'} \ldots$ arbitrary ; $J \simeq K \simeq 1$

Contributions	k a	b	c
$\frac{E_v}{hc}$	0	2	4
$+B_v^{xx} J(J+1)$	2	4	6
$+(B_v^{zz} - B_v^{xx})K^2$	2	4	6
$-2(\zeta_t B^{zz})_v K\ell_t$	2	4	6
$-D_e^J J^2(J+1)^2$	6		
$-D_e^{JK} J(J+1)K^2$	6		
$-D_e^K K^4$	6		
$+\eta_t^J J(J+1)K\ell_t$	6		
$+\eta_t^K K^3\ell_t$	6		

		$-\dfrac{16(q_{12}^{t})^2(2K\ell_t+1)\left[J(J+1)-K\ell_t(K\ell_t+1)\right]}{B_e^{zz}-B_e^{xx}+2B_e^{zz}\zeta_t^{z}}$	6		
		$+\dfrac{12\left[q_{12}^{t}(2K\ell_t-1)+2d_{12}^{t,t}\right]^2\left[J(J+1)-K\ell_t(K\ell_t-1)\right]}{2K\ell_t(B_e^{zz}-B_e^{xx}+2B_e^{zz}\zeta_t^{z})-B_e^{zz}+B_e^{xx}-6B_e^{zz}\zeta_t^{z}-8x_{\ell_t\ell_t}}$	6		
		$-\dfrac{3(q_0^{t})^2\left[J(J+1)-K\ell_t(K\ell_t+1)\right]\left[J(J+1)-(K\ell_t+1)(K\ell_t+2)\right]}{K\ell_t(B_e^{zz}-B_e^{xx}-B_e^{zz}\zeta_t^{z})+B_e^{zz}-B_e^{xx}-3B_e^{zz}\zeta_t^{z}+2x_{\ell_t\ell_t}}$	6		
$K\ell_t \neq 1$		$+\dfrac{4(q_0^{t})^2\left[J(J+1)-K\ell_t(K\ell_t-1)\right]\left[J(J+1)-(K\ell_t-1)(K\ell_t-2)\right]}{(K\ell_t-1)(B_e^{zz}-B_e^{xx}-B_e^{zz}\zeta_t^{z})}$	6		
$K\ell_t = 1$	$\psi_{1,1}^{+}$ state	$+4\,q_{eff}^{t}\;J(J+1)$	4	6	
	$\psi_{1,1}^{-}$ state	$-4\,q_{eff}^{t}\;J(J+1)$	4	6	

TABLE XLVIII, 5 B . $v_t = 3$, $|\ell_t| = 1$; $v_{t'} = \ldots = 0$; $v_n, v_{n'} \ldots$ arbitrary ; $J \simeq K \simeq 30$

Contributions	h: a	b	c
$\frac{E_v}{hc}$	0	2	4
$+B_v^{xx} J(J+1)$	0	2	4
$+(B_v^{zz} - B_v^{xx})K^2$	0	2	4
$-2(\zeta_t B^{zz})_v K\ell_t$	1	3	
$-D_v^J J^2 (J+1)^2$	2	4	
$-D_v^{JK} J(J+1) K^2$	2	4	
$-D_v^K K^4$	2	4	
$+\eta_t^J J(J+1) K\ell_t$	3		
$+\eta_t^K K^3 \ell_t$	3		

$+H^{J}J^{3}(J+1)^{3}$	4		
$+H^{JJK}J^{2}(J+1)^{2}K^{2}$	4		
$+H^{JKK}J(J+1)K^{4}$	4		
$+H^{K}K^{6}$	4		
$+\dfrac{12(q_{12}^{t})^{2}(2K\ell_t-1)^{2}\left[J(J+1)-K\ell_t(K\ell_t-1)\right]}{2K\ell_t(B_e^{zz}-B_e^{xx}+2B_e^{zz}\zeta_t^{z})-B_e^{zz}+B_e^{xx}-6B_e^{zz}\zeta_t^{z}}$	4		
$-\dfrac{16(q_{12}^{t})^{2}(2K\ell_t+1)\left[J(J+1)-K\ell_t(K\ell_t+1)\right]}{B_e^{zz}-B_e^{xx}+2B_e^{zz}\zeta_t^{z}}$	4		

TABLE XVLIII, 6 A . $v_t = 3$, $|\ell_t| = 3$; $v_{t'} = \ldots = 0$; $v_n, v_{n'}, \ldots$ arbitrary ; $J \simeq K \simeq 1$

Contributions	k: a	b	c
$\frac{E_v}{hc}$	0	2	4
$+ B_v^{xx} J(J+1)$	2	4	6
$+ (B_v^{zz} - B_v^{xx}) K^2$	2	4	6
$- 2(\zeta_t B^{zz})_v K\ell_t$	2	4	6
$- D_e^J J^2 (J+1)^2$	6		
$- D_e^{JK} J(J+1) K^2$	6		
$- D_e^K K^4$	6		
$+ \eta_t^J J(J+1) K\ell_t$	6		
$+ \eta_t^K K^3 \ell_t$	6		

		$-\dfrac{4\left[q_{12}^{t}(2K\ell_t+3)+6\,d_{12}^{t,t}\right]^2\left[9J(J+1)-K\ell_t(K\ell_t+3)\right]}{18K\ell_t(B_e^{zz}-B_e^{xx}+2B_e^{zz}\zeta_t^{z})+27\left[B_e^{zz}-B_e^{xx}-2B_e^{zz}\zeta_t^{z}-8x_{\ell_t\ell_t}\right]}$	6		
		$+\dfrac{(q_0^{t})^2\left[9J(J+1)-K\ell_t(K\ell_t-3)\right]\left[9J(J+1)-(K\ell_t-3)(K\ell_t-6)\right]}{9K\ell_t(B_e^{zz}-B_e^{xx}-B_e^{zz}\zeta_t^{z})-27(B_e^{zz}-B_e^{xx}+B_e^{zz}\zeta_t^{z}-2x_{\ell_t\ell_t})}$	6		
$\lvert K\rvert \neq 0$	$-\dfrac{192(g_6^{t})^2}{B_e^{zz}\zeta_t^{z}K\ell_t}$		6		
$K = 0$	$\psi_{0,3}^{+}$ state	$+\,48g_6^{t}$	4		
	$\psi_{0,3}^{-}$ state	$-\,48g_6^{t}$	4		

TABLE XLVIII, 6 B . $v_t = 3$, $|\ell_t| = 3$; $v_{t'} = \dots = 0$; $v_n, v_{n'} \dots$ arbitrary ; $J \simeq K \simeq 30$

Contributions	k: a	b	c
$\frac{E_v}{hc}$	0	2	4
$+ B_v^{xx} J(J+1)$	0	2	4
$+ (B_v^{zz} - B_v^{xx}) K^2$	0	2	4
$- 2(\zeta_t B^{zz})_v K\ell_t$	1	3	
$- D_v^J J^2(J+1)^2$	2	4	
$- D_v^{JK} J(J+1)K^2$	2	4	
$- D_v^K K^4$	2	4	
$+ \eta_t^J J(J+1)K\ell_t$	3		
$+ \eta_t^K K^3 \ell_t$	3		

$+H^{J}J^{3}(J+1)^{3}$	4		
$+H^{JJK}J^{2}(J+1)^{2}K^{2}$	4		
$+H^{JKK}J(J+1)K^{4}$	4		
$+H^{K}K^{6}$	4		
$-\dfrac{4(q_{12}^{t})^{2}(2K\ell_{t}+3)^{2}\left[9J(J+1)-K\ell_{t}(K\ell_{t}+3)\right]}{18K\ell_{t}(B_{e}^{zz}-B_{e}^{xx}+2B_{e}^{zz}\zeta_{t}^{z})+27(B_{e}^{zz}-B_{e}^{xx}-2B_{e}^{zz}\zeta_{t}^{z})}$	4		

TABLE XLVIII, 7 A . $v_t = v_{t'} = 1$, $\ell_t + \ell_{t'} = 0$; $v_{t''} = \dots = 0$; v_n, $v_{n'} \dots$ arbitrary; $J \simeq K \simeq 1$

Contributions	h: a	b	c
$\frac{E_v}{hc}$	0	2	4
$+ B_v^{xx} J(J+1)$	2	4	6
$+ (B_v^{zz} - B_v^{xx}) K^2$	2	4	6
$- D_e^J J^2 (J+1)^2$	6		
$- D_e^{JK} J(J+1) K^2$	6		
$- D_e^K K^4$	6		
$\pm 2 \sqrt{(B_e^{zz} \zeta_t^z - B_e^{zz} \zeta_{t'}^z)^2 K^2 + 4 (r_0^{tt'})^2}$	2		
$\pm \frac{\sum_s \left[8 r_0^{tt'} g_{22}^{tt',s} - B_e^{zz} (\zeta_t^z - \zeta_{t'}^z)(\eta_{ts} - \eta_{t's}) K^2 \right] (v_s + \frac{d_s}{2})}{\left[(B_e^{zz} \zeta_t^z - B_e^{zz} \zeta_{t'}^z)^2 K^2 + 4 (r_0^{tt'})^2 \right]^{\frac{1}{2}}}$	4		

	$\pm\Big[\Big\{\sum_{ss'}\big[4g_{22}^{tt',s}g_{22}^{tt',s'}+\frac{1}{4}(\eta_{ts}-\eta_{t's})(\eta_{ts'}-\eta_{t's'})K^2\big]-2\sum_{s\leqslant s'}B_e^{zz}(\zeta_t^z-\zeta_{t'}^z)(\xi_{ss'}^t-\xi_{ss'}^{t'})K^2\Big\}(v_s+\frac{d_s}{2})(v_{s'}+\frac{d_{s'}}{2})+\Big\{8r_0^{tt'}g_{22}^{tt',J}J(J+1)$			
$\lvert K\rvert \neq 0$	$+2\rho K^2-B_e^{zz}(\zeta_t^z-\zeta_{t'}^z)(\eta_t^J-\eta_{t'}^J)J(J+1)K^2-B_e^{zz}(\zeta_t^z-\zeta_{t'}^z)(\eta_t^K-\eta_{t'}^K)K^4\Big\}\Big]$ $\times\Big[(B_e^{zz}\zeta_t^z-B_e^{zz}\zeta_{t'}^z)^2K^2+4(r_0^{tt'})^2\Big]^{-\frac{1}{2}}$	6		
	$+\frac{4(a_{2,K}^-)^2(4r_0^{tt'}q_0^t+A_K^{\pm}q_0^{t'})^2}{B_K^{\pm}\left[16(r_0^{tt'})^2+(A_K^{\pm})^2\right]}+\frac{4(a_{2,K}^+)^2(4r_0^{tt'}q_0^t+A_{-K}^{\pm}q_0^{t'})^2}{B_{-K}^{\pm}\left[16(r_0^{tt'})^2+(A_{-K}^{\pm})^2\right]}$	6		
	$+\frac{4(a_{1,K}^+)^2\left\{4r_0^{tt'}\left[q_{12}^t(2K+1)-d_{12}^{t,t'}\right]+A_K^{\pm}\left[q_{12}^{t'}(2K+1)-d_{12}^{t',t}\right]\right\}^2}{C_K^{\pm}\left[16(r_0^{tt'})^2+(A_K^{\pm})^2\right]}$	6		
	$+\frac{4(a_{1,K}^-)^2\left\{4r_0^{tt'}\left[q_{12}^t(2K-1)+d_{12}^{t,t'}\right]+A_{-K}^{\pm}\left[q_{12}^{t'}(2K-1)+d_{12}^{t',t}\right]\right\}^2}{C_{-K}^{\pm}\left[16(r_0^{tt'})^2+(A_{-K}^{\pm})^2\right]}$	6		

$K = 0$	$\psi^{+}_{0,1,-1}$ state	$+\,4r^{tt'}_{\text{eff}}$	2	4	
		$+\,4\left[g^{tt',J}_{22} - \dfrac{(q^{t}_{12} + q^{t'}_{12} - d^{t';t}_{12} - d^{t,t'}_{12})^2}{B^{zz}_e - B^{xx}_e + 2B^{zz}_e(\zeta^{z}_{t} + \zeta^{z}_{t'}) + 2x_{l_t l_{t'}} - 4r^{tt'}_{0}}\right] J(J+1)$	6		
		$-\,\dfrac{2(q^{t}_{0} + q^{t'}_{0})^2\, J(J+1)\left[J(J+1)-2\right]}{2(B^{zz}_e - B^{xx}_e) - 2B^{zz}_e(\zeta^{z}_{t} + \zeta^{z}_{t'}) + x_{l_t l_{t'}} - 2r^{tt'}_{0}}$	6		
	$\psi^{-}_{0,1,-1}$ state	$-\,4r^{tt'}_{\text{eff}}$	2	4	
		$-\,4\left[g^{tt',J}_{22} + \dfrac{(q^{t}_{12} - q^{t'}_{12} + d^{t';t}_{12} - d^{t,t'}_{12})^2}{B^{zz}_e - B^{xx}_e + 2B^{zz}_e(\zeta^{z}_{t} + \zeta^{z}_{t'}) + 2x_{l_t l_{t'}} + 4r^{tt'}_{0}}\right] J(J+1)$	6		
		$-\,\dfrac{2(q^{t}_{0} - q^{t'}_{0})^2\, J(J+1)\left[J(J+1)-2\right]}{2(B^{zz}_e - B^{xx}_e) - 2B^{zz}_e(\zeta^{z}_{t} + \zeta^{z}_{t'}) + x_{l_t l_{t'}} + 2r^{tt'}_{0}}$	6		

$$a^{\pm}_{i,K} = \sqrt{J(J+1) - K(K\pm 1)}\ \sqrt{J(J+1) - (K\pm 1)(K\pm 2)}\ \cdots\ \sqrt{J(J+1) - \left[K \pm (i-1)\right](K \pm i)}$$

$$\rho = 4 r_0^{tt'} q_{22}^{tt',K} + B_e^{zz}(\zeta_t^z - \zeta_{t'}^z)\Big[\Delta(B_e^{zz}\zeta_t^z) - \Delta(B_e^{zz}\zeta_{t'}^z)$$

$$- \xi_{l_t l_t}^{t} + \xi_{l_t l_{t'}}^{t} - \xi_{l_{t'} l_{t'}}^{t} + \xi_{l_t l_t}^{t'} - \xi_{l_t l_{t'}}^{t'} + \xi_{l_{t'} l_{t'}}^{t'}\Big]$$

$$A_K^{\pm} = 2KB_e^{zz}(\zeta_t^z - \zeta_{t'}^z) \pm 2\sqrt{(B_e^{zz}\zeta_t^z - B_e^{zz}\zeta_{t'}^z)^2 K^2 + 4(r_0^{tt'})^2}$$

$$B_K^{\pm} = 4(B_e^{zz} - B_e^{xx})(K-1) - 2B_e^{zz}(\zeta_t^z + \zeta_{t'}^z)(K-2) - 2x_{l_t l_{t'}} \pm 2\sqrt{(B_e^{zz}\zeta_t^z - B_e^{zz}\zeta_{t'}^z)^2 K^2 + 4(r_0^{tt'})^2}$$

$$C_K^{\pm} = -(B_e^{zz} - B_e^{xx})(2K+1) - 2B_e^{zz}(\zeta_t^z + \zeta_{t'}^z)(K+1) - 2x_{l_t l_{t'}} \pm 2\sqrt{(B_e^{zz}\zeta_t^z - B_e^{zz}\zeta_{t'}^z)^2 K^2 + 4(r_0^{tt'})^2}$$

TABLE XLVIII, 7 B ; $v_t = v_{t'} = 1$; $\ell_t + \ell_{t'} = 0$; $v_{t''} = \ldots = 0$; $v_n, v_{n'} \ldots$ arbitrary ; $J \simeq K \simeq 30$

Contributions	k		
	a	b	c
$\frac{E_v}{hc}$	0	2	4
$+ B_v^{xx} J(J+1)$	0	2	4
$+ (B_v^{zz} - B_v^{xx})(K\ell_t)^2$	0	2	4
$- 2\left[(\zeta_t B^{zz})_v - (\zeta_{t'} B^{zz})_v\right] K\ell_t$	1	3	
$- D_v^J J^2 (J+1)^2$	2	4	
$- D_v^{JK} J(J+1)(K\ell_t)^2$	2	4	
$- D_v^K (K\ell_t)^4$	2	4	
$+ (\eta_t^J - \eta_{t'}^J) J(J+1) K\ell_t$	3		
$+ (\eta_t^K - \eta_{t'}^K)(K\ell_t)^3$	3		

$+H^{J}J^{3}(J+1)^{3}$	4		
$+H^{JJK}J^{2}(J+1)^{2}(K\ell_t)^{2}$	4		
$+H^{JKK}J(J+1)(K\ell_t)^{4}$	4		
$+H^{K}(K\ell_t)^{6}$	4		
$-\dfrac{16(r_0^{tt'})^{2}}{4B_e^{zz}K\ell_t(\zeta_t^z-\zeta_{t'}^z)}$	3		
$-\dfrac{4(q_{12}^{t})^{2}(2K\ell_t+1)^{2}\left[J(J+1)-K\ell_t(K\ell_t+1)\right]}{(2K\ell_t+1)(B_e^{zz}-B_e^{xx}+2B_e^{zz}\zeta_t^z)+2B_e^{zz}\zeta_{t'}^z+2x_{\ell_t\ell_{t'}}}$	4		
$+\dfrac{4(q_{12}^{t'})(2K\ell_t-1)^{2}\left[J(J+1)-K\ell_t(K\ell_t-1)\right]}{(2K\ell_t-1)(B_e^{zz}-B_e^{xx}+2B_e^{zz}\zeta_{t'}^z)-2B_e^{zz}\zeta_t^z-2x_{\ell_t\ell_{t'}}}$	4		

All the contributions in this table take into account the relation $\ell_{t'}=-\ell_t$

TABLE XLVIII, 8 A . $v_t = v_{t'} = 1$; $|\ell_t + \ell_{t'}| = 2$; $v_{t''} = \ldots = 0$; $v_n, v_{n'}, \ldots$ arbitrary ; $J \simeq K \simeq 1$

Contributions	k: a	b	c
$\frac{E_v}{hc}$	0	2	4
$+B_v^{xx} J(J+1)$	2	4	6
$+(B_v^{zz} - B_v^{xx})(K\ell_t)^2$	2	4	6
$-2\left[(\zeta_t B^{zz})_v + (\zeta_{t'} B^{zz})_v\right] K\ell_t$	2	4	6
$-D_e^J J^2 (J+1)^2$	6		
$-D_e^{JK} J(J+1)(K\ell_t)^2$	6		
$-D_e^K (K\ell_t)^4$	6		
$+(\eta_t^J + \eta_{t'}^J) J(J+1) K\ell_t$	6		
$+(\eta_t^K + \eta_{t'}^K)(K\ell_t)^3$	6		

$K\ell_t \neq -1$		$+ 4\left[J(J+1) - K\ell_t(K\ell_t+1)\right] \sum_{i=+,-} \left[(B^i_{K\ell_t})^{-1} \left[16(r_0^{tt'})^2 + (A^i_{K\ell_t+1})^2\right]^{-1} \left\{A^i_{K\ell_t+1}\left[q_{12}^{t'}(2K\ell_t+1) + d_{12}^{t',t}\right] - 4r_0^{tt'}\left[q_{12}^{t}(2K\ell_t+1) + d_{12}^{t,t'}\right]\right\}^2 \right]$	6			
$K\ell_t \neq 2$		$+ 4\left[J(J+1) - K\ell_t(K\ell_t-1)\right]\left[J(J+1) - (K\ell_t-1)(K\ell_t-2)\right] \sum_{i=+,-} \left\{ (A^i_{K\ell_t-2}q_0^{t'} - 4r_0^{tt'}q_0^{t})^2 (C^i_{K\ell_t})^{-1} \left[16(r_0^{tt'})^2 + (A^i_{K\ell_t-2})^2\right]^{-1} \right\}$	6			
$K\ell_t = -1$	$\Psi^+_{1,-1,-1}$ state	$+ 4J(J+1)\left[f_{222}^{tt'} + \dfrac{(q_{12}^{t} - q_{12}^{t'} - d_{12}^{t,t'} + d_{12}^{t',t})^2}{B_e^{zz} - B_e^{xx} + 2B_e^{zz}(\zeta_t^z + \zeta_{t'}^z) + 2x_{\ell_t\ell_{t'}} + 4r_0^{tt'}} \right]$	6			
	$\Psi^-_{1,-1,-1}$ state	$- 4J(J+1)\left[f_{222}^{tt'} - \dfrac{(q_{12}^{t} + q_{12}^{t'} - d_{12}^{t,t'} - d_{12}^{t',t})^2}{B_e^{zz} - B_e^{xx} + 2B_e^{zz}(\zeta_t^z + \zeta_{t'}^z) + 2x_{\ell_t\ell_{t'}} - 4r_0^{tt'}} \right]$	6			

$K\ell_t = 2$	$\psi^{+}_{2,1,1}$ state	$+\dfrac{2J(J+1)\left[J(J+1)-2\right](q_0^t+q_0^{t'})^2}{2(B_e^{zz}-B_e^{xx})-2B_e^{zz}(\zeta_t^z+\zeta_{t'}^z)+x_{\ell_t\ell_{t'}}-2r_0^{tt'}}$	6		
	$\psi^{-}_{2,1,1}$ state	$+\dfrac{2J(J+1)\left[J(J+1)-2\right](q_0^t-q_0^{t'})^2}{2(B_e^{zz}-B_e^{xx})-2B_e^{zz}(\zeta_t^z+\zeta_{t'}^z)+x_{\ell_t\ell_{t'}}+2r_0^{tt'}}$	6		

$$A^{\pm}_{K\ell_t+i} = 2(K\ell_t+i)B_e^{zz}(\zeta_t^z-\zeta_{t'}^z) \pm 2\sqrt{(B_e^{zz}\zeta_t^z-B_e^{zz}\zeta_{t'}^z)^2(K\ell_t+i)^2+4(r_0^{tt'})^2}$$

$$B^{\pm}_{K\ell_t} = \pm 2\sqrt{(B_e^{zz}\zeta_t^z-B_e^{zz}\zeta_{t'}^z)^2(K\ell_t+1)^2+4(r_0^{tt'})^2}-(2K\ell_t+1)(B_e^{zz}-B_e^{xx})-2K\ell_t B_e^{zz}(\zeta_t^z+\zeta_{t'}^z)+2x_{\ell_t\ell_{t'}}$$

$$C^{\pm}_{K\ell_t} = \pm 2\sqrt{(B_e^{zz}\zeta_t^z-B_e^{zz}\zeta_{t'}^z)^2(K\ell_t-2)^2+4(r_0^{tt'})^2}+4(K\ell_t-1)(B_e^{zz}-B_e^{xx})-2K\ell_t B_e^{zz}(\zeta_t^z+\zeta_{t'}^z)+2x_{\ell_t\ell_{t'}}$$

All the contributions in this table take into account the relation $\ell_{t'}=\ell_t$

TABLE XLVIII, 8 B . $v_t = v_{t'} = 1$; $|\ell_t + \ell_{t'}| = 2$; $v_{t''} = \ldots = 0$; $v_n, v_{n'} \ldots$ arbitrary ; $J \simeq K \simeq 30$

Contributions	k: a	b	c
$\frac{E_v}{hc}$	0	2	4
$+ B_v^{xx} J(J+1)$	0	2	4
$+ (B_v^{zz} - B_v^{xx})(K\ell_t)^2$	0	2	4
$- 2[(\zeta_t B^{zz})_v + (\zeta_{t'} B^{zz})_v] K\ell_t$	1	3	
$- D_v^J J^2 (J+1)^2$	2	4	
$- D_v^{JK} J(J+1)(K\ell_t)^2$	2	4	
$- D_v^K (K\ell_t)^4$	2	4	
$+ (\eta_t^J + \eta_{t'}^J) J(J+1) K\ell_t$	3		
$+ (\eta_t^K + \eta_{t'}^K)(K\ell_t)^3$	3		

$+H^{J}J^{3}(J+1)^{3}$	4			
$+H^{JJK}J^{2}(J+1)^{2}(K\ell_t)^{2}$	4			
$+H^{JKK}J(J+1)(K\ell_t)^{4}$	4			
$+H^{K}(K\ell_t)^{6}$	4			
$-4(2K\ell_t+1)^{2}\left[J(J+1)-K\ell_t(K\ell_t+1)\right]\left\{\frac{(q_{12}^{t})^{2}}{(2K\ell_t+1)(B_e^{zz}-B_e^{xx}+2B_e^{zz}\zeta_t^{z})-2B_e^{zz}\zeta_{t'}^{z}-2x_{\ell_t\ell_{t'}}} + \frac{(q_{12}^{t'})^{2}}{(2K\ell_t+1)(B_e^{zz}-B_e^{xx}+2B_e^{zz}\zeta_{t'}^{z})-2B_e^{zz}\zeta_t^{z}-2x_{\ell_t\ell_{t'}}}\right\}$	4			

All the contributions in this table take into account the relation $\ell_{t'} = \ell_t$

TABLE XLIX . Coefficients appearing in the expressions of energies or frequencies

$$\frac{E_v}{hc} \begin{cases} \text{(a)} & \sum_{s} \omega_s \left(v_s + \frac{d_s}{2}\right) \\ \text{(b)} & \sum_{\substack{ss' \\ s \leq s'}} x_{ss'} \left(v_s + \frac{d_s}{2}\right)\left(v_{s'} + \frac{d_{s'}}{2}\right) + \sum_{\substack{tt' \\ t \leq t'}} x_{l_t l_{t'}} l_t l_{t'} \\ \text{(c)} & \sum_{\substack{ss's'' \\ s \leq s' \leq s''}} y_{ss's''} \left(v_s + \frac{d_s}{2}\right)\left(v_{s'} + \frac{d_{s'}}{2}\right)\left(v_{s''} + \frac{d_{s''}}{2}\right) \\ & + \sum_{\substack{stt' \\ t \leq t'}} y_{s l_t l_{t'}} \left(v_s + \frac{d_s}{2}\right) l_t l_{t'} + \sum_{s} \Delta\omega_s \left(v_s + \frac{d_s}{2}\right) \end{cases}$$

$$\begin{matrix} B_v^{ii} \\ (i = x, y, z) \end{matrix} \begin{cases} \text{(a)} & B_e^{ii} \\ \text{(b)} & -\sum_{s} \alpha_s^{i} \left(v_s + \frac{d_s}{2}\right) \\ \text{(c)} & \sum_{\substack{ss' \\ s \leq s'}} \gamma_{ss'}^{i} \left(v_s + \frac{d_s}{2}\right)\left(v_{s'} + \frac{d_{s'}}{2}\right) + \sum_{\substack{tt' \\ t \leq t'}} \gamma_{l_t l_{t'}}^{i} l_t l_{t'} + \Delta B_e^{ii} \end{cases}$$

TABLE XLIX . Continued

D_v^i $(i = J, JK, K)$	(a)	D_e^i
	(b)	$\sum_s \beta_s^i (v_s + \frac{d_s}{2})$
$(\zeta_t B^{zz})_v$	(a)	$\zeta_t^z B_e^{zz}$
	(b)	$-\frac{1}{2} \sum_s \eta_{ts} (v_s + \frac{d_s}{2})$
	(c)	$-\sum_{\substack{s s' \\ s \leqslant s'}} \xi_{ss'}^t (v_s + \frac{d_s}{2})(v_{s'} + \frac{d_{s'}}{2}) - \sum_{\substack{t' t'' \\ t \leqslant t' \leqslant t''}} \xi_{l_{t'} l_{t''}}^t l_{t'} l_{t''} + \Delta(\zeta_t^z B_e^{zz})$
q_{eff}^t	(a)	q_0^t
	(b)	$\sum_s f_{22}^{t,s} (v_s + \frac{d_s}{2})$
$r_{eff}^{tt'}$	(a)	$r_0^{tt'}$
	(b)	$\sum_s g_{22}^{tt',s} (v_s + \frac{d_s}{2})$

An index s designates any vibration (non degenerate or twofold degenerate).

An index t designates a twofold degenerate vibration.

The constant ΔB_e^{ii} is a sum of three coefficients arising from operators :

- P^4 in $h_2^\dagger$ (coefficients $b^{\alpha\alpha}$ of the terms in $J(J+1)$ or K^2 in the diagonal matrix element ; see Chapter VIII, Table XXXV).
- P^2 in $h_4^\dagger$
- $r^4 P^2$ in $h_4^\dagger$

The constant $\Delta(\zeta_t^z B_e^{zz})$ is a sum of two coefficients arising from operators :

- $r^2 P^3$ in $h_3^\dagger$ (coefficient η_t in Chapter VIII, Table XXXV)
- $r^2 P$ in $h_5^\dagger$

TABLE L . Correspondance between the symbols used in Chapter X for the coefficients of the matrix elements (first column), and the corresponding symbols used in the preceding chapters (second column)

A - Coefficients appearing in the diagonal matrix elements

y_{sss} ; $y_{nnn'}$; y_{nnt} ; y_{ttn} ; $y_{ttt'}$	μ_1^{s} ; $\mu^{nn'}$; μ_3^{nt} ; μ_1^{nt} ; $\mu_1^{tt'}$
$y_{nn'n''}$; $y_{nn't}$; $y_{ntt'}$; $y_{tt't''}$	$\mu^{nn'n''}$; $\mu^{nn't}$; $\mu^{tt'n}$; $\mu_1^{tt't''}$
$y_{nl_tl_t}$; $y_{tl_tl_t}$; $y_{t'l_tl_t}$; $y_{nl_tl_{t'}}$; $y_{tl_tl_{t'}}$	μ_2^{nt} ; μ_2^{t} ; $\mu_2^{tt'}$; $\mu_2^{tt'n}$; $\mu_3^{tt'}$
$y_{t''l_tl_{t'}}$; $\Delta\omega_n$; $\Delta\omega_t$	$\mu_2^{tt't''}$; $\mu_2^{n}+\mu_3^{n}$; $\mu_3^{t}+\mu_4^{t}$
γ_{ss}^{x} ; $\gamma_{nn'}^{x}$; γ_{nt}^{x} ; $\gamma_{tt'}^{x}$; $\gamma_{l_tl_t}^{x}$; $\gamma_{l_tl_{t'}}^{x}$	μ_3^{sJ} ; $\mu^{nn'J}$; μ^{ntJ} ; $\mu_1^{tt'J}$; μ_4^{tJ} ; $\mu_2^{tt'J}$
$\gamma_{ss}^{z}-\gamma_{ss}^{x}$; $\gamma_{nn'}^{z}-\gamma_{nn'}^{x}$; $\gamma_{nt}^{z}-\gamma_{nt}^{x}$; $\gamma_{tt'}^{z}-\gamma_{tt'}^{x}$	μ_3^{sK} ; $\mu^{nn'K}$; μ^{ntK} ; $\mu_1^{tt'K}$
$\gamma_{l_tl_t}^{z}-\gamma_{l_tl_t}^{x}$; $\gamma_{l_tl_{t'}}^{z}-\gamma_{l_tl_{t'}}^{x}$	μ_4^{tK} ; $\mu_2^{tt'K}$
ΔB_e^{xx} ; $\Delta B_e^{zz}-\Delta B_e^{xx}$	$-b_0^{xx}+\mu_4^{J}+\mu_5^{J}$; $-(b_0^{zz}-b_0^{xx})+\mu_4^{K}+\mu_5^{K}$
α_s^{x} ; α_s^{z} ; β_s^{J} ; β_s^{JK} ; β_s^{K}	α_s^{xx} ; α_s^{zz} ; $-\mu_1^{sJ}$; $-\mu^{sJK}$; $-\mu_2^{sK}$

$\Delta(\zeta_t^z B_e^{zz})$	$-\frac{\eta_t}{2}$ + coefficient appearing in the diagonal element of $r^2 P(h_5^+)$
H^J ; H^{JJK} ; H^{JKK} ; H^K	μ_1^J ; μ_1^{JK} ; μ_3^{JK} ; μ_3^K

B - Coefficients appearing in the nondiagonal matrix elements

q_0^t ; q_3 ; q_{12}^t ; $r_0^{tt'}$	$F_\pm^t$; $R_7^x \pm i R_7^y$; $E_\pm^t$; $R_\pm^{tt'}$
d_3^t ; $d_{12}^{t,t}$;	a_t'' ; $({}^x\mathcal{R}_t - {}^y\mathcal{I}_t) \pm i ({}^x\mathcal{I}_t + {}^y\mathcal{R}_t)$
$d_{12}^{t,t'}$	$({}^x\mathcal{R}_{t't} - {}^y\mathcal{I}_{t't}) \pm i ({}^x\mathcal{I}_{t't} + {}^y\mathcal{R}_{t't})$
$g_{22}^{tt';s}$; $g_{22}^{tt';J}$; $g_{22}^{tt';K}$	$g_{22-}^{tt';s}$; $g_{22-}^{tt';J}$; $g_{22-}^{tt';K}$
$f_{22}^{t,s}$; $f_{22}^{t,J}$; $f_{22}^{t,K}$	$f_{22+}^{t,s}$; $f_{22+}^{t,J}$; $f_{22+}^{t,K}$

An index s designates any vibration (non degenerate or twofold degenerate).

An index n designates a non degenerate vibration.

An index t designates a twofold degenerate vibration.

TABLE LI . Selection rules for dipole transitions of molecules C_{3v} (neglecting inversion)

		$\sum_t \Delta \ell_t + \sum_{n(A_2)} \Delta v_n$	
		even	odd
Parallel Bands $\begin{cases} \sum_t \Delta \ell_t = 3p \\ \Delta K = 0 \end{cases}$	Q Branches $\Delta J = 0$	$\Psi^+ \leftrightarrow \Psi^-$	$\Psi^+ \leftrightarrow \Psi^+$ $\Psi^- \leftrightarrow \Psi^-$
	R and P Branches $\Delta J = \pm 1$	$\Psi^+ \leftrightarrow \Psi^+$ $\Psi^- \leftrightarrow \Psi^-$	$\Psi^+ \leftrightarrow \Psi^-$
Perpendicular Bands $\begin{cases} \sum_t \Delta \ell_t - \Delta K = 3p \\ \Delta K = \pm 1 \end{cases}$	Q Branches $\Delta J = 0$	$\Psi^+ \leftrightarrow \Psi^+$ $\Psi^- \leftrightarrow \Psi^-$	$\Psi^+ \leftrightarrow \Psi^-$
	R and P Branches $\Delta J = \pm 1$	$\Psi^+ \leftrightarrow \Psi^-$	$\Psi^+ \leftrightarrow \Psi^+$ $\Psi^- \leftrightarrow \Psi^-$
Rotational Transitions $\begin{cases} \sum_t \Delta \ell_t = 0 \\ \Delta K = 0 \end{cases}$	$\Delta J = 1$	$\Psi^+ \leftrightarrow \Psi^+$ $\Psi^- \leftrightarrow \Psi^-$	
Transitions $A_1 \leftrightarrow A_2$ and $E \leftrightarrow E$		Transitions $A_1 \leftrightarrow A_2$	

TABLES LII . FREQUENCIES OF LINES IN THE ROTATION SPECTRUM OF MOLECULES C_{3v} (NEGLECTING INVERSION)

TABLE LII,1 . Rotational frequencies for $v_t = v_{t'} = \ldots = 0$; $v_n, v_{n'} \ldots$ arbitrary

$\nu_{J,\|K\|}$	h: a	b	c
$2(J+1)B_v^{xx}$	2	4	6
$-4(J+1)^3 D_e^J$	6		
$-2(J+1)K^2 D_e^{JK}$	6		

TABLE LII, 2 . Rotational frequencies for $v_t = 1$, $\|\ell_t\| = 1$; $v_{t'} = \ldots = 0$; $v_n, v_{n'} \ldots$ arbitrary

$\nu_{J,K\ell_t}$	h: a	b	c
$2(J+1)\Big[B_v^{xx}$	2	4	6
$-D_e^{JK} + \eta_t^J - \dfrac{12(q_{12}^t)^2}{B_e^{zz} - B_e^{xx} + 2B_e^{zz}\zeta_t^z}\Big]$	6		

		$-4(J+1)^3 D_e^J$	6		
		$-2(J+1)(K\ell_t-1)^2 D_e^{JK}$	6		
		$+2(J+1)(K\ell_t-1)\left(\eta_t^J - 2D_e^{JK} - \dfrac{8(q_{12}^t)^2}{B_e^{zz} - B_e^{xx} + 2B_e^{zz}\zeta_t^z} - \dfrac{2(q_o^t)^2}{B_e^{zz} - B_e^{xx} - B_e^{zz}\zeta_t^z}\right)$	6		
$K\ell_t \neq 1$	$+\dfrac{4(J+1)^3(q_o^t)^2}{(K\ell_t-1)(B_e^{zz} - B_e^{xx} - B_e^{zz}\zeta_t^z)}$		6		
$K\ell_t = 1$	$\psi^+ \longleftrightarrow \psi^+$ transition	$+4(J+1)\, q_{eff}^t$	4	6	
	$\psi^- \longleftrightarrow \psi^-$ transition	$-4(J+1)\, q_{eff}^t$	4	6	

The result obtained here for $\nu_{J,K\ell_t}$ is identical to the one previously obtained by M.L.Grenier-Besson and G.Amat [58].

TABLE LII, 3 . Rotational frequencies for $v_t = 2$, $\ell_t = 0$; $v_{t'} = \ldots = 0$; $v_n, v_{n'}, \ldots$ arbitrary

$\nu_{J,\|K\|}$	k: a	b	c
$2(J+1)B_v^{xx}$	2	4	6
$-4(J+1)^3 D_e^J$	6		
$-2(J+1)K^2 D_e^{JK}$	6		
$+\dfrac{16(q_0^t)^2(J+1)\left[(J+1)^2 B' + K^2(2A' - B') - B'\right]}{K^2 A'^2 - B'^2}$	6		
$+\dfrac{32(J+1)\left\{4K^2\left[(q_{12}^t)^2(B-2A) + 2A\, q_{12}^t\, d_{12}^{t,t}\right] + B(q_{12}^t - d_{12}^{t,t})^2\right\}}{4K^2 A^2 - B^2}$	6		

$A = B_e^{zz} - B_e^{xx} + 2B_e^{zz}\zeta_t^z$ $A' = B_e^{zz} - B_e^{xx} - B_e^{zz}\zeta_t^z$

$B = B_e^{zz} - B_e^{xx} + 4B_e^{zz}\zeta_t^z + 4x_{\ell_t \ell_t}$ $B' = B_e^{zz} - B_e^{xx} - 2B_e^{zz}\zeta_t^z + x_{\ell_t \ell_t}$

TABLE LII, 4. Rotational frequencies for $v_t = 2$, $|\ell_t| = 2$; $v_{t'} = \ldots = 0$; $v_n, v_{n'} \ldots$ arbitrary

	$\nu_{J,K\ell_t}$	k: a	b	c
	$2(J+1)B_v^{xx}$	2	4	6
	$-4(J+1)^3 D_e^J$	6		
	$-2(J+1)K^2 D_e^{JK}$	6		
	$+2(J+1)K\ell_t \eta_t^J$	6		
$K\ell_t \neq -2$	$-\dfrac{16(J+1)\left[q_{12}^t(K\ell_t+1)+d_{12}^{t,t}\right]^2}{K\ell_t(B_e^{zz}-B_e^{xx}+2B_e^{zz}\zeta_t^z)+B_e^{zz}-B_e^{xx}-4x_{\ell_t\ell_t}}$	6		
$K\ell_t \neq 4$	$+\dfrac{4(q_0^t)^2(J+1)\left[4(J+1)^2-(K\ell_t-2)^2\right]}{K\ell_t(B_e^{zz}-B_e^{xx}-B_e^{zz}\zeta_t^z)-2(B_e^{zz}-B_e^{xx}-x_{\ell_t\ell_t})}$	6		

			a	b	c
$K\ell_t = 4$	$\psi^+ \longrightarrow \psi^+$ transition	$+ \dfrac{8(q_0^t)^2 (J+1)\left[4(J+1)^2 - (K\ell_t - 2)^2\right]}{K\ell_t(B_e^{zz} - B_e^{xx} - B_e^{zz}\zeta_t^z) - 2(B_e^{zz} - B_e^{xx} - x_{\ell_t\ell_t})}$	6		
	$\psi^- \longrightarrow \psi^-$ transition	0			
$K\ell_t = -2$	$\psi^+ \longrightarrow \psi^+$ transition	$+ 16(J+1) f_{24}^t$	6		
	$\psi^- \longrightarrow \psi^-$ transition	$- 16(J+1)\left\{ f_{24}^t + \dfrac{2\left[q_{12}^t(K\ell_t + 1) + d_{12}^{t,t}\right]^2}{K\ell_t(B_e^{zz} - B_e^{xx} + 2B_e^{zz}\zeta_t^z) + B_e^{zz} - B_e^{xx} - 4x_{\ell_t\ell_t}} \right\}$	6		

TABLE LII, 5 . Rotational frequencies for $v_t = 3$, $|\ell_t| = 1$; $v_{t'} = \ldots = 0$; v_n, $v_{n'}$ ... arbitrary

$\nu_{J,K\ell_t}$	h: a	h: b	h: c
$2(J+1)\left[B_v^{xx}\right.$	2	4	6
$\left. - \dfrac{16(q_{12}^t)^2}{B_e^{zz} - B_e^{xx} + 2B_e^{zz}\zeta_t^z}\right]$	6		

		$-4(J+1)^3 D_e^J$	6		
		$-2(J+1)K^2 D_e^{JK}$	6		
		$+2(J+1)K\ell_t\left[\eta_t^J - \dfrac{32(q_{12}^t)^2}{B_e^{zz}-B_e^{xx}+2B_e^{zz}\zeta_t^z}\right]$	6		
		$+\dfrac{24(J+1)\left[(q_{12}^t(2K\ell_t-1)+2d_{12}^{t,t}\right]^2}{2K\ell_t(B_e^{zz}-B_e^{xx}+2B_e^{zz}\zeta_t^z)-(B_e^{zz}-B_e^{xx}+6B_e^{zz}\zeta_t^z+8x_{\ell_t\ell_t})}$	6		
		$-\dfrac{12(J+1)\left[(J+1)^2-(K\ell_t+1)^2\right](q_0^t)^2}{K\ell_t(B_e^{zz}-B_e^{xx}-B_e^{zz}\zeta_t^z)+B_e^{zz}-B_e^{xx}-3B_e^{zz}\zeta_t^z+2x_{\ell_t\ell_t}}$	6		
$K\ell_t \neq 1$		$+\dfrac{16(J+1)\left[(J+1)^2-(K\ell_t-1)^2\right](q_0^t)^2}{(K\ell_t-1)\left[B_e^{zz}-B_e^{xx}-B_e^{zz}\zeta_t^z\right]}$	6		
$K\ell_t = 1$	$\psi^+ \longrightarrow \psi^+$ transition	$+8(J+1)q_{eff}^t$	4	6	
	$\psi^- \longrightarrow \psi^-$ transition	$-8(J+1)q_{eff}^t$	4	6	

TABLE LII, 6 . Rotational frequencies for $v_t = 3$, $|\ell_t| = 3$; $v_{t'} = \ldots = 0$; v_n , $v_{n'}$, ... arbitrary

$\nu_{J,K\ell_t}$	k: a	b	c
$2(J+1)B_v^{xx}$	2	4	6
$-4(J+1)^3 D_e^J$	6		
$-2(J+1)K^2 D_e^{JK}$	6		
$+2(J+1)K\ell_t \eta_t^J$	6		
$-\dfrac{72(J+1)\left[q_{12}^t(2K\ell_t+3)+6d_{12}^{t,t}\right]^2}{18K\ell_t(B_e^{zz}-B_e^{xx}+2B_e^{zz}\zeta_t^z)+27(B_e^{zz}-B_e^{xx}-2B_e^{zz}\zeta_t^z-8x_{\ell_t\ell_t})}$	6		
$+\dfrac{36(J+1)\left[9(J+1)^2-(K\ell_t-3)^2\right](q_0^t)^2}{9K\ell_t(B_e^{zz}-B_e^{xx}-B_e^{zz}\zeta_t^z)-27(B_e^{zz}-B_e^{xx}+B_e^{zz}\zeta_t^z-2x_{\ell_t\ell_t})}$	6		

TABLE LII, 7a . Rotational frequencies for $v_t = v_{t'} = 1$, $\ell_t + \ell_{t'} = 0$; $v_{t''} = \ldots = 0$; $v_n , v_{n'} \ldots$ arbitrary

	$\nu^+_{J,\|K\|}$ $\nu^-_{J,\|K\|}$	k a	k b	k c
	$2(J+1)B_v^{xx}$	2	4	6
	$-4(J+1)^3 D_e^J$	6		
	$-2(J+1)K^2 D_e^{JK}$	6		
$K \neq 0$	$\pm 2(J+1)\left[\dfrac{8r_0^{tt'} g_{zz}^{tt'J} - K^2 B_e^{zz}(\zeta_t^z - \zeta_{t'}^z)(\eta_t^J - \eta_{t'}^J)}{\sqrt{(B_e^{zz}\zeta_t^z - B_e^{zz}\zeta_{t'}^z)^2 K^2 + 4(r_0^{tt'})^2}} \pm \dfrac{4\left\{4r_0^{tt'}\left[q_{12}^t(2K+1) - d_{12}^{t,t'}\right] + A_K^{\pm}\left[q_{12}^{t'}(2K+1) - d_{12}^{t',t}\right]\right\}^2}{C_K^{\pm}\left[16(r_0^{tt'})^2 + (A_K^{\pm})^2\right]} \pm \dfrac{4\left\{4r_0^{tt'}\left[q_{12}^t(2K-1) + d_{12}^{t,t'}\right] + A_{-K}^{\pm}\left[q_{12}^{t'}(2K-1) + d_{12}^{t',t}\right]\right\}^2}{C_{-K}^{\pm}\left[16(r_0^{tt'})^2 + (A_{-K}^{\pm})^2\right]}\right]$	6		

		$+16(J+1)\left[(J+1)^2-(K-1)^2\right]\dfrac{(4r_0^{tt'}q_0^t+A_K^{\pm}q_0^{t'})^2}{B_K^{\pm}\left[16(r_0^{tt'})^2+(A_K^{\pm})^2\right]}$	6		
		$+16(J+1)\left[(J+1)^2-(K+1)^2\right]\dfrac{(4r_0^{tt'}q_0^t+A_{-K}^{\pm}q_0^{t'})^2}{B_{-K}^{\pm}\left[16(r_0^{tt'})^2+(A_{-K}^{\pm})^2\right]}$	6		
K = 0	$\psi^+ \longrightarrow \psi^+$ transition	$+8(J+1)\left[g_{22}^{tt',J}-\dfrac{(q_{12}^t+q_{12}^{t'}-d_{12}^{t',t}-d_{12}^{t,t'})^2}{B_e^{zz}-B_e^{xx}+2B_e^{zz}(\zeta_t^z+\zeta_{t'}^z)+2x_{l_tl_{t'}}-4r_0^{tt'}}+\dfrac{(q_0^t+q_0^{t'})^2}{2(B_e^{zz}-B_e^{xx})-2B_e^{zz}(\zeta_t^z+\zeta_{t'}^z)+x_{l_tl_{t'}}-2r_0^{tt'}}\right]$	6		
		$-8(J+1)^3\dfrac{(q_0^t+q_0^{t'})^2}{2(B_e^{zz}-B_e^{xx})-2B_e^{zz}(\zeta_t^z+\zeta_{t'}^z)+x_{l_tl_{t'}}-2r_0^{tt'}}$	6		

	$\psi^- \longrightarrow \psi^-$ transition	$-8(J+1)\left[g_{22}^{tt',J} + \dfrac{(q_{12}^{t} - q_{12}^{t'} + d_{12}^{t;t} - d_{12}^{t,t'})^2}{B_e^{zz} - B_e^{xx} + 2B_e^{zz}(\zeta_t^z + \zeta_{t'}^z) + 2x_{l_t l_{t'}} + 4r_o^{tt'}} - \dfrac{(q_o^{t} - q_o^{t'})^2}{2(B_e^{zz} - B_e^{xx}) - 2B_e^{zz}(\zeta_t^z + \zeta_{t'}^z) + x_{l_t l_{t'}} + 2r_o^{tt'}}\right]$	6		
		$-8(J+1)^3 \dfrac{(q_o^{t} - q_o^{t'})^2}{2(B_e^{zz} - B_e^{xx}) - 2B_e^{zz}(\zeta_t^z + \zeta_{t'}^z) + x_{l_t l_{t'}} + 2r_o^{tt'}}$	6		

$$A_K^{\pm} = 2K\,B_e^{zz}(\zeta_t^z - \zeta_{t'}^z) \pm 2\sqrt{(B_e^{zz}\zeta_t^z - B_e^{zz}\zeta_{t'}^z)^2 K^2 + 4(r_o^{tt'})^2}$$

$$B_K^{\pm} = 4(B_e^{zz} - B_e^{xx})(K-1) - 2B_e^{zz}(\zeta_t^z + \zeta_{t'}^z)(K-2) - 2x_{l_t l_{t'}} \pm 2\sqrt{(B_e^{zz}\zeta_t^z - B_e^{zz}\zeta_{t'}^z)^2 K^2 + 4(r_o^{tt'})^2}$$

$$C_K^{\pm} = -(B_e^{zz} - B_e^{xx})(2K+1) - 2B_e^{zz}(\zeta_t^z + \zeta_{t'}^z)(K+1) - 2x_{l_t l_{t'}} \pm 2\sqrt{(B_e^{zz}\zeta_t^z - B_e^{zz}\zeta_{t'}^z)^2 K^2 + 4(r_o^{tt'})^2}$$

TABLE LII, 7b . Rotational frequencies for $v_t = v_{t'} = 1$, $\ell_t + \ell_{t'} = 0$; $v_{t''} = \dots = 0$; $v_n , v_{n'} \dots$ arbitrary

$\nu'^{+}_{J,\vert K\vert}$ $\nu'^{-}_{J,\vert K\vert}$ $(\vert K\vert \neq 0)$	h: a	b	c
$2(J+1)B_v^{xx}$	2	4	6
$-4(J+1)^3 D_e^J$	6		
$-2(J+1)K^2 D_e^{JK}$	6		
$\pm 4\sqrt{(B_e^{zz}\zeta_t^z - B_e^{zz}\zeta_{t'}^z)^2K^2 + 4(r_0^{tt'})^2}$	2		
$\pm \dfrac{\sum_s \left[16 r_0^{tt'} g_{22}^{tt';s} - 2K^2 B_e^{zz}(\zeta_t^z - \zeta_{t'}^z)(\eta_{ts} - \eta_{t's})\right](v_s + \frac{d_s}{2})}{\sqrt{(B_e^{zz}\zeta_t^z - B_e^{zz}\zeta_{t'}^z)^2K^2 + 4(r_0^{tt'})^2}}$	4		
$\pm \Big[\Big\{\sum_{ss'}\Big[8 g_{22}^{tt';s} g_{22}^{tt';s'} + \frac{K^2}{2}(\eta_{ts} - \eta_{t's})(\eta_{ts'} - \eta_{t's'})\Big] - 4K^2 \sum_{s\leq s'} B_e^{zz}(\zeta_t^z - \zeta_{t'}^z)(\xi^t_{ss'} - \xi^{t'}_{ss'})\Big\}(v_s + \frac{d_s}{2})(v_{s'} + \frac{d_{s'}}{2}) + 4K^2\rho - 2K^4 B_e^{zz}(\zeta_t^z - \zeta_{t'}^z)(\eta_t^K - \eta_{t'}^K)\Big]\Big[B_e^{zz}\zeta_t^z - B_e^{zz}\zeta_{t'}^z)^2K^2 + 4(r_0^{tt'})^2\Big]^{-\frac{1}{2}}$	6		

$\pm 2(J+1)^2$	$\dfrac{\left[8 r_0^{tt'} g_{22}^{tt',J} - K^2 B_e^{zz}(\zeta_t^z - \zeta_{t'}^z)(\eta_t^J - \eta_{t'}^J)\right]}{\sqrt{(B_e^{zz}\zeta_t^z - B_e^{zz}\zeta_{t'}^z)^2 K^2 + 4(r_0^{tt'})^2}}$	6			
$+4(a_{1,J+1,K}^{+})^2$	$\dfrac{\left\{4 r_0^{tt'}\left[q_{12}^{t}(2K+1) - d_{12}^{t,t'}\right] + A_K^{\pm}\left[q_{12}^{t'}(2K+1) - d_{12}^{t',t}\right]\right\}^2}{B_K^{\pm}\left[16(r_0^{tt'})^2 + (A_K^{\pm})^2\right]}$	6			
$+4(a_{1,J+1,K}^{-})^2$	$\dfrac{\left\{4 r_0^{tt'}\left[q_{12}^{t}(2K-1) + d_{12}^{t,t'}\right] + A_{-K}^{\pm}\left[q_{12}^{t'}(2K-1) + d_{12}^{t',t}\right]\right\}^2}{B_{-K}^{\pm}\left[16(r_0^{tt'})^2 + (A_{-K}^{\pm})^2\right]}$	6			
$-4(a_{1,J,K}^{+})^2$	$\dfrac{\left\{4 r_0^{tt'}\left[q_{12}^{t}(2K+1) - d_{12}^{t,t'}\right] + A_K^{\mp}\left[q_{12}^{t'}(2K+1) - d_{12}^{t',t}\right]\right\}^2}{B_K^{\mp}\left[16(r_0^{tt'})^2 + (A_K^{\mp})^2\right]}$	6			
$-4(a_{1,J,K}^{-})^2$	$\dfrac{\left\{4 r_0^{tt'}\left[q_{12}^{t}(2K-1) + d_{12}^{t,t'}\right] + A_{-K}^{\mp}\left[q_{12}^{t'}(2K-1) + d_{12}^{t',t}\right]\right\}^2}{B_{-K}^{\mp}\left[16(r_0^{tt'})^2 + (A_{-K}^{\mp})^2\right]}$	6			

$+\dfrac{4(a^-_{2,J+1,K})^2\left[4r_0^{tt'}q_0^t + A_K^{\pm} q_0^{t'}\right]^2}{C_K^{\pm}\left[16(r_0^{tt'})^2 + (A_K^{\pm})^2\right]} - \dfrac{4(a^-_{2,J,K})^2\left[4r_0^{tt'}q_0^t + A_K^{\mp} q_0^{t'}\right]^2}{C_K^{\mp}\left[16(r_0^{tt'})^2 + (A_K^{\mp})^2\right]}$	6		
$+\dfrac{4(a^+_{2,J+1,K})^2\left[4r_0^{tt'}q_0^t + A_{-K}^{\pm} q_0^{t'}\right]^2}{C_{-K}^{\pm}\left[16(r_0^{tt'})^2 + (A_{-K}^{\pm})^2\right]} - \dfrac{4(a^+_{2,J,K})^2\left[4r_0^{tt'}q_0^t + A_{-K}^{\mp} q_0^{t'}\right]^2}{C_{-K}^{\mp}\left[16(r_0^{tt'})^2 + (A_{-K}^{\mp})^2\right]}$	6		

$$a^{\pm}_{i,J,K} = \sqrt{J(J+1)-K(K\pm 1)}\,\sqrt{J(J+1)-(K\pm 1)(K\pm 2)}\,\ldots\ldots\ldots\,\sqrt{J(J+1)-[K\pm(i-1)](K\pm i)}$$

$$\rho = 4r_0^{tt'}g_{22}^{tt',K} + B_e^{zz}(\zeta_t^z - \zeta_{t'}^z)\left[\Delta(B_e^{zz}\zeta_t^z) - \Delta(B_e^{zz}\zeta_{t'}^z) - \xi^t_{\ell_t\ell_t} + \xi^t_{\ell_t\ell_{t'}} - \xi^t_{\ell_{t'}\ell_{t'}} + \xi^{t'}_{\ell_t\ell_t} - \xi^{t'}_{\ell_t\ell_{t'}} + \xi^{t'}_{\ell_{t'}\ell_{t'}}\right.$$

$$A_K^{\pm} = 2K\,B_e^{zz}(\zeta_t^z - \zeta_{t'}^z) \pm 2\sqrt{(B_e^{zz}\zeta_t^z - B_e^{zz}\zeta_{t'}^z)^2K^2 + 4(r_0^{tt'})^2}$$

$$B_K^{\pm} = -(2K+1)(B_e^{zz} - B_e^{xx}) - 2B_e^{zz}(\zeta_t^z + \zeta_{t'}^z)(K+1) - 2x_{\ell_t\ell_{t'}} \pm 2\sqrt{(B_e^{zz}\zeta_t^z - B_e^{zz}\zeta_{t'}^z)^2K^2 + 4(r_0^{tt'})^2}$$

$$C_K^{\pm} = 4(K-1)(B_e^{zz} - B_e^{xx}) - 2B_e^{zz}(\zeta_t^z + \zeta_{t'}^z)(K-2) - 2x_{\ell_t\ell_{t'}} \pm 2\sqrt{(B_e^{zz}\zeta_t^z - B_e^{zz}\zeta_{t'}^z)^2K^2 + 4(r_0^{tt'})^2}$$

TABLE LII, 8 . Rotational frequencies for $v_t = v_{t'} = 1$, $|\ell_t + \ell_{t'}| = 2$; $v_{t''} = \ldots = 0$; v_n , $v_{n'}$ ... arbitrary

	$\nu_{J,K\ell_t}$	k: a	b	c
	$2(J+1)B_v^{xx}$	2	4	6
	$-4(J+1)^3 D_e^J$	6		
	$-2(J+1)K^2 D_e^{JK}$	6		
	$+2(J+1)K\ell_t(\eta_t^J + \eta_{t'}^J)$	6		
$K\ell_t \neq -1$	$+8(J+1)\sum_{i=+,-} \dfrac{\left\{A^i_{K\ell_t+1}\left[q_{12}^{t'}(2K\ell_t+1)+d_{12}^{t',t}\right]-4r_0^{tt'}\left[q_{12}^{t}(2K\ell_t+1)+d_{12}^{t,t'}\right]\right\}^2}{B^i_{K\ell_t}\left[16(r_0^{tt'})^2+(A^i_{K\ell_t+1})^2\right]}$	6		
$K\ell_t \neq 2$	$+16(J+1)\left[(J+1)^2-(K\ell_t-1)^2\right]\sum_{i=+,-} \dfrac{(A^i_{K\ell_t-2}q_0^{t'}-4r_0^{tt'}q_0^{t})^2}{C^i_{K\ell_t}\left[16(r_0^{tt'})^2+(A^i_{K\ell_t-2})^2\right]}$	6		

$K\ell_t = 2$	$\psi^+ \longrightarrow \psi^+$ transition	$+\dfrac{8(J+1)\left[(J+1)^2-1\right](q_0^t+q_0^{t'})^2}{2(B_e^{zz}-B_e^{xx})-2B_e^{zz}(\zeta_t^z+\zeta_{t'}^z)+x_{\ell_t\ell_{t'}}-2r_0^{tt'}}$	6		
	$\psi^- \longrightarrow \psi^-$ transition	$+\dfrac{8(J+1)\left[(J+1)^2-1\right](q_0^t-q_0^{t'})^2}{2(B_e^{zz}-B_e^{xx})-2B_e^{zz}(\zeta_t^z+\zeta_{t'}^z)+x_{\ell_t\ell_{t'}}+2r_0^{tt'}}$	6		
$K\ell_t = -1$	$\psi^+ \longrightarrow \psi^+$ transition	$+\ 8(J+1)\left[f_{222}^{tt'}+\dfrac{(q_{12}^t-q_{12}^{t'}-d_{12}^{t,t'}+d_{12}^{t';t})^2}{B_e^{zz}-B_e^{xx}+2B_e^{zz}(\zeta_t^z+\zeta_{t'}^z)+2x_{\ell_t\ell_{t'}}+4r_0^{tt'}}\right]$	6		
	$\psi^- \longrightarrow \psi^-$ transition	$-\ 8(J+1)\left[f_{222}^{tt'}-\dfrac{(q_{12}^t+q_{12}^{t'}-d_{12}^{t,t'}-d_{12}^{t';t})^2}{B_e^{zz}-B_e^{xx}+2B_e^{zz}(\zeta_t^z+\zeta_{t'}^z)+2x_{\ell_t\ell_{t'}}-4r_0^{tt'}}\right]$	6		

$$A^{\pm}_{K\ell_t+i} = 2(K\ell_t+i)B_e^{zz}(\zeta_t^z-\zeta_{t'}^z) \pm 2\sqrt{(B_e^{zz}\zeta_t^z-B_e^{zz}\zeta_{t'}^z)^2(K\ell_t+i)^2+4(r_0^{tt'})^2}$$

$$B^{\pm}_{K\ell_t} = -(2K\ell_t+1)(B_e^{zz}-B_e^{xx})-2K\ell_t B_e^{zz}(\zeta_t^z+\zeta_{t'}^z)+2x_{\ell_t\ell_{t'}} \pm 2\sqrt{(B_e^{zz}\zeta_t^z-B_e^{zz}\zeta_{t'}^z)^2(K\ell_t+1)^2+4r_0^{tt'})^2}$$

$$C^{\pm}_{K\ell_t} = 4(K\ell_t-1)(B_e^{zz}-B_e^{xx})-2K\ell_t B_e^{zz}(\zeta_t^z+\zeta_{t'}^z)+2x_{\ell_t\ell_{t'}} \pm 2\sqrt{(B_e^{zz}\zeta_t^z-B_e^{zz}\zeta_{t'}^z)^2(K\ell_t-2)^2+4(r_0^{tt'})^2}$$

All the contributions in this table take into account the relation $\ell_{t'} = \ell_t$

TABLES LIII . FREQUENCIES OF LINES IN THE ROTATION-VIBRATION SPECTRUM

OF MOLECULES C_{3v} (NEGLECTING INVERSION)

TABLE LIII,1 Aα . Transition from v''($v_t = v_{t'} = \ldots = 0$; v_n , $v_{n'}$ ... arbitrary) to v' ($v_t = v_{t'} = \ldots = 0$; v_n, $v_{n'}$ ... arbitrary) ; $J \simeq K \simeq 1$

${}^{Q}Q_{J,\|K\|}$	k a	b	c
$\dfrac{E_{v'} - E_{v''}}{hc}$	0	2	4
$+ J(J+1)(B^{xx}_{v'} - B^{xx}_{v''})$	–	4	
$+ K^2\left[(B^{zz}_{v'} - B^{zz}_{v''}) - (B^{xx}_{v'} - B^{xx}_{v''})\right]$	–	4	

TABLE LIII, 1 A β . Transition from $v''(v_t = v_{t'} = \ldots = 0\,;\ v_n, v_{n'} \ldots$ arbitrary) to $v'(v_t = v_{t'} = \ldots = 0\,;\ v_n, v_{n'} \ldots$ arbitrary) ; $J \simeq K \simeq 1$

${}^{Q}R_{J,\|K\|}$ ${}^{Q}P_{J,\|K\|}$	k: a	b	c
$\dfrac{E_{v'} - E_{v''}}{hc}$	0	2	4
$+ m\,(B^{xx}_{v'} + B^{xx}_{v''})$	2	4	
$+ m^2 (B^{xx}_{v'} - B^{xx}_{v''})$	–	4	
$+ K^2\left[(B^{zz}_{v'} - B^{zz}_{v''}) - (B^{xx}_{v'} - B^{xx}_{v''})\right]$	–	4	

TABLE LIII, 1 B α . Transition from $v''(v_t = v_{t'} = \ldots = 0 ; v_n , v_{n'} \ldots$ arbitrary) to $v'(v_t = v_{t'} = \ldots = 0 ; v_n , v_{n'} , \ldots$ arbitrary) ; $J \simeq K \simeq 30$

${}^{Q}Q_{J,\|K\|}$	k: a	b	c
$\frac{E_{v'} - E_{v''}}{hc}$	0	2	4
$+ J(J+1)(B^{xx}_{v'} - B^{xx}_{v''})$	–	2	4
$+ K^2\left[(B^{zz}_{v'} - B^{zz}_{v''}) - (B^{xx}_{v'} - B^{xx}_{v''})\right]$	–	2	4
$- J^2(J+1)^2(D^{J}_{v'} - D^{J}_{v''})$	–	4	
$- J(J+1)K^2(D^{JK}_{v'} - D^{JK}_{v''})$	–	4	
$- K^4(D^{K}_{v'} - D^{K}_{v''})$	–	4	

TABLE LIII, 1 Bβ . Transition from v'' ($v_t = v_{t'} = \ldots = 0$; $v_n , v_{n'} \ldots$ arbitrary) to v' ($v_t = v_{t'} = \ldots = 0$; $v_n , v_{n'} \ldots$ arbitrary) ; $J \simeq K \simeq 30$

${}^{Q}R_{J,\,\lvert K\rvert}$ ${}^{Q}P_{J,\,\lvert K\rvert}$	k: a	b	c
$\frac{E_{v'} - E_{v''}}{hc}$	0	2	4
$+ m (B^{xx}_{v'} + B^{xx}_{v''})$	1	3	
$+ m^2 (B^{xx}_{v'} - B^{xx}_{v''})$	–	2	4
$+ K^2 \left[(B^{zz}_{v'} - B^{zz}_{v''}) - (B^{xx}_{v'} - B^{xx}_{v''}) \right]$	–	2	4
$- 4 m^3 D^{J}_{e}$	3		
$- 2 m K^2 D^{JK}_{e}$	3		
$- m^4 (D^{J}_{v'} - D^{J}_{v''})$	–	4	
$- m^2 K^2 (D^{JK}_{v'} - D^{JK}_{v''})$	–	4	
$- K^4 (D^{K}_{v'} - D^{K}_{v''})$	–	4	

TABLE LIII, 1 C α. Transition from $v''(v_t = v_{t'} = \ldots = 0 ; v_n, v_{n'} \ldots$ arbitrary) to v' $(v_t = v_{t'} = \ldots = 0 ; v_n, v_{n'} \ldots$ arbitrary) ; $J \simeq 30$, $K \simeq 1$

$^QQ_{J\|K\|}$			h: a	h: b	h: c
$\frac{E_{v'} - E_{v''}}{hc}$			0	2	4
$+ J(J+1)(B^{xx}_{v'} - B^{xx}_{v''})$			–	2	4
$+ K^2 [(B^{zz}_{v'} - B^{zz}_{v''}) - (B^{xx}_{v'} - B^{xx}_{v''})]$			–	4	
$- J^2(J+1)^2(D^J_{v'} - D^J_{v''})$			–	4	
$\|K\| = 3$	$\psi^- \longrightarrow \psi^+$ transition	$+2J(J+1)[J(J+1)-2][J(J+1)-6](f_6 - \frac{(q_3)^2}{B^{zz}_e - B^{xx}_e})$	4		
	$\psi^+ \longrightarrow \psi^-$ transition	$-2J(J+1)[J(J+1)-2][J(J+1)-6](f_6 - \frac{(q_3)^2}{B^{zz}_e - B^{xx}_e})$	4		

TABLE LIII, 1 Cβ . Transition from v'' ($v_t = v_{t'} = \ldots = 0$; v_n , $v_{n'} \ldots$ arbitrary) to v' ($v_t = v_{t'} = \ldots = 0$; v_n , $v_{n'} \ldots$ arbitrary) ; $J \simeq 30$, $K \simeq 1$

${}^QR_{J,\|K\|}$ ${}^QP_{J,\|K\|}$	k a	k b	k c
$\frac{E_{v'} - E_{v''}}{hc}$	0	2	4
$+ m(B^{xx}_{v'} - B^{xx}_{v''})$	1	3	
$+ m^2(B^{xx}_{v'} - B^{xx}_{v''})$	–	2	4
$- 4m^3 D^J_e$	3		
$- m^4(D^J_{v'} - D^J_{v''})$	–	4	
$+ K^2[(B^{zz}_{v'} - B^{zz}_{v''}) - (B^{xx}_{v'} - B^{xx}_{v''})]$	–	4	

TABLE LIII, 2 A α . Transition from $v''(v_t = v_{t'} = \ldots = 0 ; v_n , v_{n'} \ldots$ arbitrary) to v' ($v_t = 1$, $|\ell_t| = 1$; $v_{t'} = \ldots = 0$; $v_n , v_{n'} \ldots$ arbitrary) ; $J \simeq K \simeq 1$

	${}^{R}Q_{J,\,|K|}$ ${}^{P}Q_{J,\,|K|}$	k: a	b	c
$\frac{E_{v'} - E_{v''}}{hc}$		0	2	4
$+ B^{zz}_{v'} - B^{xx}_{v'} - 2(\zeta_t B^{zz})_{v'}$		2	4	
$\pm 2|K|[B^{zz}_{v'} - B^{xx}_{v'} - (\zeta_t B^{zz})_{v'}]$		2	4	
$+ K^2[(B^{zz}_{v'} - B^{zz}_{v''}) - (B^{xx}_{v'} - B^{xx}_{v''})]$		–	4	
$|K| \neq 0$	$+J(J+1)(B^{xx}_{v'} - B^{xx}_{v''})$	–	4	
$K = 0$	$+J(J+1)[B^{xx}_{v'} - B^{xx}_{v''}$	–	4	
	$-2q^t_0]$	4		

TABLE LIII, 2 Aβ . Transition from $v''(v_t = v_{t'} = \ldots = 0\ ; v_n, v_{n'} \ldots$ arbitrary) to $v'(v_t = 1\ , |\ell_t| = 1\ ; v_{t'} = \ldots = 0\ ; v_n, v_{n'} \ldots$ arbitrary) ; $J \simeq K \simeq 1$

	$^RR_{J,\|K\|}$ $^RP_{J,\|K\|}$ $^PR_{J,\|K\|}$ $^PP_{J,\|K\|}$	k: a	b	c
$\dfrac{E_{v'} - E_{v''}}{hc}$		0	2	4
	$+ B^{zz}_{v'} - B^{xx}_{v'} - 2(\zeta_t B^{zz})_{v'}$	2	4	
	$\pm 2\|K\|[B^{zz}_{v'} - B^{xx}_{v'} - (\zeta_t B^{zz})_{v'}]$	2	4	
	$+ K^2[(B^{zz}_{v'} - B^{zz}_{v''}) - (B^{xx}_{v'} - B^{xx}_{v''})]$	-	4	
$\|K\| \neq 0$	$+ m(B^{xx}_{v'} + B^{xx}_{v''})$	2	4	
	$+ m^2(B^{xx}_{v'} - B^{xx}_{v''})$	-	4	
$K = 0$	$+ m[B^{xx}_{v'} + B^{xx}_{v''}$	2	4	
	$+ 2q^t_0]$	4		
	$+ m^2[B^{xx}_{v'} - B^{xx}_{v''}$	-	4	
	$+ 2q^t_0]$	4		

TABLE LIII, 2 Bα . Transition from v''($v_t = v_{t'} = \ldots = 0$; $v_n, v_{n'} \ldots$ arbitrary) to v'($v_t = 1, |\ell_t| = 1$; $v_{t'} = \ldots = 0$; $v_n, v_{n'} \ldots$ arbitrary) ; $J \simeq K \simeq 30$

${}^RQ_{J,\,|K|}$ ${}^PQ_{J,\,|K|}$	h: a	b	c
$\dfrac{E_{v'} - E_{v''}}{hc}$	0	2	4
$+ B^{zz}_{v'} - B^{xx}_{v'} - 2(\zeta_t B^{zz})_{v'}$	2	4	
$\pm 2\,|K| \left[B^{zz}_{v'} - B^{xx}_{v'} - (\zeta_t B^{zz})_{v'} \right]$	1	3	
$+ J(J+1) \left[B^{xx}_{v'} - B^{xx}_{v''} \right.$	–	2	4
$\left. - D^{JK}_e + \eta^J_t - \dfrac{12 (q^t_{12})^2}{B^{zz}_e - B^{xx}_e + 2 B^{zz}_e \zeta^z_t} \right]$	4		
$+ K^2 \left[B^{zz}_{v'} - B^{zz}_{v''} - (B^{xx}_{v'} - B^{xx}_{v''}) \right.$	–	2	4
$\left. - 6 D^K_e + 3 \eta^K_t + \dfrac{36 (q^t_{12})^2}{B^{zz}_e - B^{xx}_e + 2 B^{zz}_e \zeta^z_t} \right]$	4		
$\mp J(J+1)\,|K|\, \left(2 D^{JK}_e - \eta^J_t + \dfrac{8 (q^t_{12})^2}{B^{zz}_e - B^{xx}_e + 2 B^{zz}_e \zeta^z_t} \right)$	3		

$\mp \lvert K \rvert^3 \left(4D_e^K - \eta_t^K - \dfrac{8(q_{12}^t)^2}{B_e^{zz} - B_e^{xx} + 2B_e^{zz}\zeta_t^z}\right)$	3		
$-J^2(J+1)^2(D_{v'}^J - D_{v''}^J)$	–	4	
$-J(J+1)K^2(D_{v'}^{JK} - D_{v''}^{JK})$	–	4	
$-K^4(D_{v'}^K - D_{v''}^K)$	–	4	

TABLE LIII, 2 B β. Transition from $v''(v_t = v_{t'} = \dots = 0 ; v_n , v_{n'} \dots$ arbitrary) to $v'(v_t = 1 , |\ell_t| = 1 ; v_{t'} = \dots = 0 ; v_n , v_{n'} \dots$ arbitrary) ; $J \simeq K \simeq 30$

$^{R}R_{J,\|K\|}$ $^{R}P_{J,\|K\|}$ $^{P}R_{J,\|K\|}$ $^{P}P_{J,\|K\|}$	h: a	b	c
$\frac{E_{v'} - E_{v''}}{hc}$	0	2	4
$+ B^{zz}_{v'} - B^{xx}_{v'} - 2(\zeta_t B^{zz})_{v'}$	2	4	
$+ m(B^{xx}_{v'} + B^{xx}_{v''})$	1	3	
$\pm 2\|K\|[B^{zz}_{v'} - B^{xx}_{v'} - (\zeta_t B^{zz})_{v'}]$	1	3	
$+ m^2 [B^{xx}_{v'} - B^{xx}_{v''}$	–	2	4
$- D^{JK}_e + \eta^J_t + \frac{12(q^t_{12})^2}{B^{zz}_e - B^{xx}_e + 2B^{zz}_e \zeta^z_t}]$	4		
$\mp m\|K\|(2D^{JK}_e - \eta^J_t + \frac{8(q^t_{12})^2}{B^{zz}_e - B^{xx}_e + 2B^{zz}_e \zeta^z_t})$	4		

$+ K^2 \left[B^{zz}_{v'} - B^{zz}_{v''} - (B^{xx}_{v'} - B^{xx}_{v''}) \right.$	–	2	4
$\left. - 6 D^K_e + 3 \eta^K_t + \dfrac{36 (q^t_{12})^2}{B^{zz}_e - B^{xx}_e + 2 B^{zz}_e \zeta^z_t} \right]$	4		
$- 4 m^3 D^J_e$	3		
$\mp m^2 \lvert K \rvert \left(2 D^{JK}_e - \eta^J_t + \dfrac{8 (q^t_{12})^2}{B^{zz}_e - B^{xx}_e + 2 B^{zz}_e \zeta^z_t} \right)$	3		
$- 2 m K^2 D^{JK}_e$	3		
$\mp \lvert K \rvert^3 \left(4 D^K_e - \eta^K_t - \dfrac{8 (q^t_{12})^2}{B^{zz}_e - B^{xx}_e + 2 B^{zz}_e \zeta^z_t} \right.$	3		
$- m^4 (D^J_{v'} - D^J_{v''})$	–	4	
$- m^2 K^2 (D^{JK}_{v'} - D^{JK}_{v''})$	–	4	
$- K^4 (D^K_{v'} - D^K_{v''})$	–	4	

TABLE LIII, 3 Aα. Transition from $v''(v_t = v_{t'} = \ldots = 0; v_n, v_{n'} \ldots$ arbitrary) to $v'(v_t = 2, \ell_t = 0; v_{t'} = \ldots = 0; v_n, v_{n'} \ldots$ arbitrary); $J \simeq K \simeq 1$

${}^{Q}Q_{J\,|K|}$	h a	h b	h c
$\dfrac{E_{v'} - E_{v''}}{hc}$	0	2	4
$+ J(J+1)(B^{xx}_{v'} - B^{xx}_{v''})$	–	4	
$+ K^2[(B^{zz}_{v'} - B^{zz}_{v''}) - (B^{xx}_{v'} - B^{xx}_{v''})]$	–	4	

TABLE LIII, 3 Aβ . Transition from $v''(v_t = v_{t'} = \ldots = 0 ; v_n, v_{n'} \ldots$ arbitrary) to v' $(v_t = 2, \ell_t = 0 ; v_{t'} = \ldots = 0 ; v_n, v_{n'} \ldots$ arbitrary) ; $J \simeq K \simeq 1$

${}^{Q}R_{J,\|K\|}$ ${}^{Q}P_{J,\|K\|}$	h		
	a	b	c
$\frac{E_{v'} - E_{v''}}{hc}$	0	2	4
$+ m(B^{xx}_{v'} + B^{xx}_{v''})$	1	3	
$+ m^2(B^{xx}_{v'} - B^{xx}_{v''})$	–	4	
$+ K^2[(B^{zz}_{v'} - B^{zz}_{v''}) - (B^{xx}_{v'} - B^{xx}_{v''})]$	–	4	

TABLE LIII, 3 Bα . Transition from $v''(v_t = v_{t'} = \ldots = 0\ ; v_n, v_{n'} \ldots$ arbitrary) to $v'\ (v_t = 2, \ell_t = 0\ ; v_{t'} = \ldots = 0\ ; v_n, v_{n'} \ldots$ arbitrary); $J \simeq K \simeq 30$

${}^{Q}Q_{J,\,\lvert K \rvert}$	k: a	b	c
$\dfrac{E_{v'} - E_{v''}}{hc}$	0	2	4
$+ J(J+1)(B^{xx}_{v'} - B^{xx}_{v''})$	–	2	4
$+ K^2\left[(B^{zz}_{v'} - B^{zz}_{v''}) - (B^{xx}_{v'} - B^{xx}_{v''})\right]$	–	2	4
$- J^2(J+1)^2(D^J_{v'} - D^J_{v''})$	–	4	
$- J(J+1)K^2(D^{JK}_{v'} - D^{JK}_{v''})$	–	4	
$- K^4(D^K_{v'} - D^K_{v''})$	–	4	
$+ 16(q^t_{12})^2 \Big\{ J(J+1)(B^{zz}_e - B^{xx}_e + 4B^{zz}_e \zeta^z_t + 4x_{\ell_t\ell_t}) - K^2(3B^{zz}_e - 3B^{xx}_e + 16B^{zz}_e\zeta^z_t + 20x_{\ell_t\ell_t})$			
$- 4J(J+1)K^2(B^{zz}_e - B^{xx}_e - 4x_{\ell_t\ell_t}) + 4K^4(3B^{zz}_o - 3B^{xx}_e + 4B^{zz}_e\zeta^z_t - 4x_{\ell_t\ell_t})\Big\}$			
$\left[4K^2(B^{zz}_e - B^{xx}_e + 2B^{zz}_e\zeta^z_t)^2 - (B^{zz}_e - B^{xx}_e + 4B^{zz}_e\zeta^z_t + 4x_{\ell_t\ell_t})^2\right]^{-1}$	4		

TABLE LIII, $3B\beta$. Transition from $v''(v_t = v_{t'} = \ldots = 0; v_n, v_{n'} \ldots$ arbitrary) to $v'(v_t = 2, l_t = 0; v_{t'} = \ldots = 0; v_n, v_{n'} \ldots$ arbitrary); $J \simeq K \simeq 30$

${}^{Q}R_{J,\|K\|}$ ${}^{Q}P_{J,\|K\|}$	k: a	b	c
$\frac{E_{v'} - E_{v''}}{hc}$	0	2	4
$+ m(B^{xx}_{v'} + B^{xx}_{v''})$	1	3	
$+ m^2(B^{xx}_{v'} - B^{xx}_{v''})$	–	2	4
$+ K^2[(B^{zz}_{v'} - B^{zz}_{v''}) - (B^{xx}_{v'} - B^{xx}_{v''})]$	–	2	4
$- 4m^3 D^J_e - 2mK^2 D^{JK}_e$	3		
$- m^4(D^J_{v'} - D^J_{v''}) - m^2K^2(D^{JK}_{v'} - D^{JK}_{v''}) - K^4(D^K_{v'} - D^K_{v''})$	–	4	
$+ 16(q^t_{12})^2 \{ m(m+1)(B^{zz}_e - B^{xx}_e + 4B^{zz}_e\zeta^z_t + 4x_{l_tl_t}) - K^2(3B^{zz}_e - 3B^{xx}_e + 16B^{zz}_e\zeta^z_t + 20x_{l_tl_t})$ $- 4m(m+1)K^2(B^{zz}_e - B^{xx}_e - 4x_{l_tl_t}) + 4K^4(3B^{zz}_e - 3B^{xx}_e + 4B^{zz}_e\zeta^z_t - 4x_{l_tl_t}) \} [4K^2(B^{zz}_e$ $- B^{xx}_e + 2B^{zz}_e\zeta^z_t)^2 - (B^{zz}_e - B^{xx}_e + 4B^{zz}_e\zeta^z_t + 4x_{l_tl_t})^2]^{-1}$	4		

TABLE LIII, 4 A α. Transition from $v''(v_t = v_{t'} = \ldots = 0; v_n, v_{n'} \ldots$ arbitrary) to $v'(v_t = 2, |\ell_t| = 2; v_{t'} = \ldots = 0; v_n, v_{n'} \ldots$ arbitrary); $J \simeq K \simeq 1$

${}^RQ_{J,|K|}$ ${}^PQ_{J,|K|}$	h a	b	c
$\frac{E_{v'} - E_{v''}}{hc}$	0	2	4
$+B^{zz}_{v'} - B^{xx}_{v'} + 4(\zeta_t B^{zz})_{v'}$	2	4	
$\pm 2|K|\left[B^{zz}_{v'} - B^{xx}_{v'} + 2(\zeta_t B^{zz})_{v'}\right]$	2	4	
$+ J(J+1)(B^{xx}_{v'} - B^{xx}_{v''})$	–	4	
$+ K^2\left[(B^{zz}_{v'} - B^{zz}_{v''}) - (B^{xx}_{v'} - B^{xx}_{v''})\right]$	–	4	

TABLE LIII, $4A\beta$. Transition from v'' ($v_t = v_{t'} = \ldots = 0$; $v_n, v_{n'} \ldots$ arbitrary) to v' ($v_t = 2, |l_t| = 2$; $v_{t'} = \ldots = 0$; $v_n, v_{n'} \ldots$ arbitrary) ; $J \simeq K \simeq 1$

${}^{R}R_{J,\,|K|}$ ${}^{R}P_{J,\,|K|}$ ${}^{P}R_{J,\,|K|}$ ${}^{P}P_{J,\,|K|}$	h: a	b	c
$\dfrac{E_{v'} - E_{v''}}{hc}$	0	2	4
$+ B^{zz}_{v'} - B^{xx}_{v'} + 4(\zeta_t B^{zz})_{v'}$	2	4	
$+ m(B^{xx}_{v'} + B^{xx}_{v''})$	2	4	
$\pm 2|K|\left[B^{zz}_{v'} - B^{xx}_{v'} + 2(\zeta_t B^{zz})_{v'}\right]$	2	4	
$+ m^2(B^{xx}_{v'} - B^{xx}_{v''})$	–	4	
$+ K^2\left[(B^{zz}_{v'} - B^{zz}_{v''}) - (B^{xx}_{v'} - B^{xx}_{v''})\right]$	–	4	

TABLE LIII, 4 B α. Transition from v'' ($v_t = v_{t'} = \ldots = 0$; $v_n, v_{n'} \ldots$ arbitrary) to v' ($v_t = 2$, $|\ell_t| = 2$; $v_{t'} = \ldots = 0$; $v_n, v_{n'} \ldots$ arbitrary); $J \simeq K \simeq 30$

${}^{R}Q_{J,\|K\|}$ ${}^{P}Q_{J,\|K\|}$	k: a	b	c
$\dfrac{E_{v'} - E_{v''}}{hc}$	0	2	4
$+ B^{zz}_{v'} - B^{xx}_{v'} + 4(\zeta_t B^{zz})_{v'}$	2	4	
$\pm 2\|K\|\left[B^{zz}_{v'} - B^{xx}_{v'} + 2(\zeta_t B^{zz})_{v'}\right]$	1	3	
$+ J(J+1)\left[B^{xx}_{v'} - B^{xx}_{v''}\right.$	–	2	4
$\left. - D^{JK}_e - 2\eta^J_t\right]$	4		
$+ K^2\left[B^{zz}_{v'} - B^{zz}_{v''} - (B^{xx}_{v'} - B^{xx}_{v''})\right.$	–	2	4
$\left. - 6D^K_e - 6\eta^K_t\right]$	4		
$\mp 2J(J+1)\|K\|(D^{JK}_e + \eta^J_t)$	3		
$\mp 2\|K\|^3(2D^K_e + \eta^K_t)$	3		

$-J^2(J+1)^2(D^J_{v'}-D^J_{v''})$	–	4	
$-J(J+1)K^2(D^{JK}_{v'}-D^{JK}_{v''})$	–	4	
$-K^4(D^K_{v'}-D^K_{v''})$	–	4	
$\pm\dfrac{8(q^t_{12})^2(2\lvert K\rvert\pm 1)^2\left[J(J+1)-\lvert K\rvert(\lvert K\rvert\pm 1)\right]}{2\lvert K\rvert(B^{zz}_e-B^{xx}_e+2B^{zz}_e\zeta_t)\pm(B^{zz}_e-B^{xx}_e+4B^{zz}_e\zeta^z_t+4x_{l_t l_t})}$	4		

TABLE LIII, 4 B β. Transition from v'' ($v_t = v_{t'} = \ldots = 0$; $v_n, v_{n'} \ldots$ arbitrary) to v' ($v_t = 2$, $|l_t| = 2$; $v_{t'} = \ldots = 0$; $v_n, v_{n'} \ldots$ arbitrary); $J \simeq K \simeq 30$

${}^{R}R_{J,\|K\|}$ ${}^{R}P_{J,\|K\|}$ ${}^{P}R_{J,\|K\|}$ ${}^{P}P_{J,\|K\|}$	k: a	b	c
$\frac{E_{v'} - E_{v''}}{hc}$	0	2	4
$+ B^{zz}_{v'} - B^{xx}_{v'} + 4(\zeta_t B^{zz})_{v'}$	2	4	
$+ m(B^{xx}_{v'} + B^{xx}_{v''})$	1	3	
$\pm 2\|K\|[B^{zz}_{v'} - B^{xx}_{v'} + 2(\zeta_t B^{zz})_{v'}]$	1	3	
$+ m^2[B^{xx}_{v'} - B^{xx}_{v''}$	–	2	4
$- D^{JK}_e - 2\eta^J_t]$	4		
$\mp 2m\|K\|(D^{JK}_e + \eta^J_t)$	4		
$+ K^2[B^{zz}_{v'} - B^{zz}_{v''} - (B^{xx}_{v'} - B^{xx}_{v''})$	–	2	4
$- 6D^K_e - 6\eta^K_t]$	4		

$- 4 m^3 D_e^J$	3		
$\mp 2 m^2 \lvert K \rvert (D_e^{JK} + \eta_t^J)$	3		
$- 2 m K^2 D_e^{JK}$	3		
$\mp 2 \lvert K \rvert^3 (2 D_e^K + \eta_t^K)$	3		
$- m^4 (D_{v'}^J - D_{v''}^J)$	-	4	
$- m^2 K^2 (D_{v'}^{JK} - D_{v''}^{JK})$	-	4	
$- K^4 (D_{v'}^K - D_{v''}^K)$	-	4	
$\pm \dfrac{8(q_{12}^t)^2 (2\lvert K \rvert \pm 1)^2 [m(m+1) - \lvert K \rvert(\lvert K \rvert \pm 1)]}{2\lvert K \rvert (B_e^{zz} - B_e^{xx} + 2B_e^{zz}\zeta_t^z) \pm (B_e^{zz} - B_e^{xx} + 4B_e^{zz}\zeta_t^z + 4x_{l_t l_t})}$	4		

TABLE LIII, 5 A α . Transition from $v''(v_t = v_{t'} = \ldots = 0 ; v_n, v_{n'} \ldots$ arbitrary) to $v'(v_t = 3, |\ell_t| = 1 ; v_{t'} = \ldots = 0 ; v_n, v_{n'} \ldots$ arbitrary) ; $J \simeq K \simeq 1$

	$^RQ_{J,\|K\|}$ $^PQ_{J,\|K\|}$	k: a	k: b	k: c
	$\dfrac{E_{v'} - E_{v''}}{hc}$	0	2	4
	$+ B^{zz}_{v'} - B^{xx}_{v'} - 2(\zeta_t B^{zz})_{v'}$	2	4	
	$\pm 2\|K\| \left[B^{zz}_{v'} - B^{xx}_{v'} - (\zeta_t B^{zz})_{v'} \right]$	2	4	
	$+ K^2 \left[(B^{zz}_{v'} - B^{zz}_{v''}) - (B^{xx}_{v'} - B^{xx}_{v''}) \right]$	-	4	
$\|K\| \neq 0$	$+ J(J+1)(B^{xx}_{v'} - B^{xx}_{v''})$	-	4	
$K = 0$	$+ J(J+1)\left[B^{xx}_{v'} - B^{xx}_{v''} \right.$	-	4	
	$\left. - 4 q^t_0 \right]$	4		

TABLE LIII, 5 Aβ. Transition from v'' ($v_t = v_{t'} = \ldots = 0$; $v_n, v_{n'} \ldots$ arbitrary) to v' ($v_t = 3$, $|\ell_t| = 1$; $v_{t'} = \ldots = 0$; $v_n, v_{n'} \ldots$ arbitrary); $J \simeq K \simeq 1$

	${}^{R}R_{J,\,\lvert K\rvert}$ ${}^{R}P_{J,\,\lvert K\rvert}$ ${}^{P}R_{J,\,\lvert K\rvert}$ ${}^{P}P_{J,\,\lvert K\rvert}$	k: a	b	c
	$\dfrac{E_{v'} - E_{v''}}{hc}$	0	2	4
	$+ B^{zz}_{v'} - B^{xx}_{v'} - 2(\zeta_t B^{zz})_{v'}$	2	4	
	$\pm 2\lvert K\rvert [B^{zz}_{v'} - B^{xx}_{v'} - (\zeta_t B^{zz})_{v'}]$	2	4	
	$+ K^2 [B^{zz}_{v'} - B^{zz}_{v''} - (B^{xx}_{v'} - B^{xx}_{v''})]$	–	4	
$\lvert K\rvert \neq 0$	$+ m(B^{xx}_{v'} + B^{xx}_{v''})$	2	4	
	$+ m^2(B^{xx}_{v'} - B^{xx}_{v''})$	–	4	
$K = 0$	$+ m[B^{xx}_{v'} + B^{xx}_{v''}$	2	4	
	$+ 4q^t_0]$	4		
	$+ m^2[B^{xx}_{v'} - B^{xx}_{v''}$	–	4	
	$+ 4q^t_0]$	4		

TABLE LIII, 5 Bα . Transition from $v''(v_t = v_{t'} = \ldots = 0 ; v_n , v_{n'} \ldots$ arbitrary) to $v'(v_t = 3, |\ell_t| = 1 ; v_{t'} = \ldots = 0 ; v_n , v_{n'} \ldots$ arbitrary) ; $J \simeq K \simeq 30$

${}^{R}Q_{J,\,|K|}$ ${}^{P}Q_{J,\,|K|}$	k: a	b	c
$\dfrac{E_{v'} - E_{v''}}{hc}$	0	2	4
$+ B^{zz}_{v'} - B^{xx}_{v'} - 2(\zeta_t B^{zz})_{v'}$	2	4	
$\pm 2\,|K|\left[B^{zz}_{v'} - B^{xx}_{v'} - (\zeta_t B^{zz})_{v'}\right]$	1	3	
$+ J(J+1)\left[B^{xx}_{v'} - B^{xx}_{v''}\right.$	-	2	4
$\left. + \eta^{J}_{t} - D^{JK}_{e} - \dfrac{48(q^{t}_{12})^2}{B^{zz}_{e} - B^{xx}_{e} + 2B^{zz}_{e}\zeta^{z}_{t}}\right]$	4		
$+ K^2\left[B^{zz}_{v'} - B^{zz}_{v''} - (B^{xx}_{v'} - B^{xx}_{v''})\right.$	-	2	4
$\left. + 3\eta^{K}_{t} - 6D^{K}_{e} + \dfrac{14{\cdot}4(q^{t}_{12})^2}{B^{zz}_{e} - B^{xx}_{e} + 2B^{zz}_{e}\zeta^{z}_{t}}\right]$	4		

$\mp J(J+1)\lvert K\rvert\left(2D_e^{JK} - \eta_t^J + \dfrac{32(q_{12}^t)^2}{B_e^{zz} - B_e^{xx} + 2B_e^{zz}\zeta_t^z}\right)$	3		
$\mp \lvert K\rvert^3\left(4D_e^K - \eta_t^K - \dfrac{32(q_{12}^t)^2}{B_e^{zz} - B_e^{xx} + 2B_e^{zz}\zeta_t^z}\right)$	3		
$-J^2(J+1)^2(D_{v'}^J - D_{v''}^J)$	–	4	
$-J(J+1)K^2(D_{v'}^{JK} - D_{v''}^{JK})$	–	4	
$-K^4(D_{v'}^K - D_{v''}^K)$	–	4	
$\pm \dfrac{12(q_{12}^t)^2(2\lvert K\rvert \pm 1)^2\left[J(J+1) - \lvert K\rvert(\lvert K\rvert \pm 1)\right]}{2\lvert K\rvert(B_e^{zz} - B_e^{xx} + 2B_e^{zz}\zeta_t^z) \pm (B_e^{zz} - B_e^{xx} - 2B_e^{zz}\zeta_t^z)}$	4		

TABLE LIII, $5B\beta$. Transition from $v''(v_t = v_{t'} = \dots = 0; v_n, v_{n'} \dots$ arbitrary) to $v'(v_t = 3, |\ell_t| = 1; v_{t'} = \dots = 0; v_n, v_{n'} \dots$ arbitrary); $J \simeq K \simeq 30$

$^{R}R_{J,\|K\|}$ $^{R}P_{J,\|K\|}$ $^{P}R_{J,\|K\|}$ $^{P}P_{J,\|K\|}$	h: a	b	c
$\dfrac{E_{v'} - E_{v''}}{hc}$	0	2	4
$+ B^{zz}_{v'} - B^{xx}_{v'} - 2(\zeta_t B^{zz})_{v'}$	2	4	
$+ m(B^{xx}_{v'} + B^{xx}_{v''})$	1	3	
$\pm 2\|K\|\left[B^{zz}_{v'} - B^{xx}_{v'} - (\zeta_t B^{zz})_{v'}\right]$	1	3	
$+ m^2\left[B^{xx}_{v'} - B^{xx}_{v''}\right.$	–	2	4
$\left. - D^{JK}_e + \eta^J_t - \dfrac{48(q^t_{12})^2}{B^{zz}_e - B^{xx}_e + 2B^{zz}_e \zeta^z_t}\right]$	4		
$+ K^2\left[B^{zz}_{v'} - B^{zz}_{v''} - (B^{xx}_{v'} - B^{xx}_{v''})\right.$	–	2	4
$\left. - 6D^K_e + 3\eta^K_t + \dfrac{144(q^t_{12})^2}{B^{zz}_e - B^{xx}_e + 2B^{zz}_e \zeta^z_t}\right]$	4		

$\mp m\,\lvert K\rvert\left(2D_e^{JK}-\eta_t^J+\dfrac{32(q_{12}^t)^2}{B_e^{zz}-B_e^{xx}+2B_e^{zz}\zeta_t^z}\right)$	4		
$-4m^3 D_e^J$	3		
$\mp m^2\,\lvert K\rvert\left(2D_e^{JK}-\eta_t^J+\dfrac{32(q_{12}^t)^2}{B_e^{zz}-B_e^{xx}+2B_e^{zz}\zeta_t^z}\right)$	3		
$-2mK^2 D_e^{JK}$	3		
$\mp \lvert K\rvert^3\left(4D_e^K-\eta_t^K-\dfrac{32(q_{12}^t)^2}{B_e^{zz}-B_e^{xx}+2B_e^{zz}\zeta_t^z}\right)$	3		
$-m^4(D_{v'}^J-D_{v''}^J)$	-	4	
$-m^2K^2(D_{v'}^{JK}-D_{v''}^{JK})$	-	4	
$-K^4(D_{v'}^K-D_{v''}^K)$	-	4	
$\pm\dfrac{12(q_{12}^t)^2(2\lvert K\rvert\pm1)^2\left[m(m+1)-\lvert K\rvert(\lvert K\rvert\pm1)\right]}{2\lvert K\rvert(B_e^{zz}-B_e^{xx}+2B_e^{zz}\zeta_t^z)\pm(B_e^{zz}-B_e^{xx}-2B_e^{zz}\zeta_t^z)}$	4		

TABLE LIII, 6 A α. Transition from $v''(v_t = v_{t'} = \ldots = 0\,;\, v_n, v_{n'} \ldots$ arbitrary) to $v'(v_t = 3, |\ell_t| = 3\,;\, v_{t'} = \ldots = 0\,;\, v_n, v_{n'} \ldots$ arbitrary) ; $J \simeq K \simeq 1$

${}^{Q}Q_{J,\,K\ell_t}$		k: a	b	c
$\dfrac{E_{v'} - E_{v''}}{hc}$		0	2	4
$-2K\ell_t(\zeta_t B^{zz})_{v'}$		2	4	
$+J(J+1)(B^{xx}_{v'} - B^{xx}_{v''})$		–	4	
$+K^2\left[(B^{zz}_{v'} - B^{zz}_{v''}) - (B^{xx}_{v'} - B^{xx}_{v''})\right]$		–	4	
$K = 0$	$+48\,g^t_6$	4		

TABLE LIII, 6 A β. Transition from $v''(v_t = v_{t'} = \ldots = 0; v_n, v_{n'} \ldots$ arbitrary) to v' ($v_t = 3, |\ell_t| = 3$; $v_{t'} = \ldots = 0; v_n, v_{n'} \ldots$ arbitrary); $J \simeq K \simeq 1$

	${}^{Q}R_{J,K\ell_t}$ ${}^{Q}P_{J,K\ell_t}$	k: a	b	c
$\frac{E_{v'} - E_{v''}}{hc}$		0	2	4
$+ m(B^{xx}_{v'} + B^{xx}_{v''})$		2	4	
$- 2K\ell_t(\zeta_t B^{zz})_{v'}$		2	4	
$+ m^2(B^{xx}_{v'} - B^{xx}_{v''})$		–	4	
$+ K^2[(B^{zz}_{v'} - B^{zz}_{v''}) - (B^{xx}_{v'} - B^{xx}_{v''})]$		–	4	
$K = 0$	$-48g^t_6$	4		

TABLE LIII, 6 B α. Transition from v'' ($v_t = v_{t'} = \ldots = 0$; $v_n, v_{n'} \ldots$ arbitrary) to v' ($v_t = 3, |\ell_t| = 3$; $v_{t'} = \ldots = 0$; $v_n, v_{n'} \ldots$ arbitrary) ; $J \simeq K \simeq 30$

${}^{Q}Q_{J,K\ell_t}$	k: a	b	c
$\frac{E_{v'} - E_{v''}}{hc}$	0	2	4
$-2K\ell_t(\zeta_t B^{zz})_{v'}$	1	3	
$+J(J+1)(B^{xx}_{v'} - B^{xx}_{v''})$	–	2	4
$+K^2[(B^{zz}_{v'} - B^{zz}_{v''}) - (B^{xx}_{v'} - B^{xx}_{v''})]$	–	2	4
$+J(J+1)K\ell_t\eta^J_t$	3		
$+K^3\ell_t\eta^K_t$	3		
$-J^2(J+1)^2(D^J_{v'} - D^J_{v''})$	–	4	
$-J(J+1)K^2(D^{JK}_{v'} - D^{JK}_{v''})$	–	4	

$- K^4 \left(D^K_{v'} - D^K_{v''}\right)$	–	4	
$- \dfrac{4\left(q^t_{12}\right)^2 \left(2K\ell_t + 3\right)^2 \left[9J(J+1) - K\ell_t \left(K\ell_t + 3\right)\right]}{18K\ell_t \left(B^{zz}_e - B^{xx}_e + 2B^{zz}_e \zeta^z_t\right) + 27\left(B^{zz}_e - B^{xx}_e - 2B^{zz}_e \zeta^z_t\right)}$	4		

TABLE LIII, 6 B β . Transition from $v''(v_t = v_{t'} = \ldots = 0 ; v_n , v_{n'} \ldots$ arbitrary) to v' $(v_t = 3, |\ell_t| = 3 ;$ $v_{t'} = \ldots = 0 ; v_n = v_{n'} \ldots$ arbitrary) ; $J \simeq K \simeq 30$

${}^{Q}R_{J,K\ell_t}$ ${}^{Q}P_{J,K\ell_t}$	k		
	a	b	c
$\frac{E_{v'} - E_{v''}}{hc}$	0	2	4
$+ m(B^{xx}_{v'} + B^{xx}_{v''})$	1	3	
$- 2K\ell_t(\zeta_t B^{zz})_{v'}$	1	3	
$+ m^2(B^{xx}_{v'} - B^{xx}_{v''})$	–	2	4
$+ K^2\left[(B^{zz}_{v'} - B^{zz}_{v''}) - (B^{xx}_{v'} - B^{xx}_{v''})\right]$	–	2	4
$- 4m^3 D^J_e$	3		
$+ m^2 K\ell_t \eta^J_t$	3		
$- 2mK^2 D^{JK}_e$	3		
$+ K^3 \ell_t \eta^K_t$	3		

$- m^4(D^J_{v'} - D^J_{v''})$	–	4	
$- m^2 K^2(D^{JK}_{v'} - D^{JK}_{v''})$	–	4	
$- K^4(D^K_{v'} - D^K_{v''})$	–	4	
$+ m K\ell_t \eta^J_t$	4		
$- \dfrac{4(q^t_{12})^2(2K\ell_t+3)^2[9m(m+1) - K\ell_t(K\ell_t+3)]}{18K\ell_t(B^{zz}_e - B^{xx}_e + 2B^{zz}_e\zeta^z_t) + 27(B^{zz}_e - B^{xx}_e - 2B^{zz}_e\zeta^z_t)}$	4		

TABLE LIII, 7 A α . Transition from v'' ($v_t = v_{t'} = \ldots = 0$; $v_n, v_{n'} \ldots$ arbitrary) to v' ($v_t = v_{t'} = 1$, $\ell_t + \ell_{t'} = 0$; $v_{t''} = \ldots = 0$; $v_n, v_{n'} \ldots$ arbitrary) ; $J \simeq K \simeq 1$

	${}^{Q}Q^{+}_{J,\lvert K\rvert}$ ${}^{Q}Q^{-}_{J,\lvert K\rvert}$	k: a	b	c
	$\dfrac{E_{v'} - E_{v''}}{hc}$	0	2	4
	$+ J(J+1)(B^{xx}_{v'} - B^{xx}_{v''})$	–	4	
	$+ K^2\left[(B^{zz}_{v'} - B^{zz}_{v''}) - (B^{xx}_{v'} - B^{xx}_{v''})\right]$	–	4	
$\lvert K\rvert \neq 0$	$\pm 2\sqrt{4(r^{tt'}_0)^2 + K^2(B^{zz}_e \zeta^z_t - B^{zz}_e \zeta^z_{t'})^2}$	2		
	$\pm \dfrac{\sum_s \left[8 r^{tt'}_0 g^{tt',s}_{22} - K^2 B^{zz}_e(\zeta^z_t - \zeta^z_{t'})(\eta_{ts} - \eta_{t's})\right](v_s + \frac{d_s}{2})}{\sqrt{4(r^{tt'}_0)^2 + K^2(B^{zz}_e \zeta^z_t - B^{zz}_e \zeta^z_{t'})^2}}$	4		
$K = 0$	$- 4 r^{tt'}_{eff}$	2	4	

TABLE LIII, 7 Aβ. Transition from v'' ($v_t = v_{t'} = \ldots = 0$; $v_n, v_{n'} \ldots$ arbitrary) to v' ($v_t = v_{t'} = 1$, $\ell_t + \ell_{t'} = 0$; $v_{t''} = \ldots = 0$; $v_n, v_{n'} \ldots$ arbitrary); $J \simeq K \simeq 1$

	$Q_{R^+_{J,\|K\|}}$ $Q_{R^-_{J,\|K\|}}$ $Q_{P^+_{J,\|K\|}}$ $Q_{P^-_{J,\|K\|}}$	k: a	b	c
$\dfrac{E_{v'} - E_{v''}}{hc}$		0	2	4
$+ m(B^{xx}_{v'} + B^{xx}_{v''})$		2	4	
$+ m^2(B^{xx}_{v'} - B^{xx}_{v''})$		-	4	
$+ K^2\left[(B^{zz}_{v'} - B^{zz}_{v''}) - (B^{xx}_{v'} - B^{xx}_{v''})\right]$		-	4	
$\|K\| \neq 0$	$\pm 2\sqrt{4(r_0^{tt'})^2 + K^2(B_e^{zz}\zeta_t^z - B_e^{zz}\zeta_{t'}^z)^2}$	2		
	$\pm \dfrac{\sum_s \left[8 r_0^{tt'} g_{22}^{tt',s} - K^2 B_e^{zz}(\zeta_t^z - \zeta_{t'}^z)(\eta_{ts} - \eta_{t's})\right](v_s + \frac{d_s}{2})}{\sqrt{4(r_0^{tt'})^2 + K^2(B_e^{zz}\zeta_t^z - B_e^{zz}\zeta_{t'}^z)^2}}$	4		
$K = 0$	$+ 4 r_{eff}^{tt'}$	2	4	

TABLE LIII, 7 Bα. Transition from v'' ($v_t = v_{t'} = \ldots = 0$; $v_n, v_{n'} \ldots$ arbitrary) to v' ($v_t = v_{t'} = 1$, $\ell_t + \ell_{t'} = 0$; $v_{t''} = \ldots = 0$; $v_n, v_{n'} \ldots$ arbitrary) ; $J \simeq K \simeq 30$

${}^{Q}Q_{J,K\ell_t}$	h: a	b	c
$\dfrac{E_{v'} - E_{v''}}{hc}$	0	2	4
$- 2K\ell_t \left[(\zeta_t B^{zz})_{v'} - (\zeta_{t'} B^{zz})_{v'} \right]$	1	3	
$+ J(J+1)(B^{xx}_{v'} - B^{xx}_{v''})$	–	2	4
$+ (K\ell_t)^2 \left[(B^{zz}_{v'} - B^{zz}_{v''}) - (B^{xx}_{v'} - B^{xx}_{v''}) \right]$	–	2	4
$+ J(J+1)K\ell_t(\eta^J_t - \eta^J_{t'})$	3		
$+ (K\ell_t)^3(\eta^K_t - \eta^K_{t'})$	3		
$- \dfrac{4(r^{tt'}_0)^2}{K\ell_t B^{zz}_e (\zeta^z_t - \zeta^z_{t'})}$	3		
$- J^2(J+1)^2(D^J_{v'} - D^J_{v''})$	–	4	

$-J(J+1)(K\ell_t)^2(D^{JK}_{v'} - D^{JK}_{v''})$	–	4	
$-(K\ell_t)^4(D^{K}_{v'} - D^{K}_{v''})$	–	4	
$-\dfrac{4(q^{t}_{12})^2(2K\ell_t+1)^2\left[J(J+1)-K\ell_t(K\ell_t+1)\right]}{(2K\ell_t+1)(B^{zz}_e - B^{xx}_e + 2B^{zz}_e\zeta^{z}_{t}) + 2B^{zz}_e\zeta^{z}_{t'} + 2x_{\ell_t\ell_{t'}}}$	4		
$+\dfrac{4(q^{t'}_{12})^2(2K\ell_t-1)^2\left[J(J+1)-K\ell_t(K\ell_t-1)\right]}{(2K\ell_t-1)(B^{zz}_e - B^{xx}_e + 2B^{zz}_e\zeta^{z}_{t'}) - 2B^{zz}_e\zeta^{z}_{t} - 2x_{\ell_t\ell_{t'}}}$	4		

The contributions in this table take into account the relation $\ell_{t'} = -\ell_t$

TABLE LIII, 7 Bβ . Transition from v'' ($v_t = v_{t'} = \ldots = 0$; $v_n, v_{n'} \ldots$ arbitrary) to v' ($v_t = v_{t'} = 1$, $\ell_t + \ell_{t'} = 0$; $v_{t''} = \ldots = 0$; $v_n, v_{n'} \ldots$ arbitrary) ; $J \simeq K \simeq 30$

${}^{Q}R_{J,K\ell_t}$ ${}^{Q}P_{J,K\ell_t}$	k: a	b	c
$\frac{E_{v'} - E_{v''}}{hc}$	0	2	4
$+ m(B^{xx}_{v'} + B^{xx}_{v''})$	1	3	
$- 2K\ell_t \left[(\zeta_t B^{zz})_{v'} - (\zeta_{t'} B^{zz})_{v'}\right]$	1	3	
$+ m^2 (B^{xx}_{v'} - B^{xx}_{v''})$	–	2	4
$+ (K\ell_t)^2 \left[(B^{zz}_{v'} - B^{zz}_{v''}) - (B^{xx}_{v'} - B^{xx}_{v''})\right]$	–	2	4
$- 4m^3 D^J_e$	3		
$- 2m (K\ell_t)^2 D^{JK}_e$	3		
$+ m^2 K\ell_t (\eta^J_t - \eta^J_{t'})$	3		
$+ (K\ell_t)^3 (\eta^K_t - \eta^K_{t'})$	3		

$-\dfrac{4(r_0^{tt'})^2}{K\ell_t B_e^{zz}(\zeta_t^z - \zeta_{t'}^z)}$	3		
$-m^4(D_{v'}^J - D_{v''}^J)$	–	4	
$-m^2(K\ell_t)^2(D_{v'}^{JK} - D_{v''}^{JK})$	–	4	
$-(K\ell_t)^4(D_{v'}^K - D_{v''}^K)$	–	4	
$+m K\ell_t(\eta_t^J - \eta_{t'}^J)$	4		
$-\dfrac{4(q_{12}^t)^2(2K\ell_t+1)^2\left[m(m+1) - K\ell_t(K\ell_t+1)\right]}{(2K\ell_t+1)(B_e^{zz} - B_e^{xx} + 2B_e^{zz}\zeta_t^z) + 2B_e^{zz}\zeta_{t'}^z + 2x_{\ell_t\ell_{t'}}}$	4		
$+\dfrac{4(q_{12}^{t'})^2(2K\ell_t-1)^2\left[m(m+1) - K\ell_t(K\ell_t-1)\right]}{(2K\ell_t-1)(B_e^{zz} - B_e^{xx} + 2B_e^{zz}\zeta_{t'}^z) - 2B_e^{zz}\zeta_t^z - 2x_{\ell_t\ell_{t'}}}$	4		

The contributions in this table take into account the relation $\ell_{t'} = -\ell_t$

TABLE LIII, 8 A α. Transition from v'' ($v_t = v_{t'} = \ldots = 0$; $v_n, v_{n'} \ldots$ arbitrary) to v' ($v_t = v_{t'} = 1$, $|\ell_t + \ell_{t'}| = 2$; $v_{t''} = \ldots = 0$; $v_n, v_{n'} \ldots$ arbitrary) ; $J \simeq K \simeq 1$

${}^{R}Q_{J,\|K\|}$ ${}^{P}Q_{J,\|K\|}$	k: a	b	c
$\dfrac{E_{v'} - E_{v''}}{hc}$	0	2	4
$+ B^{zz}_{v'} - B^{xx}_{v'} + 2(\zeta_t B^{zz})_{v'} + 2(\zeta_{t'} B^{zz})_{v'}$	2	4	
$\pm 2\|K\| \left[B^{zz}_{v'} - B^{xx}_{v'} + (\zeta_t B^{zz})_{v'} + (\zeta_{t'} B^{zz})_{v'} \right]$	2	4	
$+ J(J+1)(B^{xx}_{v'} - B^{xx}_{v''})$	-	4	
$+ K^2 \left[(B^{zz}_{v'} - B^{zz}_{v''}) - (B^{xx}_{v'} - B^{xx}_{v''}) \right]$	-	4	

TABLE LIII, 8 Aβ . Transition from v'' ($v_t = v_{t'} = \ldots = 0$; $v_n , v_{n'} \ldots$ arbitrary) to v' ($v_t = v_{t'} = 1$, $|\ell_t + \ell_{t'}| = 2$; $v_{t''} = \ldots = 0$; $v_n , v_{n'} \ldots$ arbitrary) ; $J \simeq K \simeq 1$

${}^{R}R_{J,\|K\|}$ ${}^{R}P_{J,\|K\|}$ ${}^{P}R_{J,\|K\|}$ ${}^{P}P_{J,\|K\|}$	k		
	a	b	c
$\dfrac{E_{v'} - E_{v''}}{hc}$	0	2	4
$+ B^{zz}_{v'} - B^{xx}_{v'} + 2(\zeta_t B^{zz})_{v'} + 2(\zeta_{t'} B^{zz})_{v'}$	2	4	
$+ m(B^{xx}_{v'} + B^{xx}_{v''})$	2	4	
$\pm 2\|K\|\left[B^{zz}_{v'} - B^{xx}_{v'} + (\zeta_t B^{zz})_{v'} + (\zeta_{t'} B^{zz})_{v'}\right]$	2	4	
$+ m^2(B^{xx}_{v'} - B^{xx}_{v''})$	-	4	
$+ K^2\left[(B^{zz}_{v'} - B^{zz}_{v''}) - (B^{xx}_{v'} - B^{xx}_{v''})\right]$	-	4	

TABLE LIII, 8 B α . Transition from v'' ($v_t = v_{t'} = \ldots = 0 ; v_n , v_{n'} \ldots$ arbitrary) to v' ($v_t = v_{t'} = 1$, $|\ell_t + \ell_{t'}| = 2 ; v_{t''} = \ldots = 0 ; v_n , v_{n'} \ldots$ arbitrary) ; $J \simeq K \simeq 30$

${}^{R}Q_{J}\,\lvert K\rvert$ ${}^{P}Q_{J}\,\lvert K\rvert$	h a	h b	h c
$\dfrac{E_{v'} - E_{v''}}{hc}$	0	2	4
$+ B^{zz}_{v'} - B^{xx}_{v'} + 2\,(\zeta_t B^{zz})_{v'} + 2\,(\zeta_{t'} B^{zz})_{v'}$	2	4	
$\pm\, 2\,\lvert K\rvert \left[B^{zz}_{v'} - B^{xx}_{v'} + (\zeta_t B^{zz})_{v'} + (\zeta_{t'} B^{zz})_{v'} \right]$	1	3	
$+ J(J+1) \left[B^{xx}_{v'} - B^{xx}_{v''} \right.$	–	2	4
$\left. - D^{JK}_{e} - \eta^{J}_{t} - \eta^{J}_{t'} \right]$	4		
$+ K^2 \left[B^{zz}_{v'} - B^{zz}_{v''} - (B^{xx}_{v'} - B^{xx}_{v''}) \right.$	–	2	4
$\left. - 6\,D^{K}_{e} - 3\,\eta^{K}_{t} - 3\,\eta^{K}_{t'} \right]$	4		
$\mp J(J+1)\,\lvert K\rvert\,(2\,D^{JK}_{e} + \eta^{J}_{t} + \eta^{J}_{t'})$	3		
$\mp \lvert K\rvert^3\,(4\,D^{K}_{e} + \eta^{K}_{t} + \eta^{K}_{t'})$	3		

$-J^2(J+1)^2(D^J_{v'}-D^J_{v''})$	–	4	
$-J(J+1)K^2(D^{JK}_{v'}-D^{JK}_{v''})$	–	4	
$-K^4(D^K_{v'}-D^K_{v''})$	–	4	
$\pm\dfrac{4(2\lvert K\rvert\pm1)^2\left[J(J+1)-\lvert K\rvert(\lvert K\rvert\pm1)\right](q^t_{12})^2}{2\lvert K\rvert(B^{zz}_e-B^{xx}_e+2B^{zz}_e\zeta^z_t)\pm\left[B^{zz}_e-B^{xx}_e+2B^{zz}_e(\zeta^z_t+\zeta^z_{t'})+2x_{l_tl_{t'}}\right]}$	4		
$\pm\dfrac{4(2\lvert K\rvert\pm1)^2\left[J(J+1)-\lvert K\rvert(\lvert K\rvert\pm1)\right](q^{t'}_{12})^2}{2\lvert K\rvert(B^{zz}_e-B^{xx}_e+2B^{zz}_e\zeta^z_{t'})\pm\left[B^{zz}_e-B^{xx}_e+2B^{zz}_e(\zeta^z_t+\zeta^z_{t'})+2x_{l_tl_t}\right]}$	4		

TABLE LIII, $8B\beta$. Transition from v'' ($v_t = v_{t'} = \ldots = 0$; $v_n, v_{n'} \ldots$ arbitrary) to v' ($v_t = v_{t'} = 1$, $|\ell_t + \ell_{t'}| = 2$; $v_{t''} = \ldots = 0$; $v_n, v_{n'} \ldots$ arbitrary); $J \simeq K \simeq 30$

$^RR_{J,\|K\|}$ $^RP_{J,\|K\|}$ $^PR_{J,\|K\|}$ $^PP_{J,\|K\|}$	k: a	b	c
$\frac{E_{v'} - E_{v''}}{hc}$	0	2	4
$+B^{zz}_{v'} - B^{xx}_{v'} + 2(\zeta_t B^{zz})_{v'} + 2(\zeta_{t'} B^{zz})_{v'}$	2	4	
$+m(B^{xx}_{v'} + B^{xx}_{v''})$	1	3	
$\pm 2\|K\|\left[B^{zz}_{v'} - B^{xx}_{v'} + (\zeta_t B^{zz})_{v'} + (\zeta_{t'} B^{zz})_{v'}\right]$	1	3	
$+m^2\left[B^{xx}_{v'} - B^{xx}_{v''}\right.$	–	2	4
$\left. - D^{JK}_e - \eta^J_t - \eta^J_{t'}\right]$	4		
$+K^2\left[B^{zz}_{v'} - B^{zz}_{v''} - (B^{xx}_{v'} - B^{xx}_{v''})\right.$	–	2	4
$\left. - 6D^K_e - 3\eta^K_t - 3\eta^K_{t'}\right]$	4		
$-4m^3 D^J_e$	3		

$\mp m^2 \lvert K\rvert (2D_e^{JK} + \eta_t^J + \eta_{t'}^J)$	3		
$-2m K^2 D_e^{JK}$	3		
$\mp \lvert K\rvert^3 (4D_e^K + \eta_t^K + \eta_{t'}^K)$	3		
$\mp m \lvert K\rvert (2D_e^{JK} + \eta_t^J + \eta_{t'}^J)$	4		
$-m^4 (D_{v'}^J - D_{v''}^J)$	–	4	
$-m^2 K^2 (D_{v'}^{JK} - D_{v''}^{JK})$	–	4	
$-K^4 (D_{v'}^K - D_{v''}^K)$	–	4	
$\pm \dfrac{4(2\lvert K\rvert \pm 1)^2 [m(m+1) - \lvert K\rvert(\lvert K\rvert \pm 1)] (q_{12}^t)^2}{2\lvert K\rvert (B_e^{zz} - B_e^{xx} + 2B_e^{zz}\zeta_t^z) \pm [B_e^{zz} - B_e^{xx} + 2B_e^{zz}(\zeta_t^z + \zeta_{t'}^z) + 2x_{l_t l_{t'}}]}$	4		
$\pm \dfrac{4(2\lvert K\rvert \pm 1)^2 [m(m+1) - \lvert K\rvert(\lvert K\rvert \pm 1)] (q_{12}^{t'})^2}{2\lvert K\rvert (B_e^{zz} - B_e^{xx} + 2B_e^{zz}\zeta_{t'}^z) \pm [B_e^{zz} - B_e^{xx} + 2B_e^{zz}(\zeta_t^z + \zeta_{t'}^z) + 2x_{l_t l_{t'}}]}$	4		

For $L = 0$, E^+ and E^- are the two components of a $A_1 A_2$ doublet. When $L \neq 0$ and $|L| \neq 3$, E^+ is twofold degenerate and E^- is twofold degenerate. When $|L| = 3$, each one of the levels E^+ and E^- splits into two sublevels belonging to species A_1 and A_2 .
In the case of Tables XLVIII ; 7 A and 8 A ($v_t = v_{t'} = 1$; $J \simeq K \simeq 1$), the ℓ-type vibrational resonance is strong and makes resonance submatrices of degree two appear in the energy matrix. When $\ell_t = \ell_{t'} = \pm 1$ (Table XLVIII, 8 A), the energy can be expressed in terms of K, ℓ_t and $\ell_{t'}$, but when $\ell_{t'} = - \ell_t = \pm 1$ (Table XLVIII, 7 A), the energy is given by the sum or the difference of two expressions F'_1 and F'_2 as follows :

$$E^{\pm}_{(v_n \ldots, v_t, v_{t'}, J, |K|)} = F'_1(v_n \ldots, v_t, v_{t'}, J, |K|)$$

$$\pm F'_2(v_n \ldots, v_t, v_{t'}, J, |K|)$$

For $K = 0$, E^+ and E^- are the two components of a A_1 and A_2 doublet. For any other value of $|K|$, both levels E^+ and E^- are independently twofold degenerate.

3 . The expansions of the coefficients are not given in Table XLIX when only the first term is needed in the computation.

4 . φ^+_a and φ^+_b are two linear combinations of the functions $\psi^+_{2,-1}$ and $\psi^+_{4,1}$, φ^-_a and φ^-_b are the same linear combinations of the functions $\psi^-_{2,-1}$ and $\psi^-_{4,1}$:

$$\varphi^{\pm}_a = c_{a_1} \psi^{\pm}_{2,-1} + c_{a_2} \psi^{\pm}_{4,1}$$

$$\varphi^{\pm}_b = c_{b_1} \psi^{\pm}_{2,-1} + c_{b_2} \psi^{\pm}_{4,1}$$

where

$$c_{a_1} = \frac{-\lambda}{\sqrt{\lambda^2 + (\mu_-)^2}} \quad ; \quad c_{a_2} = \frac{\mu_-}{\sqrt{\lambda^2 + (\mu_-)^2}}$$

$$c_{b_1} = \frac{-\lambda}{\sqrt{\lambda^2 + (\mu_+)^2}} \quad ; \quad c_{b_2} = \frac{\mu_+}{\sqrt{\lambda^2 + (\mu_+)^2}}$$

with

$$\lambda = 2 q_0^t \sqrt{J(J+1)-6} \quad \sqrt{J(J+1)-12}$$

$$\mu_\pm = -6(B_e^{zz} - B_e^{xx} - B_e^{zz} \zeta_t^z) \pm \sqrt{36(B_e^{zz} - B_e^{xx} - B_e^{zz} \zeta_t^z)^2 + \lambda^2}$$

5 . In reference [25] (see Chapter VI), non-vanishing matrix elements of totally symmetric operators H (or H', or $H^\dagger$) have been studied. Similar arguments can be used for components of the electric dipole moment along space fixed axis M_X , M_Y , M_Z. For all molecules with a threefold axis they lead to a selection rule expressed by equation (X,4).

6 . Molecules with normal vibrations belonging to species A_2 are not very common. Results given in Tables LIII would however hold in such molecules for transitions corresponding to an even value of $\sum_n \Delta v_n$
(species A_2)

CHAPTER XI

MOLECULES WITH A FOURFOLD SYMMETRY

INTRODUCTION

In the present chapter, we shall compute, for molecules with a fourfold axis, the wave numbers of transitions belonging to rotation and rotation-vibration spectra. Calculations are carried out to fourth order in the latter case and to sixth order in the former. The more common symmetry groups for molecules with a fourfold axis are C_{4v} and D_{2d}, therefore we restrict ourselves to these two groups.

The calculation is very similar to that of Chapter X. For this reason, we shall only emphasize the particularities arising from the symmetry of the problem. Doing so, we shall deal essentially with the non-vanishing off-diagonal matrix elements and with selection rules. The non-vanishing off-diagonal matrix elements for C_{4v} and D_{2d} groups obey the general rule of Chapter VI. They are given in Tables XXXVI and XXXVII of Chapter VIII and in Tables XL and XLIV of Chapter IX. Selection rules are given in the present chapter.

ROTATION-VIBRATION ENERGIES FOR MOLECULES BELONGING TO SYMMETRY GROUPS C_{4v} AND D_{2d}.

We have computed the rotation-vibration levels in the following vibrational states, numbered from 1 to 6 :

1 $v_t = v_{t'} = \ldots = 0$

2 $v_t = 1$; $\ell_t = \pm 1$

3 $v_t = 2$; $\ell_t = 0$ $\Big\}$ $v_{t'} = \ldots = 0$ (cases 2, 3, 4)

4 $v_t = 2$; $\ell_t = \pm 2$

5 $v_t = v_{t'} = 1$; $\ell_t = - \ell_{t'} = \pm 1$

6 $v_t = v_{t'} = 1$; $\ell_t = \ell_{t'} = \pm 1$

$\Big\}$ $v_n, v_{n'} \ldots$ arbitrary (cases 1 to 6)

Now, let us point out that the case $J \simeq 30$, $K \simeq 1$ will not be studied here. As a matter of fact, one is led to solve secular equations of degree larger than two and explicit formulae cannot be found for the energy levels. Therefore, we have considered only the two cases :

A $J \simeq K \simeq 1$

B $J \simeq K \simeq 30$

The energies given in Tables LIV, 1 A, 1 B, 2 A, 2 B, 3 A, 3 B, 4 B, 5 B, and 6 B have been obtained by using a perturbation method to diagonalize the Hamiltonian submatrices. The energies given in Tables LIV, 4 A, 5 A and 6 A have been obtained by solving secular equations of degree two, which lead to expressions much more complicated than in the preceding case. In Tables LIV, the rotation-vibration energies are given in terms of rotational quantum numbers and coefficients which depend upon molecular parameters and vibrational quantum numbers. When it is necessary, these coefficients can be expanded in power series with respect to vibrational quantum numbers as shown in Table LV.

The presentation of Tables LIV is very similar[1] to the one used in Chapter X (although strong resonances do not occur necessarily in the same vibrational states). For this reason, we shall not describe it again. The symbols used for diagonal contributions are the same as in Chapter X. For off-diagonal contributions, there are some differences with the notations of Chapters I to IX. We give in Table LVI the correspondance between these two notations.

FREQUENCIES OF ROTATION AND ROTATION-VIBRATION LINES FOR C_{4v} AND D_{2d} MOLECULES

A . Selection rules

As seen in Chapter X, the ordinary selection rules need to be completed when the basic wave functions are symmetric or antisymmetric linear combinations $\psi^{\pm}$ of the functions ψ_0 . For C_{4v} and D_{2d} symmetry groups, ψ^+ and ψ^- functions can be labeled as A_1 , A_2 , B_1 , B_2 or E. By using symmetry properties of these functions, it has been possible to establish, for dipole transitions of C_{4v} and D_{2d} molecules, the complete selection rules given in Tables LVII. In these tables[2],

$$\sum_n \Delta v_n^{(B)} = \Delta v_{n_1}^{(B)} + \Delta v_{n_2}^{(B)} + \dots + \Delta v_{n_i}^{(B)} + \dots$$

where $v_{n_i}^{(B)}$ is the vibrational quantum number for a non degenerate vibration of symmetry species B_1 or B_2.

$$\sum_n \Delta v_n^{(2)} = \Delta v_{n_1}^{(2)} + \Delta v_{n_2}^{(2)} + \dots + \Delta v_{n_i}^{(2)} + \dots$$

where $v_{n_i}^{(2)}$ is the vibrational quantum number for a non degenerate vibration of symmetry species A_2 or B_2 .

B . Frequencies of Rotation Lines

Obviously, molecules D_{2d} have no pure rotation spectrum[3]. Line frequencies have been computed for C_{4v} molecules to sixth order, in two types of vibrational states, and results are given in Tables LVIII where the values of J and K have been assumed to be small (case A)

- In Table LVIII,1 , are given rotational transitions in a vibrational state where only non degenerate vibrations are excited.

- In Table LVIII,2 , are given rotational transitions in a vibrational state where, besides non degenerate vibrations, one twofold degenerate vibration is excited by one quantum.

Frequencies are expressed in terms of rotational quantum numbers J and K associated with the lower level of the transition. A symbol $\nu_{J,|K|}$ is used for the frequency in states 1, where all ℓ_t are equal to zero ; a symbol $\nu_{J,K\ell_t}$ is used in the states 2, where $v_t = 1, \ell_t = \pm 1$

The frequencies are obtained from Table LVIII by adding up the various contributions written one under the other, taking into account the value of $|K|$ or $K\ell_t$ when it is necessary.

Let us describe briefly the two tables.

Table LVIII, 1

The various contributions listed in this table can be numbered from I to IV as shown in the scheme hereafter.

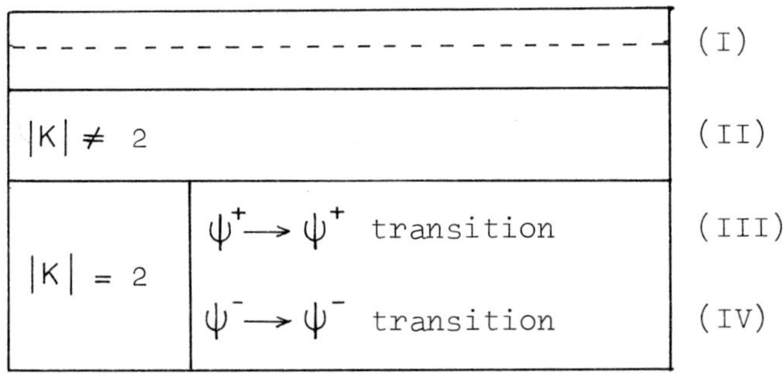

The frequency of a transition is obtained by adding up two out of the four contributions as follows :

If $|K| \neq 2$: $\nu_{J,|K|} = (I) + (II)$

If $|K| = 2$: two transitions are possible

$\psi^+ \to \psi^+$: $\nu_{J,|2|,++} = (I) + (III)$

$\psi^- \to \psi^-$: $\nu_{J,|2|,--} = (I) + (IV)$

Table LVIII, 2

This table is more complicated. In a vibrational state where one twofold degenerate vibration is excited by one quantum, the frequencies of rotational lines depend on the product $K\ell_t$. Table LVIII, 2 can be summarized as follows :

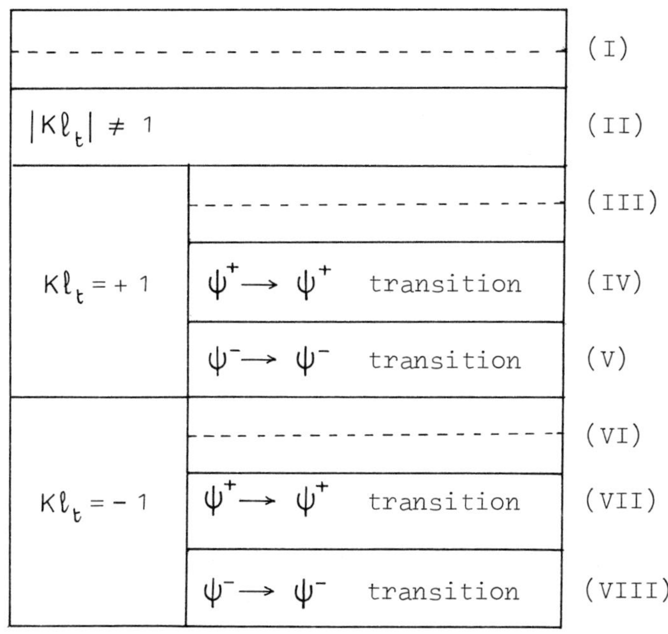

The frequencies are given by the following expressions :

If $|K\ell_t| \neq 1$: $\nu_{J,K\ell_t} = (I) + (II)$

If $K\ell_t = +1$ $\begin{cases} \psi^+ \longrightarrow \psi^+ : & \nu_{J,1,++} = (I) + (III) + (IV) \\ \psi^- \longrightarrow \psi^- : & \nu_{J,1,--} = (I) + (III) + (V) \end{cases}$

If $K\ell_t = -1$ $\begin{cases} \psi^+ \longrightarrow \psi^+ : & \nu_{J,-1,++} = (I) + (VI) + (VII) \\ \psi^- \longrightarrow \psi^- : & \nu_{J,-1,--} = (I) + (VI) + (VIII) \end{cases}$

C . Frequencies of Rotation-Vibration Lines

The Tables LIX give line frequencies for transitions between a level where only non degenerate vibrations are excited and various upper levels, for C_{4v} and D_{2d} molecules. Molecular vibrations may be of species A_1, A_2, B_1, B_2 or E and we must make a distinction between the transitions according to the parities of $\Sigma_n \Delta v_n^{(B)}$ and $\Sigma_n \Delta v_n^{(2)}$. In the labeling of Tables LIX, the number from 1 to 6, characterizes the upper vibrational level, with the same conventions as for Tables LIV. The capital letter A or B refers to the range of rotational quantum numbers.The greek letters α and β refer to lines belonging to Q branches and R or P branches respectively. At last, for perpendicular bands (Tables LIX, 2) we consider two cases : Tables (R), refering to

$$ {}^RP_{J,|K|} \quad {}^RQ_{J,|K|} \quad {}^RR_{J,|K|} \text{ lines} $$

and Tables (P) , refering to

$$ {}^PP_{J,|K|} \quad {}^PQ_{J,|K|} \quad {}^PR_{J,|K|} \text{ lines.} $$

The symbols v' and v'' characterize respectively the upper and the lower level of the transitions. Quantum numbers J and K are those associated with the lower level. The notation used to characterize the lines follows closely that of Chapter X.

In parallel bands for which the upper level corresponds to zero values of all quantum numbers ℓ_t symbols :

$$ {}^QQ_{J,|K|} \quad {}^QR_{J,|K|} \quad {}^QP_{J,|K|} $$

are used for the lines (Tables LIX , 1 and 3).

In parallel bands where at least one quantum number ℓ_t is different from zero in the upper level,symbols :

$$ {}^QQ_{J,K\ell_t} \quad {}^QR_{J,K\ell_t} \quad {}^QP_{J,K\ell_t} $$

are used for the lines when $J \simeq K \simeq 30$ (Tables LIX , 4 B, 5 B and 6 B).

When J and K are small, symbols :

$$ {}^{Q}Q^{+}_{J,|K|} \qquad {}^{Q}R^{+}_{J,|K|} \qquad {}^{Q}P^{+}_{J,|K|} $$

or

$$ {}^{Q}Q^{-}_{J,|K|} \qquad {}^{Q}R^{-}_{J,|K|} \qquad {}^{Q}P^{-}_{J,|K|} $$

are used (Tables LIX, 4 A , 5 A and 6 A) respectively when the upper level of the transition is $E^{+}_{(|K|)}$ (energy level obtained taking the upper signs in Tables LIV) or $E^{-}_{(|K|)}$ (energy level obtained taking the lower signs in Tables LIV).

In perpendicular bands, symbols

$$ {}^{R}Q_{J,|K|} \qquad {}^{R}R_{J,|K|} \qquad {}^{R}P_{J,|K|} $$

are used for lines with $\Delta|K| = +1$ and K =0, 1, 2, 3 ...; symbols

$$ {}^{P}Q_{J,|K|} \qquad {}^{P}R_{J,|K|} \qquad {}^{P}P_{J,|K|} $$

are used for lines with $\Delta|K| = -1$ and K = 1, 2, 3 ...

Furthermore, the following conventions have been used in tables giving R and P branches (Tablesβ) :

$$ m = J+1 \quad \text{for} \quad {}^{R}R_{J,|K|} \quad {}^{Q}R_{J,|K|} \quad {}^{P}R_{J,|K|} \quad \text{and} \quad {}^{Q}R_{J,K\ell_t} $$

$$ m = -J \quad \text{for} \quad {}^{R}P_{J,|K|} \quad {}^{Q}P_{J,|K|} \quad {}^{P}P_{J,|K|} \quad \text{and} \quad {}^{Q}P_{J,K\ell_t} $$

Let us give with some details the directions for use of Tables LIX.

Tables LIX,1

These tables give wave numbers for transitions occurring only if

$$ \Sigma\Delta v_n^{(B)} \quad \text{is} \quad \begin{cases} \text{even for } C_{4v} \text{ group} \\ \text{odd for } D_{2d} \text{ group} \end{cases} $$

The wave numbers are obtained by adding up the various contributions written one under the other.

Tables LIX,2

These tables are more complicated. Table LIX, 2 A α(R) appears as shown hereunder and the wave number $\nu_{J,\lvert K\rvert}$ is given by adding up three out of the contributions numbered from (I) to (IX)

TABLE LIX, 2 A α (R)

- - - - - - - - - -			(I)
$\Sigma \Delta v_n^{(B)}$ even	- - - - - - - - - -		(II)
	$\lvert K\rvert \neq 0$		(III)
	K = 0	$\Sigma \Delta v_n^{(2)}$ even $(\psi^+ \to \psi^-)$	(IV)
		$\Sigma \Delta v_n^{(2)}$ odd $(\psi^+ \to \psi^+)$	(V)
$\Sigma \Delta v_n^{(B)}$ odd	- - - - - - - - - -		(VI)
	$\lvert K\rvert \neq 0$		(VII)
	K = 0	$\Sigma \Delta v_n^{(2)}$ even $(\psi^+ \to \psi^-)$	(VIII)
		$\Sigma \Delta v_n^{(2)}$ odd $(\psi^+ \to \psi^+)$	(IX)

If $\Sigma \Delta v_n^{(B)}$ is even

- for $\lvert K\rvert \neq 0$: $\nu_{J,\lvert K\rvert}$ = (I) + (II) + (III)

- for K = 0 : $\nu_{J,0} = \begin{cases} \text{(I) + (II) + (IV) if } \Sigma \Delta v_n^{(2)} \text{ is even} \\ \text{(I) + (II) + (V) if } \Sigma \Delta v_n^{(2)} \text{ is odd} \end{cases}$

If $\Sigma \Delta v_n^{(B)}$ is odd

- for $\lvert K\rvert \neq 0$: $\nu_{J,\lvert K\rvert}$ = (I) + (VI) + (VII)

- for K = 0 : $\nu_{J,0} = \begin{cases} \text{(I) + (VI) + (VIII) if } \Sigma \Delta v_n^{(2)} \text{ is even} \\ \text{(I) + (VI) + (IX) if } \Sigma \Delta v_n^{(2)} \text{ is odd} \end{cases}$

Table LIX, 2 A β (R) is very similar and must be used in the same way.

Tables LIX, 2 A α (P) and LIX, 2 A β (P) appear as follows :

TABLES LIX , 2 A α (P) and 2 A β (P)

- - - - - - - - - - - - - - - - - - - -		(I)
$\Sigma \Delta v_n^{(B)}$ even	- - - - - - - - - - - - - - - -	(II)
	$\lvert K \rvert \neq 2$	(III)
	$\lvert K \rvert = 2$	(IV)
$\Sigma \Delta v_n^{(B)}$ odd	- - - - - - - - - - - - - - - -	(V)
	$\lvert K \rvert \neq 2$	(VI)
	$\lvert K \rvert = 2$	(VII)

If $\Sigma \Delta v_n^{(B)}$ is even

- for $|K| \neq 2$: $\nu_{J,\,|K|}$ = (I) + (II) + (III)
- for $|K| = 2$: $\nu_{J,\,|2|}$ = (I) + (II) + (IV)

There are two transitions corresponding to $\nu_{J,|2|}$: one for which the upper level is $\psi^{+}_{1,-1}$ (its wave number is obtained taking the upper signs in contribution IV), the second for which the upper level is $\psi^{-}_{1,-1}$ (its wave number is obtained taking the lower signs in contribution IV).

If $\Sigma \Delta v_n^{(B)}$ is odd

- for $|K| \neq 2$: $\nu_{J,\,|K|}$ = (I) + (V) + (VI)
- for $|K| = 2$: $\nu_{J,\,|2|}$ = (I) + (V) + (VII)

There are also two transitions corresponding to $\nu_{J,|2|}$ as when $\Sigma \Delta v_n^{(B)}$ is even.

Tables LIX, 2 B α(R) , 2 B α(P) , 2 B β(R) , 2 B β(P) have a more simple form, which can be summarized as follows:

TABLES LIX, 2 Bα(R) , 2 Bα (P) , 2 Bβ(R) , 2 Bβ (P)

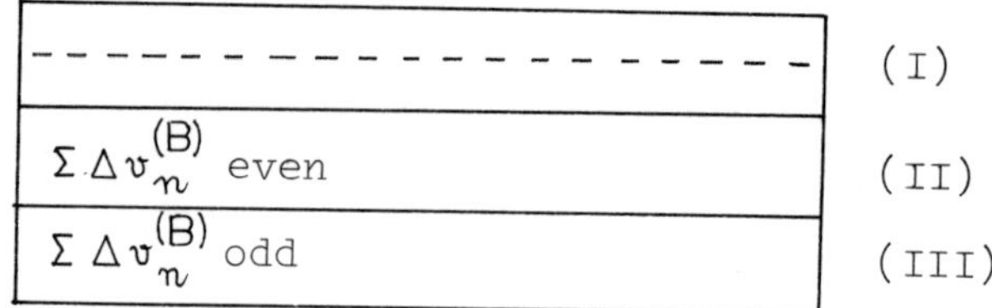

- - - - - - - - - - - - - - - - - - - -	(I)
$\Sigma \Delta v_n^{(B)}$ even	(II)
$\Sigma \Delta v_n^{(B)}$ odd	(III)

If $\Sigma \Delta v_n^{(B)}$ is even : $\nu_{J,|K|}$ = (I) + (II)

If $\Sigma \Delta v_n^{(B)}$ is odd : $\nu_{J,|K|}$ = (I) + (III)

Tables LIX,3 :

These tables give wave numbers for the transitions occurring only

if $\Sigma \Delta v_n^{(B)}$ is $\begin{cases} \text{even for } C_{4v} \text{ group} \\ \text{odd for } D_{2d} \text{ group} \end{cases}$

In addition, when K = 0 in Tables LIX , 3 Aβ and 3 Bβ , transitions indicated occur only if

$$\Sigma \Delta v_n^{(2)} \text{ is } \begin{cases} \text{even for } C_{4v} \text{ group} \\ \text{odd for } D_{2d} \text{ group} \end{cases}$$

Tables LIX,4

These tables give wave numbers for transitions occurring only if

$$\Sigma \Delta v_n^{(B)} \text{ is } \begin{cases} \text{odd for } C_{4v} \text{ group} \\ \text{even for } D_{2d} \text{ group} \end{cases}$$

When $J \simeq K \simeq 1$, the upper energy level (v_t = 2 , $\ell_t = \pm 2$) has been computed by solving second order determinants, and one obtains two levels, $E^+_{(|K|)}$ and $E^-_{(|K|)}$, for each value of $|K|$, corresponding respectively to upper signs and lower signs in Tables LIV, 4 A . This has the following effect in Tables LIX, 4 A :

Table LIX, 4 Aα gives two wave numbers values for a Q line, according to the upper level which may be $E^+_{(|K|)}$ or $E^-_{(|K|)}$. The two transitions ${}^{Q}Q^{+}_{J,|K|}$ and ${}^{Q}Q^{-}_{J,|K|}$ are obtained taking respectively upper signs and lower signs.

Table LIX, 4 Aβ has the following form :

TABLE LIX, 4 Aβ

<table>
<tr><td colspan="3">- -</td><td>(I)</td></tr>
<tr><td colspan="3">$|K| \neq 0$</td><td>(II)</td></tr>
<tr><td rowspan="2">$K = 0$</td><td>C_{4v} group, $\Sigma_n \Delta v^{(2)}$ even
D_{2d} group, $\Sigma \Delta v_n^{(2)}$ odd</td><td>$(\psi^+ \rightarrow \psi^+)$</td><td>(III)</td></tr>
<tr><td>C_{4v} group, $\Sigma \Delta v_n^{(2)}$ odd
D_{2d} group, $\Sigma \Delta v_n^{(2)}$ even</td><td>$(\psi^+ \rightarrow \psi^-)$</td><td>(IV)</td></tr>
</table>

If $|K| \neq 0$: $\nu_{J,|K|}$ = (I) + (II) with the upper sign for ${}^{Q}Q^{+}_{J,|K|}$ line, the lower sign for ${}^{Q}Q^{-}_{J,|K|}$ line.

If $K = 0$: $\nu_{J,0}$ =
- (I) + (III) when $\Sigma \Delta v_n^{(2)}$ is { even for C_{4v} ; odd for D_{2d} }
- (I) + (IV) when $\Sigma \Delta v_n^{(2)}$ is { odd for C_{4v} ; even for D_{2d} }

Tables LIX, 4 Bα and 4 Bβ present no difficulties and give the wave number, not in terms of $|K|$ but of $K\ell_t$.

<u>Tables LIX,5</u>

These tables give wave numbers for transitions occurring only if

$$\Sigma \Delta v_n^{(B)} \quad \text{is} \quad \begin{cases} \text{even for } C_{4v} \text{ group} \\ \text{odd for } D_{2d} \text{ group} \end{cases}$$

They must be used in the same manner as Tables LIX, 4.

Tables LIX,6

These tables give wave numbers for transitions occurring only if

$$\Sigma \Delta v_n^{(B)} \quad \text{is} \quad \begin{cases} \text{odd for } C_{4v} \text{ group} \\ \\ \text{even for } D_{2d} \text{ group} \end{cases}$$

They must be used in the same manner as Tables LIX, 4.

NOTES TO CHAPTER XI

1 . The computation, however, is carried out to sixth order only for levels 1 and 2 with $J \simeq K \simeq 1$ (Tables LIV,1 A and LIV,2 A) , since the frequencies of rotational lines are given only for these two vibrational states (Tables LVIII,1 and LVIII,2).

2 . The same convention is used throughout the chapter.

3 . However, rotational transitions can be expected to occur in vibrational excited states if a dipole moment is induced by vibration.

4 . The values of the quantum numbers $v_n \, v_{n'} \ldots$ associated with the lower vibrational state v'' and with the upper vibrational state v' are not stated in Tables LIX,1 , LIX,3 , LIX,4 , LIX,5 and LIX,6 . Actually they cannot be chosen arbitrarily in both the lower and the upper vibrational states, since the following selection rules have to be taken into account :

Tables LIX,1 , LIX,3 and LIX,5 :

group C_{4v} : $\Delta v_n^{(B)}$ even

group D_{2d} : $\Delta v_n^{(B)}$ odd

Tables LIX,4 and LIX,6 ;

group C_{4v} : $\Delta v_n^{(B)}$ odd

group D_{2d} : $\Delta v_n^{(B)}$ even

TABLES LIV . ROTATION-VIBRATION ENERGIES FOR MOLECULES BELONGING TO SYMMETRY GROUPS C_{4v} AND D_{2d}

TABLE LIV, 1 A . $v_t = v_{t'} = \ldots = 0$; $v_n, v_{n'} \ldots$ arbitrary ; $J \simeq K \simeq 1$

Contributions			k: a	b	c
$\frac{E_v}{hc}$			0	2	4
$+B_v^{xx} J(J+1)$			2	4	6
$+(B_v^{zz} - B_v^{xx}) K^2$			2	4	6
$-D_e^J J^2 (J+1)^2$			6		
$-D_e^{JK} J(J+1) K^2$			6		
$-D_e^K K^4$			6		
$\lvert K \rvert \neq 2$	0				
$\lvert K \rvert = 2$	ψ_2^+ state	$+R_6 J(J+1)[J(J+1)-2]$	6		
	ψ_2^- state	$-R_6 J(J+1)[J(J+1)-2]$	6		

TABLE LIV, 1 B . $v_t = v_{t'} = \ldots 0$; $v_n, v_{n'} \ldots$ arbitrary ; $J \simeq K \simeq 30$

Contributions	h: a	h: b	h: c
$\frac{E_v}{hc}$	0	2	4
$+B_v^{xx} J(J+1)$	0	2	4
$+(B_v^{zz} - B_v^{xx})K^2$	0	2	4
$-D_v^J J^2(J+1)^2$	2	4	
$-D_v^{JK} J(J+1)K^2$	2	4	
$-D_v^K K^4$	2	4	
$+H^J J^3(J+1)^3$	4		
$+H^{JJK} J^2(J+1)^2 K^2$	4		
$+H^{JKK} J(J+1)K^4$	4		
$+H^K K^6$	4		

TABLE LIV, 2 A . $v_t = 1$, $|\ell_t| = 1$; $v_{t'} = \ldots = 0$; $v_n, v_{n'}, \ldots$ arbitrary ; $J \simeq K \simeq 1$

Contributions	k: a	b	c
$\frac{E_v}{hc}$	0	2	4
$+B_v^{xx} J(J+1)$	2	4	6
$+(B_v^{zz} - B_v^{xx})K^2$	2	4	6
$-2(\zeta_t B^{zz})_v K\ell_t$	2	4	6
$-D_e^J J^2(J+1)^2$	6		
$-D_e^{JK} J(J+1)K^2$	6		
$-D_e^K K^4$	6		
$+\eta_t^J J(J+1)K\ell_t$	6		
$+\eta_t^K K^3 \ell_t$	6		

$\lvert K\rvert \neq 1$		$-\dfrac{(a^{+}_{2,J,K\ell_t})^2 (G^t)^2}{(K\ell_t+1)[B^{zz}_e - B^{xx}_e + \zeta^z_t B^{zz}_e]} + \dfrac{(a^{-}_{2,J,K\ell_t})^2 (F^t)^2}{(K\ell_t-1)[B^{zz}_e - B^{xx}_e - \zeta^z_t B^{zz}_e]}$	6		
$K\ell_t = 1$	$\psi^{+}_{1,1}$ state	$+ 2J(J+1) F^t_{eff}$	4	6	
		$-(a^{+}_{2,J,1})^2 \left\{(G^t)^2/2(B^{zz}_e - B^{xx}_e + \zeta^z_t B^{zz}_e)\right\}$	6		
	$\psi^{-}_{1,1}$ state	$- 2J(J+1) F^t_{eff}$	4	6	
		$-(a^{+}_{2,J,1})^2 \left\{(G^t)^2/2(B^{zz}_e - B^{xx}_e + \zeta^z_t B^{zz}_e\right\}$	6		
$K\ell_t = -1$	$\psi^{+}_{1,-1}$ state	$+ 2J(J+1) G^t_{eff}$	4	6	
		$-(a^{+}_{2,J,1})^2 \left\{(F^t)^2/2(B^{zz}_e - B^{xx}_e - \zeta^z_t B^{zz}_e)\right\}$	6		
	$\psi^{-}_{1,-1}$ state	$- 2J(J+1) G^t_{eff}$	4	6	
		$-(a^{+}_{2,J,1})^2 \left\{(F^t)^2/2(B^{zz}_e - B^{xx}_e - \zeta^z_t B^{zz}_e)\right\}$	6		

$$a^{\pm}_{2,J,K\ell_t} = \sqrt{J(J+1) - K\ell_t(K\ell_t \pm 1)}\ \sqrt{J(J+1) - (K\ell_t \pm 1)(K\ell_t \pm 2)}$$

TABLE LIV,2 B. $v_t = 1, |\ell_t| = 1$; $v_{t'} = \ldots = 0$; $v_n, v_{n'} \ldots$ arbitrary ; $J \simeq K \simeq 30$

Contributions	k: a	b	c
$\frac{E_v}{hc}$	0	2	4
$+B_v^{xx} J(J+1)$	0	2	4
$+(B_v^{zz} - B_v^{xx})K^2$	0	2	4
$-2(\zeta_t B^{zz})_v K\ell_t$	1	3	
$-D_v^J J^2(J+1)^2$	2	4	
$-D_v^{JK} J(J+1)K^2$	2	4	
$-D_v^K K^4$	2	4	
$+\eta_t^J J(J+1)K\ell_t$	3		
$+\eta_t^K K^3 \ell_t$	3		
$+H^J J^3(J+1)^3$	4		
$+H^{JJK} J^2(J+1)^2 K^2$	4		
$+H^{JKK} J(J+1)K^4$	4		
$+H^K K^6$	4		

TABLE LIV, 3 A . $v_t = 2$, $\ell_t = 0$; $v_{t'} = \ldots = 0$; $v_n, v_{n'} \ldots$ arbitrary ; $J \simeq K \simeq 1$

Contributions	k a	k b	k c
$\frac{E_v}{hv}$	0	2	4
$+B_v^{xx} J(J+1)$	2	4	
$+(B_v^{zz} - B_v^{xx}) K^2$	2	4	

TABLE LIV, 3 B . $v_t = 2$, $\ell_t = 0$; $v_{t'} = \ldots = 0$; $v_n, v_{n'} \ldots$ arbitrary ; $J \simeq K \simeq 30$

Contributions	k a	k b	k c
$\frac{E_v}{hc}$	0	2	4
$+B_v^{xx} J(J+1)$	0	2	4
$+(B_v^{zz} - B_v^{xx}) K^2$	0	2	4
$-D_v^J J^2 (J+1)^2$	2	4	
$-D_v^{JK} J(J+1) K^2$	2	4	
$-D_v^K K^4$	2	4	
$+H^J J^3 (J+1)^3 + H^{JJK} J^2 (J+1)^2 K^2$	4		
$+H^{JKK} J(J+1) K^4 + H^K K^6$	4		

TABLE LIV, 4 A . $v_t = 2, |\ell_t| = 2 ; v_{t'} = \ldots = 0 ; v_n, v_{n'} \ldots$ arbitrary; $J \simeq K \simeq 1$

Contributions			k a	k b	k c
$\frac{E_v}{hc}$			0	2	4
$+ B_v^{xx} J(J+1)$			2	4	
$+ (B_v^{zz} - B_v^{xx}) K^2$			2	4	
$\lvert K \rvert \neq 0$	$\pm 4\sqrt{(\zeta_t B^{zz})_v^2 K^2 + 4(U^t)^2}$		2		
	$\mp \frac{2\sum_s [K^2 \zeta_t^z B_e^{zz} \eta_{ts} - 8U^t g_4^{t,s}](v_s + \frac{d_s}{2})}{[(\zeta_t^z B_e^{zz})^2 K^2 + 4(U^t)^2]^{\frac{1}{2}}}$		4		
$K = 0$	$\psi_{0,2}^{+}$ state	$+ 8U_{eff}^t$	2	4	
	$\psi_{0,2}^{-}$ state	$- 8U_{eff}^t$	2	4	

TABLE LIV, 4 B . $v_t = 2, |\ell_t| = 2$; $v_{t'} = \ldots = 0; v_n, v_{n'} \ldots$ arbitrary; $J \simeq K \simeq 30$

Contributions	k: a	b	c
$\frac{E_v}{hc}$	0	2	4
$+ B_v^{xx} J(J+1)$	0	2	4
$+ (B_v^{zz} - B_v^{xx}) K^2$	0	2	4
$- 2 (\zeta_t B^{zz})_v K\ell_t$	1	3	
$- D_v^J J^2 (J+1)^2$	2	4	
$- D_v^{JK} J(J+1) K^2$	2	4	
$- D_v^K K^4$	2	4	
$+ \eta_t^J J(J+1) K\ell_t$	3		
$+ \eta_t^K K^3 \ell_t$	3		
$- 16 \left\{ (U^t)^2 / \zeta_t^3 B_e^{zz} K\ell_t \right\}$	3		
$+ H^J J^3 (J+1)^3$	4		
$+ H^{JJK} J^2 (J+1)^2 K^2$	4		
$+ H^{JKK} J (J+1) K^4$	4		
$+ H^K K^6$	4		

TABLE LIV, 5 A . $v_t = v_{t'} = 1$; $\ell_t + \ell_{t'} = 0$; $v_{t''} = \ldots = 0$; $v_n, v_{n'} \ldots$ arbitrary ; $J \simeq K \simeq 1$

Contributions			h: a	h: b	h: c
$\frac{E_v}{hc}$			0	2	4
$+B_v^{xx} J(J+1)$			2	4	
$+(B_v^{zz} - B_v^{xx})K^2$			2	4	
$\lvert K \rvert \neq 0$	$\pm \sqrt{4K^2(\zeta_t^z - \zeta_{t'}^z)^2 (B_e^{zz})^2 + (r^{tt'})^2}$		2		
	$\mp \dfrac{2\sum_s \left[K^2(\zeta_t^z - \zeta_{t'}^z) B_e^{zz}(\eta_{ts} - \eta_{t's}) - 2r^{tt'} g_{22-}^{tt',s}\right](v_s + \frac{d_s}{2})}{\left[4K^2(\zeta_t^z - \zeta_{t'}^z)^2 (B_e^{zz})^2 + (r^{tt'})^2\right]^{\frac{1}{2}}}$		4		
$K = 0$	$\Psi_{0,1,-1}^{+}$ state	$+ r_{eff}^{tt'}$	2	4	
	$\Psi_{0,1,-1}^{-}$ state	$- r_{eff}^{tt'}$	2	4	

TABLE LIV, 5 B . $v_t = v_{t'} = 1$, $\ell_t + \ell_{t'} = 0$; $v_{t''} = \dots = 0$; $v_n, v_{n'} \dots$ arbitrary; $J \simeq K \simeq 30$

Contributions	k a	k b	k c
$\frac{E_v}{hc}$	0	2	4
$+B_v^{xx} J(J+1)$	0	2	4
$+(B_v^{zz} - B_v^{xx}) K^2$	0	2	4
$-2[(\zeta_t B^{zz})_v - (\zeta_{t'} B^{zz})_v] K\ell_t$	1	3	
$-D_v^J J^2 (J+1)^2$	2	4	
$-D_v^{JK} J(J+1) K^2$	2	4	
$-D_v^K K^4$	2	4	
$-\{(r^{tt'})^2 / 4 B_e^{zz} (\zeta_t^z - \zeta_{t'}^z) K\ell_t\}$	3		
$+(\eta_t^J - \eta_{t'}^J) J(J+1) K\ell_t$	3		
$+(\eta_t^K - \eta_{t'}^K) K^3 \ell_t$	3		
$+H^J J^3 (J+1)^3$	4		
$+H^{JJK} J^2 (J+1)^2 K^2$	4		
$+H^{JKK} J(J+1) K^4$	4		
$+H^K K^6$	4		

The contributions of this table take into account the relation $\ell_{t'} = -\ell_t$

TABLE LIV, 6 A . $v_t = v_{t'} = 1, |l_t + l_{t'}| = 2; v_{t''} = \dots = 0; v_n, v_{n'} \dots$ arbitrary ; $J \simeq K \simeq 1$

Contributions			k: a	k: b	k: c
$\frac{E_v}{hc}$			0	2	4
$+B_v^{xx} J(J+1)$			2	4	
$+(B_v^{zz} - B_v^{xx})K^2$			2	4	
$\lvert K\rvert \neq 0$	$\pm \sqrt{4K^2(\zeta_t^z + \zeta_{t'}^z)^2 (B_e^{zz})^2 + (S^{tt'})^2}$		2		
	$\mp \frac{2\sum_s \left[K^2(\zeta_t^z + \zeta_{t'}^z) B_e^{zz}(\eta_{ts} + \eta_{t's}) - 2S^{tt'} g_{22+}^{tt',s}\right](v_s + \frac{d_s}{2})}{\left[4K^2(\zeta_t^z + \zeta_{t'}^z)^2 B_e^{zz} + (S^{tt'})^2\right]^{\frac{1}{2}}}$		4		
$K = 0$	$\Psi_{0,1,1}^{+}$ state	$+S_{eff}^{tt'}$	2	4	
	$\Psi_{0,1,1}^{-}$ state	$-S_{eff}^{tt'}$	2	4	

TABLE LIV, 6 B . $v_t = v_{t'} = 1, |\ell_t + \ell_{t'}| = 2$; $v_{t''} = \ldots = 0$; $v_n, v_{n'} \ldots$ arbitrary ; $J \simeq K \simeq 30$

Contributions	k a	k b	k c
$\frac{E_v}{hc}$	0	2	4
$+B_v^{xx} J(J+1)$	0	2	4
$+(B_v^{zz} - B_v^{xx})K^2$	0	2	4
$-2K[(\zeta_t B^{zz})_v + (\zeta_{t'} B^{zz})_v]\ell_t$	1	3	
$-D_v^J J^2(J+1)^2$	2	4	
$-D_v^{JK} J(J+1)K^2$	2	4	
$-D_v^K K^4$	2	4	
$-\{(s^{tt'})^2/4B_e^{zz}(\zeta_t^z + \zeta_{t'}^z)K\ell_t\}$	3		
$+(\eta_t^J + \eta_{t'}^J)J(J+1)K\ell_t$	3		
$+(\eta_t^K + \eta_{t'}^K)K^3\ell_t$	3		
$+H^J J^3(J+1)^3$	4		
$+H^{JJK} J^2(J+1)^2K^2$	4		
$+H^{JKK} J(J+1)K^4$	4		
$+H^K K^6$	4		

The contributions of this table take into account the relation $\ell_{t'} = \ell_t$

TABLE LV . Coefficients appearing in the expressions of energies or frequencies

$\frac{E_v}{hc}$	(a)	$\sum_{s} \omega_s \left(v_s + \frac{d_s}{2}\right)$
	(b)	$\sum_{\substack{ss' \\ s \leq s'}} x_{ss'} \left(v_s + \frac{d_s}{2}\right)\left(v_{s'} + \frac{d_{s'}}{2}\right) + \sum_{\substack{t,t' \\ t \leq t'}} x_{l_t l_{t'}} l_t l_{t'}$
	(c)	$\sum_{\substack{ss's'' \\ s \leq s' \leq s''}} y_{ss's''} \left(v_s + \frac{d_s}{2}\right)\left(v_{s'} + \frac{d_{s'}}{2}\right)\left(v_{s''} + \frac{d_{s''}}{2}\right) + \sum_{\substack{stt' \\ t \leq t'}} y_{s l_t l_{t'}} \left(v_s + \frac{d_s}{2}\right) l_t l_{t'} + \sum_{s} \Delta\omega_s \left(v_s + \frac{d_s}{2}\right)$
B_e^{ii} $(i = x, y, z)$	(a)	B_e^{ii}
	(b)	$-\sum_{s} \alpha_s^{i} \left(v_s + \frac{d_s}{2}\right)$
	(c)	$\sum_{\substack{ss' \\ s \leq s'}} \gamma_{ss'}^{i} \left(v_s + \frac{d_s}{2}\right)\left(v_{s'} + \frac{d_{s'}}{2}\right) + \sum_{\substack{t,t' \\ t \leq t'}} \gamma_{l_t l_{t'}}^{i} l_t l_{t'} + \Delta B_e^{ii}$

TABLE LV . Continued

D^i_v $(i = J, JK, K)$	(a)	D^i_e
	(b)	$\sum_s \beta^i_s (v_s + \frac{d_s}{2})$
$(\zeta_t B^{zz})_v$	(a)	$\zeta^z_t B^{zz}_e$
	(b)	$-\frac{1}{2} \sum_s \eta_{ts} (v_s + \frac{d_s}{2})$
	(c)	$-\sum_{\substack{ss' \\ s \leq s'}} \xi^t_{ss'} (v_s + \frac{d_s}{2})(v_{s'} + \frac{d_{s'}}{2}) - \sum_{\substack{t't'' \\ t \leq t' \leq t''}} \xi^t_{l_{t'} l_{t''}} l_{t'} l_{t''} + \Delta(\zeta^z_t B^{zz}_e)$
F^t_{eff}	(a)	F^t
	(b)	$\sum_s f^{t,s}_{22+} (v_s + \frac{d_s}{2})$
G^t_{eff}	(a)	G^t
	(b)	$\sum_s f^{t,s}_{22-} (v_s + \frac{d_s}{2})$

U^t_{eff}	(a)	U^t
	(b)	$\sum_s g_4^{t,s}\left(v_s + \frac{d_s}{2}\right)$
$r^{tt'}_{eff}$	(a)	$r^{tt'}$
	(b)	$4\sum_s g_{22-}^{tt',s}\left(v_s + \frac{d_s}{2}\right)$
$S^{tt'}_{eff}$	(a)	$S^{tt'}$
	(b)	$4\sum_s g_{22+}^{tt',s}\left(v_s + \frac{d_s}{2}\right)$

See footnote to Table XLIX (Chapter X, p.257).

TABLE LVI . Correspondance between the symbols used in Chapter XI for the coefficients of the off-diagonal matrix elements (first column), and the corresponding symbols used in the Chapters I to IX (second column).

R_6	$R_6 \pm i R'_6$
F^t	$F^t_\pm$
G^t	$G^t_\pm$
U^t	$U^t_\pm$
$\mathscr{r}^{tt'}$	$4 R^{tt'}_\pm$
$S^{tt'}$	$4 S^{tt'}_\pm$

An index t designates a twofold degenerate vibration.

Coefficients of the diagonal matrix elements are the same as in Chapter X (See Table L, p.258).

TABLE LVII a. Selection rules for dipole transitions of molecules C_{4v} (Rotation and Rotation-Vibration).

	$\sum_n \Delta v_n^{(B)}$ even	$\sum_n \Delta v_n^{(B)}$ odd		$\sum_n \Delta v_n^{(2)}$ even	$\sum_n \Delta v_n^{(2)}$ odd
Parallel bands $\Delta K = 0$	$\sum_t \Delta \ell_t = 4p$	$\sum_t \Delta \ell_t = 2(2p+1)$	Q branches $\Delta J = 0$	$\psi^+ \longleftrightarrow \psi^-$	$\psi^+ \longleftrightarrow \psi^+$ $\psi^- \longleftrightarrow \psi^-$
			R and P branches $\Delta J = \pm 1$	$\psi^+ \longleftrightarrow \psi^+$ $\psi^- \longleftrightarrow \psi^-$	$\psi^+ \longleftrightarrow \psi^-$
Perpendicular bands $\Delta K = \pm 1$	$\sum_t \Delta \ell_t = \Delta K + 4p$	$\sum_t \Delta \ell_t = \Delta K + 2(2p+1)$	Q branches $\Delta J = 0$	$\psi^+ \longleftrightarrow \psi^-$	$\psi^+ \longleftrightarrow \psi^+$ $\psi^- \longleftrightarrow \psi^-$
			R and P branches $\Delta J = \pm 1$	$\psi^+ \longleftrightarrow \psi^+$ $\psi^- \longleftrightarrow \psi^-$	$\psi^+ \longleftrightarrow \psi^-$
Rotational transitions $\Delta K = 0$	$\sum_t \Delta \ell_t = 0$		$\Delta J = 1$	$\psi^+ \longleftrightarrow \psi^+$ $\psi^- \longleftrightarrow \psi^-$	
Transitions $A_1 \longleftrightarrow A_2$, $B_1 \longleftrightarrow B_2$, $E \longleftrightarrow E$				Transitions $A_1 \longleftrightarrow A_2$, $B_1 \longleftrightarrow B_2$	

TABLE LVII b . Selection rules for dipole transitions of molecules D_{2d} (Rotation-Vibration)

	$\sum_n \Delta v_n^{(B)}$			$\sum_n \Delta v_n^{(2)}$	
	even	odd		even	odd
Parallel bands $\Delta K = 0$	$\sum_t \Delta \ell_t = 2(2p+1)$	$\sum_t \Delta \ell_t = 4p$	Q branches $\Delta J = 0$	$\psi^+ \longleftrightarrow \psi^+$ $\psi^- \longleftrightarrow \psi^-$	$\psi^+ \longleftrightarrow \psi^-$
			R and P branches $\Delta J = \pm 1$	$\psi^+ \longleftrightarrow \psi^-$	$\psi^+ \longleftrightarrow \psi^+$ $\psi^- \longleftrightarrow \psi^-$
Perpendicular bands $\Delta K = \pm 1$	$\sum_t \Delta \ell_t = \Delta K + 4p$	$\sum_t \Delta \ell_t = \Delta K + 2(2p+1)$	Q branches $\Delta J = 0$	$\psi^+ \longleftrightarrow \psi^-$	$\psi^+ \longleftrightarrow \psi^+$ $\psi^- \longleftrightarrow \psi^-$
			R and P branches $\Delta J = \pm 1$	$\psi^+ \longleftrightarrow \psi^+$ $\psi^- \longleftrightarrow \psi^-$	$\psi^+ \longleftrightarrow \psi^-$
Transitions $A_1 \longleftrightarrow B_1$, $A_2 \longleftrightarrow B_2$, $E \longleftrightarrow E$				Transitions $A_1 \longleftrightarrow B_1$, $A_2 \longleftrightarrow B_2$	

TABLES LVIII . FREQUENCIES OF LINES IN THE ROTATION SPECTRUM OF MOLECULES C_{4v}

TABLE LVIII, 1 . Rotational frequencies for $v_t = v_{t'} = \ldots = 0$; $v_n, v_{n'} \ldots$ arbitrary

$\nu_{J,\|K\|}$			k: a	k: b	k: c
$2(J+1)B_v^{xx}$			2	4	6
$-4(J+1)^3 D_e^J$			6		
$-2(J+1)K^2 D_e^{JK}$			6		
$\|K\| \neq 2$	0				
$\|K\| = 2$	$\psi^+ \longrightarrow \psi^+$ transition	$+4J(J+1)(J+2)R_6$	6		
	$\psi^- \longrightarrow \psi^-$ transition	$-4J(J+1)(J+2)R_6$	6		

TABLE LVIII, 2 . Rotational frequencies for $v_t = 1, |\ell_t| = 1$; $v_{t'} = \ldots = 0$; $v_n, v_{n'} \ldots$ arbitrary

	$\nu_{J, K\ell_t}$	k: a	b	c
	$2(J+1)B_v^{xx}$	2	4	6
	$-4(J+1)^3 D_e^J$	6		
	$-2(J+1)(K\ell_t)^2 D_e^{JK}$	6		
	$+2(J+1)\eta_t^J K\ell_t$	6		
$\lvert K\ell_t \rvert \neq 1$	$-4(J+1)^3 \left\{ \frac{(G^t)^2}{(K\ell_t+1)(B_e^{zz}-B_e^{xx}+\zeta_t^z B_e^{zz})} - \frac{(F^t)^2}{(K\ell_t-1)(B_e^{zz}-B_e^{xx}-\zeta_t^z B_e^{zz})} \right\}$	6		
	$+4(J+1) \left\{ \frac{(G^t)^2 (K\ell_t+1)}{(B_e^{zz}-B_e^{xx}+\zeta_t^z B_e^{zz})} - \frac{(F^t)^2 (K\ell_t-1)}{(B_e^{zz}-B_e^{xx}-\zeta_t^z B_e^{zz})} \right\}$	6		

$K\ell_t = +1$	$-2(J+1)^3 \dfrac{(G^t)^2}{(B_e^{zz} - B_e^{xx} + \zeta_t^z B_e^{zz})} + 4(J+1) \dfrac{2(G^t)^2}{(B_e^{zz} - B_e^{xx} + \zeta_t^z B_e^{zz})}$		6		
	$\psi^+ \longrightarrow \psi^+$ transition	$+4(J+1)F_{eff}^t$	4	6	
	$\psi^- \longrightarrow \psi^-$ transition	$-4(J+1)F_{eff}^t$	4	6	
$K\ell_t = -1$	$-2(J+1)^3 \dfrac{(F^t)^2}{(B_e^{zz} - B_e^{xx} - \zeta_t^z B_e^{zz})} + 4(J+1) \dfrac{2(F^t)^2}{(B_e^{zz} - B_e^{xx} - \zeta_t^z B_e^{zz})}$		6		
	$\psi^+ \longrightarrow \psi^+$ transition	$+4(J+1)G_{eff}^t$	4	6	
	$\psi^- \longrightarrow \psi^-$ transition	$-4(J+1)G_{eff}^t$	4	6	

TABLES LIX . FREQUENCIES OF LINES IN THE ROTATION-VIBRATION SPECTRUM OF MOLECULES C_{4v} AND D_{2d} (*)

TABLE LIX, 1 Aα. Transition from $v''(v_t = v_{t'} = \dots = 0)$ to $v'(v_t = v_{t'} = \dots = 0)$; $J \simeq K \simeq 1$

${}^{Q}Q_{J,\|K\|}$	k a	k b	k c
$\dfrac{E_{v'} - E_{v''}}{hc}$	0	2	4
$+ J(J+1)(B^{xx}_{v'} - B^{xx}_{v''})$	–	4	
$+ K^2[(B^{zz}_{v'} - B^{zz}_{v''}) - (B^{xx}_{v'} - B^{xx}_{v''})]$	–	4	

TABLE LIX, 1 Aβ. Transition from $v''(v_t = v_{t'} = \dots = 0)$ to $v'(v_t = v_{t'} = \dots = 0)$; $J \simeq K \simeq 1$

${}^{Q}R_{J,\|K\|}$ ${}^{Q}P_{J,\|K\|}$	k a	k b	k c
$\dfrac{E_{v'} - E_{v''}}{hc}$	0	2	4
$+ m(B^{xx}_{v'} + B^{xx}_{v''})$	2	4	
$+ m^2(B^{xx}_{v'} - B^{xx}_{v''})$	–	4	
$+ K^2[(B^{zz}_{v'} - B^{zz}_{v''}) - (B^{xx}_{v'} - B^{xx}_{v''})]$	–	4	

(*) See note 4, page 336

TABLE LIX, 1 Bα. Transition from $v''(v_t = v_{t'} = \dots = 0)$ to $v'(v_t = v_{t'} = \dots = 0)$; $J \simeq K \simeq 30$

${}^{Q}Q_{J,\|K\|}$	k a	k b	k c
$\frac{E_{v'} - E_{v''}}{hc}$	0	2	4
$+ J(J+1)(B^{xx}_{v'} - B^{xx}_{v''})$	–	2	4
$+ K^2[(B^{zz}_{v'} - B^{zz}_{v''}) - (B^{xx}_{v'} - B^{xx}_{v''})]$	–	2	4
$- J^2(J+1)^2(D^{J}_{v'} - D^{J}_{v''}) - J(J+1)K^2(D^{JK}_{v'} - D^{JK}_{v''}) - K^4(D^{K}_{v'} - D^{K}_{v''})$	–	4	

TABLE LIX, 1 Bβ. Transition from $v''(v_t = v_{t'} = \dots = 0)$ to $v'(v_t = v_{t'} = \dots = 0)$; $J \simeq K \simeq 30$

${}^{Q}R_{J,\|K\|}$ ${}^{Q}P_{J,\|K\|}$	k a	k b	k c
$\frac{E_{v'} - E_{v''}}{hc}$	0	2	4
$+ m(B^{xx}_{v'} + B^{xx}_{v''})$	1	3	
$+ m^2(B^{xx}_{v'} - B^{xx}_{v''})$	–	2	4
$+ K^2[(B^{zz}_{v'} - B^{zz}_{v''}) - (B^{xx}_{v'} - B^{xx}_{v''})]$	–	2	4
$- 4m^3 D^{J}_{e} - 2mK^2 D^{JK}_{e}$	3		
$- m^4(D^{J}_{v'} - D^{J}_{v''}) - m^2K^2(D^{JK}_{v'} - D^{JK}_{v''}) - K^4(D^{K}_{v'} - D^{K}_{v''})$	–	4	

TABLE LIX, 2 $A\alpha(R)$. Transition from $v''(v_t = v_{t'} = \ldots = 0\,; v_n, v_{n'} \ldots$ arbitrary) to $v'(v_t = 1, |\ell_t| = 1\,; v_{t'} = \ldots = 0\,; v_n, v_{n'} \ldots$ arbitrary) ; $J \simeq K \simeq 1$

Condition		${}^{R}Q_{J,\,|K|}$	k: a	k: b	k: c
		$\dfrac{E_{v'} - E_{v''}}{hc}$	0	2	4
		$+K^2\left[(B^{zz}_{v'} - B^{zz}_{v''}) - (B^{xx}_{v'} - B^{xx}_{v''})\right]$	–	4	
$\Sigma\Delta v_n^{(B)}$ even		$+B^{zz}_{v'} - B^{xx}_{v'} - 2(\zeta_t B^{zz})_{v'}$	2	4	
		$+2|K|\left[(B^{zz}_{v'} - B^{xx}_{v'}) - (\zeta_t B^{zz})_{v'}\right]$	2	4	
	$|K| \neq 0$	$+J(J+1)(B^{xx}_{v'} - B^{xx}_{v''})$	–	4	
	$K = 0$	$\Sigma\Delta v_n^{(2)}$ even: $+J(J+1)\left[(B^{xx}_{v'} - B^{xx}_{v''})\right.$	–	4	
		($\psi^+ \longrightarrow \psi^-$ transition): $\left.-2F^t\right]$	4		
		$\Sigma\Delta v_n^{(2)}$ odd: $+J(J+1)\left[(B^{xx}_{v'} - B^{xx}_{v''})\right.$	–	4	
		($\psi^+ \longrightarrow \psi^+$ transition): $\left.+2F^t\right]$	4		

$\Sigma \Delta v_n^{(B)}$ odd	$+B_{v'}^{zz} - B_{v'}^{xx} + 2(\zeta_t B^{zz})_{v'}$			2	4	
	$+2\lvert K \rvert \left[(B_{v'}^{zz} - B_{v'}^{xx}) + (\zeta_t B^{zz})_{v'} \right]$			2	4	
	$\lvert K \rvert \neq 0$	$+ J(J+1)(B_{v'}^{xx} - B_{v''}^{xx})$		–	4	
	$K = 0$	$\Sigma \Delta v_n^{(2)}$ even	$+ J(J+1)\left[(B_{v'}^{xx} - B_{v''}^{xx} \right.$	–	4	
		($\psi^+ \longrightarrow \psi^-$ transition)	$\left. - G^t \right]$	4		
		$\Sigma \Delta v_n^{(2)}$ odd	$+ J(J+1)\left[(B_{v'}^{xx} - B_{v''}^{xx}) \right.$	–	4	
		($\psi^+ \longrightarrow \psi^+$ transition)	$\left. + G^t \right]$	4		

TABLE LIX, 2 Aα (P). Transition from $v''(v_t = v_{t'} = \dots = 0 ; v_n, v_{n'} \dots$ arbitrary) to $v'(v_t = 1, |\ell_t| = 1 ; v_{t'} = \dots = 0 ; v_n, v_{n'} \dots$ arbitrary) ; $J \simeq K \simeq 1$

${}^{P}Q_{J,\|K\|}$			k: a	b	c
$\frac{E_{v'} - E_{v''}}{hc}$			0	2	4
$+K^2\left[(B^{zz}_{v'} - B^{zz}_{v''}) - (B^{xx}_{v'} - B^{xx}_{v''})\right]$			–	4	
$\Sigma\Delta v_n^{(B)}$ even	$+B^{zz}_{v'} - B^{xx}_{v'} - 2(\zeta_t B^{zz})_{v'}$		2	4	
	$-2\|K\|\left[(B^{zz}_{v'} - B^{xx}_{v'}) - (\zeta_t B^{zz})_{v'}\right]$		2	4	
	$\|K\| \neq 2$	$+J(J+1)(B^{xx}_{v'} - B^{xx}_{v''})$	–	4	
	$\|K\| = 2$	$+J(J+1)\left[(B^{xx}_{v'} - B^{xx}_{v''})\right.$	–	4	
		$\left.\pm 2G^t\right]$	4		

$\Sigma \Delta v_n^{(B)}$ odd	$+ B^{zz}_{v'} - B^{xx}_{v'} + 2(\zeta_t B^{zz})_{v'}$		2	4	
	$-2\,\lvert K\rvert \left[(B^{zz}_{v'} - B^{xx}_{v'}) + (\zeta_t B^{zz})_{v'} \right]$		2	4	
	$\lvert K\rvert \neq 2$	$+J(J+1)(B^{xx}_{v'} - B^{xx}_{v''})$	–	4	
	$\lvert K\rvert = 2$	$+J(J+1)\left[(B^{xx}_{v'} - B^{xx}_{v''}) \right.$	–	4	
		$\left. \pm 2F^{t} \right]$	4		

TABLE LIX, 2 Aβ(R). Transition from v'' ($v_t = v_{t'} = \dots = 0$; $v_n, v_{n'} \dots$ arbitrary) to v' ($v_t = 1, |\ell_t| = 1$; $v_{t'} = \dots = 0$; $v_n, v_{n'} \dots$ arbitrary); $J \simeq K \simeq 1$

			${}^{R}R_{J,\|K\|}$ ${}^{R}P_{J,\|K\|}$	k: a	b	c
			$\frac{E_{v'} - E_{v''}}{hc}$	0	2	4
			$+ K^2\left[(B^{zz}_{v'} - B^{zz}_{v''}) - (B^{xx}_{v'} - B^{xx}_{v''})\right]$	–	4	
$\Sigma\Delta v_n^{(B)}$ even			$+ B^{zz}_{v'} - B^{xx}_{v'} - 2(\zeta_t B^{zz})_{v'}$	2	4	
			$+ 2\|K\|\left[(B^{zz}_{v'} - B^{xx}_{v'}) - (\zeta_t B^{zz})_{v'}\right]$	2	4	
	$\|K\| \neq 0$		$+ m(B^{xx}_{v'} + B^{xx}_{v''})$	2	4	
			$+ m^2(B^{xx}_{v'} - B^{xx}_{v''})$	–	4	
	$K = 0$	$\Sigma\Delta v_n^{(2)}$ even ($\psi^+ \longrightarrow \psi^+$ transition)	$+ m\left[(B^{xx}_{v'} + B^{xx}_{v''})\right.$	2	4	
			$\left. + 2F^t\right]$	4		
			$+ m^2\left[(B^{xx}_{v'} - B^{xx}_{v''})\right.$	–	4	
			$\left. + 2F^t\right]$	4		

		$\Sigma\Delta v_n^{(2)}$ odd	$+ m\left[(B_{v'}^{xx} + B_{v''}^{xx})\right.$	2	4	
			$\left.-2F^t\right]$	4		
		($\psi^+ \longrightarrow \psi^-$ transition)	$+ m^2\left[(B_{v'}^{xx} - B_{v''}^{xx})\right.$	–	4	
			$\left.-2F^t\right]$	4		
	$+ B_{v'}^{zz} - B_{v'}^{xx} + 2(\zeta_t B^{zz})_{v'}$			2	4	
	$+ 2\lvert K\rvert\left[(B_{v'}^{zz} - B_{v'}^{xx}) + (\zeta_t B^{zz})_{v'}\right]$			2	4	
	$\lvert K\rvert \neq 0$	$+ m(B_{v'}^{xx} + B_{v''}^{xx})$		2	4	
		$+ m^2(B_{v'}^{xx} - B_{v''}^{xx})$		–	4	
$\Sigma\Delta v_n^{(B)}$ odd	$K = 0$	$\Sigma\Delta v_n^{(2)}$ even	$+ m\left[(B_{v'}^{xx} + B_{v''}^{xx})\right.$	2	4	
			$\left.+2G^t\right]$	4		
		($\psi^+ \longrightarrow \psi^+$ transition)	$+ m^2\left[(B_{v'}^{xx} - B_{v''}^{xx})\right.$	–	4	
			$\left.+2G^t\right]$	4		
		$\Sigma\Delta v_n^{(2)}$ odd	$+ m\left[(B_{v'}^{xx} + B_{v''}^{xx})\right.$	2	4	
			$\left.-2G^t\right]$	4		
		($\psi^+ \longrightarrow \psi^-$ transition)	$+ m^2\left[(B_{v'}^{xx} - B_{v''}^{xx})\right.$	–	4	
			$\left.-2G^t\right]$	4		

TABLE LIX, 2 Aβ(P). Transition from v'' ($v_t = v_{t'} = \dots = 0$; $v_n, v_{n'} \dots$ arbitrary) to v' ($v_t = 1$, $|l_t| = 1$; $v_{t'} = \dots = 0$; $v_n, v_{n'} \dots$ arbitrary); $J \simeq K \simeq 1$

		$^P R_{J,\|K\|}$ $^P P_{J,\|K\|}$	k: a	b	c
$\frac{E_{v'} - E_{v''}}{hc}$			0	2	4
$+K^2[(B^{zz}_{v'} - B^{zz}_{v''}) - (B^{xx}_{v'} - B^{xx}_{v''})]$			–	4	
$\Sigma \Delta v^{(B)}_n$ even		$+B^{zz}_{v'} - B^{xx}_{v'} - 2(\zeta_t B^{zz})_{v'}$	2	4	
		$-2\|K\|[(B^{zz}_{v'} - B^{xx}_{v'}) - (\zeta_t B^{zz})_{v'}]$	–	4	
	$\|K\| \neq 2$	$+m(B^{xx}_{v'} + B^{xx}_{v''})$	2	4	
		$+m^2(B^{xx}_{v'} - B^{xx}_{v''})$	–	4	
	$\|K\| = 2$	$+m[(B^{xx}_{v'} + B^{xx}_{v''})$	2	4	
		$\pm 2G^t]$	4		
		$+m^2[(B^{xx}_{v'} - B^{xx}_{v''})$	–	4	
		$\pm 2G^t]$	4		

$\sum \Delta v_n^{(B)}$ odd		$+B_{v'}^{zz} - B_{v'}^{xx} + 2(\zeta_t B^{zz})_{v'}$	2	4	
		$-2\lvert K\rvert\left[(B_{v'}^{zz} - B_{v'}^{xx}) + (\zeta_t B^{zz})_{v'}\right]$	–	4	
	$\lvert K\rvert \neq 2$	$+m(B_{v'}^{xx} + B_{v''}^{xx})$	2	4	
		$+m^2(B_{v'}^{xx} - B_{v''}^{xx})$	–	4	
	$\lvert K\rvert = 2$	$+m\left[(B_{v'}^{xx} + B_{v''}^{xx})\right.$	2	4	
		$\left.\pm 2F^t\right]$	4		
		$+m^2\left[(B_{v'}^{xx} - B_{v''}^{xx})\right.$	–	4	
		$\left.\pm 2F^t\right]$	4		

TABLE LIX, 2 Bα (R). Transition from v'' ($v_t = v_{t'} = \ldots = 0$; v_n, $v_{n'} \ldots$ arbitrary) to v' ($v_t = 1, |\ell_t| = 1$; $v_t{}' = \ldots = 0$; $v_n, v_{n'} \ldots$ arbitrary) ; $J \simeq K \simeq 30$

	${}^{R}Q_{J,\|K\|}$	k: a	b	c
	$\frac{E_{v'} - E_{v''}}{hc}$	0	2	4
	$- J^2(J+1)^2(D^J_{v'} - D^J_{v''})$	–	4	
	$- J(J+1)\, K^2 (D^{JK}_{v'} - D^{JK}_{v''})$	–	4	
	$- K^4 (D^K_{v'} - D^K_{v''})$	–	4	
$\Sigma \Delta v_n^{(B)}$ even	$+ B^{zz}_{v'} - B^{xx}_{v'} - 2(\zeta_t B^{zz})_{v'}$	2	4	
	$+ 2\,\|K\| [B^{zz}_{v'} - B^{xx}_{v'} - (\zeta_t B^{zz})_{v'}]$	1	3	
	$+ J(J+1) [B^{xx}_{v'} - B^{xx}_{v''}$	–	2	4
	$- D^{JK}_e + \eta^J_t]$	4		
	$+ K^2 [(B^{zz}_{v'} - B^{zz}_{v''}) - (B^{xx}_{v'} - B^{xx}_{v''})$	–	2	4
	$- 6 D^K_e + 3 \eta^K_t]$	4		

	$-J(J+1)\lvert K\rvert(2D_e^{JK}-\eta_t^J)$	3		
	$-\lvert K\rvert^3(4D_e^K-\eta_t^K)$	3		
$\Sigma\Delta v_n^{(B)}$ odd	$+B_{v'}^{zz}-B_{v'}^{xx}+2(\zeta_t B^{zz})_{v'}$	2	4	
	$+2\lvert K\rvert\left[B_{v'}^{zz}-B_{v'}^{xx}+(\zeta_t B^{zz})_{v'}\right]$	1	3	
	$+J(J+1)\left[B_{v'}^{xx}-B_{v''}^{xx}\right.$	–	2	4
	$\left.-D_e^{JK}-\eta_t^J\right]$	4		
	$+K^2\left[(B_{v'}^{zz}-B_{v''}^{zz})-(B_{v'}^{xx}-B_{v''}^{xx})\right.$	–	2	4
	$\left.-6D_e^K-3\eta_t^K\right]$	4		
	$-J(J+1)\lvert K\rvert(2D_e^{JK}+\eta_t^J)$	3		
	$-\lvert K\rvert^3(4D_e^K+\eta_t^K)$	3		

TABLE LIX, 2 Bα (P). Transition from v'' ($v_t = v_{t'} = \ldots = 0$; v_n, $v_{n'}$ ... arbitrary) to v' ($v_t = 1$, $|\ell_t| = 1$; $v_{t'} = \ldots = 0$; v_n, $v_{n'}$ ... arbitrary) ; $J \simeq K \simeq 30$

	${}^{P}Q_{J,|K|}$	k: a	b	c
	$\dfrac{E_{v'} - E_{v''}}{hc}$	0	2	4
	$-J^2(J+1)^2(D^J_{v'} + D^J_{v''})$	–	4	
	$-J(J+1)\ K^2(D^{JK}_{v'} - D^{JK}_{v''})$	–	4	
	$-K^4(D^K_{v'} - D^K_{v''})$	–	4	
	$+B^{zz}_{v'} - B^{xx}_{v'} - 2(\zeta_t B^{zz})_{v'}$	2	4	
	$-2|K|\left[B^{zz}_{v'} - B^{xx}_{v'} - (\zeta_t B^{zz})_{v'}\right]$	1	3	
	$+J(J+1)\left[B^{xx}_{v'} - B^{xx}_{v''}\right.$	–	2	4
$\Sigma \Delta v_n^{(B)}$	$\left. - D^{JK}_e + \eta^J_t\right]$	4		
even	$+K^2\left[(B^{zz}_{v'} - B^{zz}_{v''}) - (B^{xx}_{v'} - B^{xx}_{v''})\right.$	–	2	4
	$\left. - 6D^K_e + 3\eta^K_t\right]$	4		

	$+J(J+1)\|K\|(2D_e^{JK}-\eta_t^J)$	3		
	$+\|K\|^3(4D_e^K-\eta_t^K)$	3		
$\Sigma\Delta v_n^{(B)}$ odd	$+B_{v'}^{zz}-B_{v'}^{xx}+2(\zeta_t B^{zz})_{v'}$	2	4	
	$-2\|K\|\left[B_{v'}^{zz}-B_{v'}^{xx}+(\zeta_t B^{zz})_{v'}\right]$	1	3	
	$+J(J+1)\left[B_{v'}^{xx}-B_{v''}^{xx}\right.$	–	2	4
	$\left.-D_e^{JK}-\eta_t^J\right]$	4		
	$+K^2\left[(B_{v'}^{zz}-B_{v''}^{zz})-(B_{v'}^{xx}-B_{v''}^{xx})\right.$	–	2	4
	$\left.-6D_e^K-3\eta_t^K\right]$	4		
	$+J(J+1)\|K\|(2D_e^{JK}+\eta_t^J)$	3		
	$+\|K\|^3(4D_e^K+\eta_t^K)$	3		

TABLE LIX, $2B\beta(R)$. Transition from v'' ($v_t = v_{t'} = \ldots = 0$; $v_n, v_{n'} \ldots$ arbitrary) to v' ($v_t = 1, |\ell_t| = 1$; $v_{t'} = \ldots = 0$; $v_n, v_{n'} \ldots$ arbitrary); $J \simeq K \simeq 30$

	${}^R R_{J,|K|}$ ${}^R P_{J,|K|}$	k: a	b	c
	$\dfrac{E_{v'} - E_{v''}}{hc}$	0	2	4
	$+m(B^{xx}_{v'} + B^{xx}_{v''})$	1	3	
	$-4m^3 D^J_e - 2m K^2 D^{JK}_e$	3		
	$-m^4(D^J_{v'} - D^J_{v''})$	–	4	
	$-m^2 K^2(D^{JK}_{v'} - D^{JK}_{v''})$	–	4	
	$-K^4(D^K_{v'} - D^K_{v''})$	–	4	
$\Sigma \Delta v_n^{(B)}$ even	$+2|K|\left[B^{zz}_{v'} - B^{xx}_{v'} - (\zeta_t B^{zz})_{v'}\right]$	1	3	
	$+B^{zz}_{v'} - B^{xx}_{v'} - 2(\zeta_t B^{zz})_{v'}$	2	4	
	$+m^2\left[B^{xx}_{v'} - B^{xx}_{v''}\right.$	–	2	4
	$\left.- D^{JK}_e + \eta^J_t\right]$	4		

	$-m\|K\|(2D_e^{JK} - \eta_t^J)$	4		
	$+K^2[(B_{v'}^{zz} - B_{v''}^{zz}) - (B_{v'}^{xx} - B_{v''}^{xx})$	-	2	4
	$-6D_e^K + 3\eta_t^K]$	4		
	$-m^2\|K\|(2D_e^{JK} - \eta_t^J) - \|K\|^3(4D_e^K - \eta_t^K)$	3		
$\Sigma\Delta v_n^{(B)}$ odd	$+2\|K\|[B_{v'}^{zz} - B_{v'}^{xx} + (\zeta_t B^{zz})_{v'}]$	1	3	
	$+B_{v'}^{zz} - B_{v'}^{xx} + 2(\zeta_t B^{zz})_{v'}$	2	4	
	$+m^2[B_{v'}^{xx} - B_{v''}^{xx}$	-	2	4
	$-D_e^{JK} - \eta_t^J]$	4		
	$-m\|K\|(2D_e^{JK} + \eta_t^J)$	4		
	$+K^2[(B_{v'}^{zz} - B_{v''}^{zz}) - (B_{v'}^{xx} - B_{v''}^{xx})$	-	2	4
	$-6D_e^K - 3\eta_t^K]$	4		
	$-m^2\|K\|(2D_e^{JK} + \eta_t^J) - \|K\|^3[4D_e^K + \eta_t^K]$	3		

TABLE LIX, $2B\beta(P)$. Transition from $v''(v_t = v_{t'} = \ldots = 0\,;\, v_n, v_{n'} \ldots$ arbitrary) to $v'(v_t = 1, |\ell_t| = 1\,;\, v_{t'} = \ldots = 0\,;\, v_n, v_{n'} \ldots$ arbitrary) ; $J \simeq K \simeq 30$

	${}^{P}R_{J,|K|}$ ${}^{P}P_{J,|K|}$	k: a	b	c
	$\frac{E_{v'} - E_{v''}}{hc}$	0	2	4
	$+ m\,(B^{xx}_{v'} + B^{xx}_{v''})$	1	3	
	$- 4m^3 D^J_e - 2m\,K^2 D^{JK}_e$	3		
	$- m^4 (D^J_{v'} - D^J_{v''})$	–	4	
	$- m^2 K^2 (D^{JK}_{v'} - D^{JK}_{v''})$	–	4	
	$- K^4 (D^K_{v'} - D^K_{v''})$	–	4	
$\Sigma \Delta v^{(B)}_n$ even	$- 2\,|K|\left[B^{zz}_{v'} - B^{xx}_{v'} - (\zeta_t B^{zz})_{v'}\right]$	1	3	
	$+ B^{zz}_{v'} - B^{xx}_{v'} - 2\,(\zeta_t B^{zz})_{v'}$	2	4	
	$+ m^2 \left[B^{xx}_{v'} - B^{xx}_{v''}\right.$	–	2	4
	$\left. - D^{JK}_e + \eta^J_t\right]$	4		

	$+m\,\lvert K\rvert(2D_e^{JK}-\eta_t^J)$	4		
	$+K^2\Big[(B_{v'}^{zz}-B_{v''}^{zz})-(B_{v'}^{xx}-B_{v''}^{xx})$	–	2	4
	$-6D_e^K+3\eta_t^K\Big]$	4		
	$+m^2\lvert K\rvert(2D_e^{JK}-\eta_t^J)+\lvert K\rvert^3(4D_e^K-\eta_t^K)$	3		
$\Sigma\Delta v_n^{(B)}$ odd	$-2\lvert K\rvert\Big[B_{v'}^{zz}-B_{v'}^{xx}+(\zeta_t B^{zz})_{v'}\Big]$	1	3	
	$+B_{v'}^{zz}-B_{v'}^{xx}+2(\zeta_t B^{zz})_{v'}$	2	4	
	$+m^2\Big[B_{v'}^{xx}-B_{v''}^{xx}$	–	2	4
	$-D_e^{JK}-\eta_t^J\Big]$	4		
	$+m\,\lvert K\rvert(2D_e^{JK}+\eta_t^J)$	4		
	$+K^2\Big[(B_{v'}^{zz}-B_{v''}^{zz})-(B_{v'}^{xx}-B_{v''}^{xx})$	–	2	4
	$-6D_e^K-3\eta_t^K\Big]$	4		
	$+m^2\lvert K\rvert(2D_e^{JK}+\eta_t^J)+\lvert K\rvert^3(4D_e^K+\eta_t^K)$	3		

TABLE LIX, 3 Aα. Transition from $v''(v_t = v_{t'} = \dots = 0)$ to $v'(v_t = 2, \ell_t = 0; v_{t'} = \dots = 0)$; $J \simeq K \simeq 1$

${}^{Q}Q_{J,\|K\|}$	k: a	b	c
$\dfrac{E_{v'} - E_{v''}}{hc}$	0	2	4
$+ J(J+1)(B^{xx}_{v'} - B^{xx}_{v''})$	–	4	
$+ K^2[(B^{zz}_{v'} - B^{zz}_{v''}) - (B^{xx}_{v'} - B^{xx}_{v''})]$	–	4	

TABLE LIX, 3 Aβ. Transition from $v''(v_t = v_{t'} = \dots = 0)$ to $v'(v_t = 2, \ell_t = 0; v_{t'} = \dots = 0)$; $J \simeq K \simeq 1$

${}^{Q}R_{J,\|K\|}$ ${}^{Q}P_{J,\|K\|}$	k: a	b	c
$\dfrac{E_{v'} - E_{v''}}{hc}$	0	2	4
$+ m(B^{xx}_{v'} + B^{xx}_{v''})$	2	4	
$+ m^2(B^{xx}_{v'} - B^{xx}_{v''})$	–	4	
$+ K^2[(B^{zz}_{v'} - B^{zz}_{v''}) - (B^{xx}_{v'} - B^{xx}_{v''})]$	–	4	

TABLE LIX, 3 Bα. Transition from $v''(v_t = v_{t'} = \dots = 0)$ to $v'(v_t = 2, \ell_t = 0 ; v_{t'} = \dots = 0)$; $J \simeq K \simeq 30$

${}^{Q}Q_{J,\,|K|}$	k: a	b	c
$\frac{E_{v'} - E_{v''}}{hc}$	0	2	4
$+ J(J+1)(B^{xx}_{v'} - B^{xx}_{v''})$	–	2	4
$+ K^2\left[(B^{zz}_{v'} - B^{zz}_{v''}) - (B^{xx}_{v'} - B^{xx}_{v''})\right]$	–	2	4
$- J^2(J+1)^2(D^J_{v'} - D^J_{v''})$	–	4	
$- J(J+1)K^2(D^{JK}_{v'} - D^{JK}_{v''})$	–	4	
$- K^4(D^K_{v'} - D^K_{v''})$	–	4	

TABLE LIX, 3 Bβ. Transition from $v''(v_t = v_{t'} = \dots = 0)$ to $v'(v_t = 2, \ell_t = 0 ; v_{t'} = \dots = 0)$; $J \simeq K \simeq 30$

${}^{Q}R_{J,\|K\|}$ ${}^{Q}P_{J,\|K\|}$	k: a	b	c
$\dfrac{E_{v'} - E_{v''}}{hc}$	0	2	4
$+ m(B^{xx}_{v'} + B^{xx}_{v''})$	1	3	
$+ m^2(B^{xx}_{v'} - B^{xx}_{v''})$	–	2	4
$+ K^2\left[(B^{zz}_{v'} - B^{zz}_{v''}) - (B^{xx}_{v'} - B^{xx}_{v''})\right]$	–	2	4
$- 4m^3 D^J_e$	3		
$- 2mK^2 D^{JK}_e$	3		
$- m^4(D^J_{v'} - D^J_{v''})$	–	4	
$- m^2K^2(D^{JK}_{v'} - D^{JK}_{v''})$	–	4	
$- K^4(D^K_{v'} - D^K_{v''})$	–	4	

TABLE LIX, 4Aα. Transition from v'' ($v_t = v_{t'} = \dots = 0$) to v' ($v_t = 2$, $|\ell_t| = 2$; $v_{t'} = \dots = 0$); $J \simeq K \simeq 1$

${}^{Q}Q^{\pm}_{J,\,|K|}$	k: a	b	c
$\dfrac{E_{v'} - E_{v''}}{hc}$	0	2	4
$+ J(J+1)(B^{xx}_{v'} - B^{xx}_{v''})$	-	4	
$+ K^2[(B^{zz}_{v'} - B^{zz}_{v''}) - (B^{xx}_{v'} - B^{xx}_{v''})]$	-	4	
$\pm 4\sqrt{(\zeta^z_t B^{zz}_e)^2 K^2 + 4(U^t)^2}$	2		
$\mp \dfrac{2\sum_s [K^2 \zeta^z_t B^{zz}_e \eta_{ts} - 8U^t g^{t,s}_{4}](v_s + \frac{d_s}{2})}{[(\zeta^z_t B^{zz}_e)^2 K^2 + 4(U^t)^2]^{\frac{1}{2}}}$	4		

TABLE LIX, $4A\beta$. Transition from $v''(v_t = v_{t'} = \dots = 0)$ to $v'(v_t = 2, |l_t| = 2; v_{t'} = \dots = 0)$; $J \simeq K \simeq 1$

$J \simeq K \simeq 1$

	${}^{Q}R^{\pm}_{J,\|K\|}$ ${}^{Q}P^{\pm}_{J,\|K\|}$		k: a	b	c
$\frac{E_{v'} - E_{v''}}{hc}$			0	2	4
$+m(B^{xx}_{v'} + B^{xx}_{v''})$			2	4	
$+m^2(B^{xx}_{v'} - B^{xx}_{v''})$			–	4	
$+K^2[(B^{zz}_{v'} - B^{zz}_{v''}) - (B^{xx}_{v'} - B^{xx}_{v''})]$			–	4	
$\|K\| \neq 0$	$\pm 4[(\zeta^z_t B^{zz}_e)^2 K^2 + 4(U^t)^2]^{\frac{1}{2}}$		2		
	$\mp\left\{2\sum_s\left[K^2 \zeta^z_t B^{zz}_e \eta_{ts} - 8U^t g^{t,s}_{44}\right]\left(v_s + \frac{d_s}{2}\right)\right\}\left[(\zeta^z_t B^{zz}_e)^2 K^2 + 4(U^t)^2\right]^{-\frac{1}{2}}$		4		
$K = 0$	$\sum\Delta v^{(2)}_n$ even (C_{4v} group) or $\sum\Delta v^{(2)}_n$ odd (D_{2d} group) ($\psi^+ \longrightarrow \psi^+$ transition)	$+8U^t_{eff}$	2	4	
	$\sum\Delta v^{(2)}_n$ odd (C_{4v} group) or $\sum\Delta v^{(2)}_n$ even (D_{2d} group) ($\psi^+ \longrightarrow \psi^-$ transition)	$-8U^t_{eff}$	2	4	

TABLE LIX, 4 Bα. Transition from $v''(v_t = v_{t'} = \dots = 0)$ to $v'(v_t = 2, |\ell_t| = 2;\ v_{t'} = \dots = 0)$; $J \simeq K \simeq 30$

${}^{Q}Q_{J,\,K\ell_t}$	k: a	b	c
$\dfrac{E_{v'} - E_{v''}}{hc}$	0	2	4
$-2(\zeta_t B^{zz})_{v'} K\ell_t$	1	3	
$+J(J+1)(B^{xx}_{v'} - B^{xx}_{v''})$	–	2	4
$+\frac{1}{4}(K\ell_t)^2[(B^{zz}_{v'} - B^{zz}_{v''}) - (B^{xx}_{v'} - B^{xx}_{v''})]$	–	2	4
$+J(J+1)K\ell_t\,\eta^J_t$	3		
$+\frac{1}{4}(K\ell_t)^3\,\eta^K_t$	3		
$-16\{(U^t)^2/\zeta^z_t B^{zz}_e K\ell_t\}$	3		
$-J^2(J+1)^2(D^J_{v'} - D^J_{v''})$	–	4	
$-\frac{1}{4}J(J+1)(K\ell_t)^2(D^{JK}_{v'} - D^{JK}_{v''})$	–	4	
$-\frac{1}{16}(K\ell_t)^4(D^K_{v'} - D^K_{v''})$	–	4	

TABLE LIX, 4 Bβ. Transition from $v''(v_t = v_{t'} = \dots = 0)$ to $v'(v_t = 2, |\ell_t| = 2; v_{t'} = \dots = 0)$; $J \simeq K \simeq 30$

${}^{Q}R_{J,K\ell_t}$ ${}^{Q}P_{J,K\ell_t}$	k		
	a	b	c
$\dfrac{E_{v'} - E_{v''}}{hc}$	0	2	4
$+ m(B^{xx}_{v'} + B^{xx}_{v''})$	1	3	
$- 2(\zeta_t B^{zz})_{v'} K\ell_t$	1	3	
$+ m^2(B^{xx}_{v'} - B^{xx}_{v''})$	–	2	4
$+ \frac{1}{4}(K\ell_t)^2[(B^{zz}_{v'} - B^{zz}_{v''}) - (B^{xx}_{v'} - B^{xx}_{v''})]$	–	2	4
$- 4m^3 D^J_e + m^2 K\ell_t \eta^J_t$	3		
$- \frac{1}{2} m (K\ell_t)^2 D^{JK}_e + \frac{1}{4}(K\ell_t)^3 \eta^K_t$	3		
$- 16\{(U^t)^2/\zeta^z_t B^{zz}_e K\ell_t\}$	3		
$+ m K\ell_t \eta^J_t$	4		
$- m^4(D^J_{v'} - D^J_{v''})$	4		
$- \frac{1}{4} m^2 (K\ell_t)^2 (D^{JK}_{v'} - D^{JK}_{v''})$	4		
$- \frac{1}{16}(K\ell_t)^4 (D^K_{v'} - D^K_{v''})$	4		

TABLE LIX, 5 Aα. Transition from $v''(v_t = v_{t'} = \dots = 0)$ to $v'(v_t = v_{t'} = 1, \ell_t + \ell_{t'} = 0); v_{t''} = \dots = 0)$; $J \simeq K \simeq 1$

${}^{Q}Q^{\pm}_{J,\,|K|}$	k a	b	c
$\frac{E_{v'} - E_{v''}}{hc}$	0	2	4
$+ J(J+1)(B^{xx}_{v'} - B^{xx}_{v''})$	–	4	
$+ K^2\left[(B^{zz}_{v'} - B^{zz}_{v''}) - (B^{xx}_{v'} - B^{xx}_{v''})\right]$	–	4	
$\pm \sqrt{4K^2(\zeta^z_t - \zeta^z_{t'})^2 (B^{zz}_e)^2 + (r^{tt'})^2}$	2		
$\mp \dfrac{2\sum_s \left[K^2(\zeta^z_t - \zeta^z_{t'}) B^{zz}_e (\eta_{ts} - \eta_{t's}) - 2r^{tt'} g^{tt',s}_{22-}\right](v_s + \frac{d_s}{2})}{\left[4K^2(\zeta^z_t - \zeta^z_{t'})^2 (B^{zz}_e)^2 + (r^{tt'})^2\right]^{\frac{1}{2}}}$	4		

TABLE LIX, 5 Aβ. Transition from $v''(v_t = v_{t'} = \dots = 0)$ to $v'(v_t = v_{t'} = 1,\ \ell_t + \ell_{t'} = 0);\ v_{t''} = \dots = 0)$; $J \simeq K \simeq 1$

	$Q_{R^{\pm}_{J,\lvert K\rvert}}$ $Q_{P^{\pm}_{J,\lvert K\rvert}}$		k: a	b	c
	$\dfrac{E_{v'} - E_{v''}}{hc}$		0	2	4
	$+ m(B^{xx}_{v'} + B^{xx}_{v''})$		2	4	
	$+ m^2(B^{xx}_{v'} - B^{xx}_{v''})$		–	4	
	$+ K^2\left[(B^{zz}_{v'} - B^{zz}_{v''}) - (B^{xx}_{v'} - B^{xx}_{v''})\right]$		–	4	
$\lvert K\rvert \neq 0$	$\pm\left\{4K^2(\zeta^z_t - \zeta^z_{t'})^2(B^{zz}_e)^2 + (r^{tt'})^2\right\}^{\frac{1}{2}}$		2		
	$\mp 2\left\{\sum_s\left[K^2(\zeta^z_t - \zeta^z_{t'})B^{zz}_e(\eta_{ts} - \eta_{t's}) - 2r^{tt'}g^{tt',s}_{22-}\right](v_s + \frac{d_s}{2})\right\}\left[4K^2(\zeta^z_t - \zeta^z_{t'})^2(B^{zz}_e)^2 + (r^{tt'})^2\right]^{-\frac{1}{2}}$		4		
$K = 0$	$\Sigma\Delta v^{(2)}_n$ even (C_{4v} group) or $\Sigma\Delta v^{(2)}_n$ odd (D_{2d} group) ($\psi^+ \longrightarrow \psi^+$ transition)	$+ r^{tt'}_{eff}$	2	4	
	$\Sigma\Delta v^{(2)}_n$ odd (C_{4v} group) or $\Sigma\Delta v^{(2)}_n$ even (D_{2d} group) ($\psi^+ \longrightarrow \psi^-$ transition)	$- r^{tt'}_{eff}$	2	4	

TABLE LIX, 5 Bα. Transition from $v''(v_t = v_{t'} = \dots = 0)$ to $v'(v_t = v_{t'} = 1$, $\ell_t + \ell_{t'} = 0$; $v_{t''} = \dots = 0)$; $J \simeq K \simeq 30$

${}^{Q}Q_{J,\ K\ell_t}$	k: a	b	c
$\dfrac{E_{v'} - E_{v''}}{hc}$	0	2	4
$-2\left[(\zeta_t B^{zz})_{v'} - (\zeta_{t'} B^{zz})_{v'}\right] K\ell_t$	1	3	
$+ J(J+1)(B^{xx}_{v'} - B^{xx}_{v''})$	-	2	4
$+ (K\ell_t)^2\left[(B^{zz}_{v'} - B^{zz}_{v''}) - (B^{xx}_{v'} - B^{xx}_{v''})\right]$	-	2	4
$+ J(J+1)(\eta^J_t - \eta^J_{t'}) K\ell_t$	3		
$+ (\eta^K_t - \eta^K_{t'})(K\ell_t)^3$	3		
$-\left\{(r^{tt'})^2/4 B^{zz}_e (\zeta^z_t - \zeta^z_{t'}) K\ell_t\right\}$	3		
$- J^2(J+1)^2(D^J_{v'} - D^J_{v''})$	-	4	
$- J(J+1)(K\ell_t)^2(D^{JK}_{v'} - D^{JK}_{v''})$	-	4	
$- (K\ell_t)^4(D^K_{v'} - D^K_{v''})$	-	4	

The contributions of this table take into account the relation $\ell_{t'} = -\ell_t$

TABLE LIX, 5 Bβ. Transition from $v''(v_t = v_{t'} = \dots = 0)$ to $v'(v_t = v_{t'} = 1,$ $\ell_t + \ell_{t'} = 0 ;\ v_{t''} = \dots = 0)$; $J \simeq K \simeq 30$

$^{Q}R_{J,\,K\ell_t}$ $\quad$ $^{Q}P_{J,\,K\ell_t}$	k a	k b	k c
$\dfrac{E_{v'} - E_{v''}}{hc}$	0	2	4
$-2\left[(\zeta_t B^{zz})_{v'} - (\zeta_{t'} B^{zz})_{v'}\right] K\ell_t$	1	3	
$+ m\,(B^{xx}_{v'} + B^{xx}_{v''})$	1	3	
$+ m^2 (B^{xx}_{v'} - B^{xx}_{v''})$	-	2	4
$+ (K\ell_t)^2 \left[(B^{zz}_{v'} - B^{xx}_{v'}) - (B^{zz}_{v''} - B^{xx}_{v''})\right]$	-	2	4
$-4\,m^3\,D^J_e$	3		
$-2m\,(K\ell_t)^2\,D^{JK}_e$	3		
$+ m^2 (\eta^J_t - \eta^J_{t'})\,K\ell_t$	3		
$+ (K\ell_t)^3 (\eta^K_t - \eta^K_{t'})$	3		
$-\left\{(r^{tt'})^2 / 4B^{zz}_e (\zeta^z_t - \zeta^z_{t'})\,K\ell_t\right\}$	3		
$+ m\,K\ell_t\,(\eta^J_t - \eta^J_{t'})$	4		
$- m^4 (D^J_{v'} - D^J_{v''})$	-	4	
$- m^2 (K\ell_t)^2 (D^{JK}_{v'} - D^{JK}_{v''})$	-	4	
$-(K\ell_t)^4 (D^K_{v'} - D^K_{v''})$	-	4	

The contributions of this table take into account the relation $\ell_{t'} = -\ell_t$

TABLE LIX, 6 Aα. Transition from $v''(v_t = v_{t'} = \dots = 0)$ to $v'(v_t = v_{t'} = 1, |\ell_t + \ell_{t'}| = 2; v_{t''} = \dots = 0)$; $J \simeq K \simeq 1$

${}^{Q}Q^{\pm}_{\lvert K \rvert}$	k: a	b	c
$\frac{E_{v'} - E_{v''}}{hc}$	0	2	4
$+J(J+1)(B^{xx}_{v'} - B^{xx}_{v''})$	–	4	
$+K^2\left[(B^{zz}_{v'} - B^{zz}_{v''}) - (B^{xx}_{v'} - B^{xx}_{v''})\right]$	–	4	
$\pm\sqrt{4K^2(\zeta^z_t + \zeta^z_{t'})^2 (B^{zz}_e)^2 + (S^{tt'})^2}$	2		
$\mp \frac{2\sum_s \left[K^2(\zeta^z_t + \zeta^z_{t'}) B^{zz}_e (\eta_{ts} + \eta_{t's}) - 2S^{tt'} g^{tt',s}_{22+}\right](v_s + \frac{d_s}{2})}{\left[4K^2(\zeta^z_t + \zeta^z_{t'})^2 (B^{zz}_e)^2 + (S^{tt'})^2\right]^{\frac{1}{2}}}$	4		

TABLE LIX, 6 Aβ. Transition from $v''(v_t = v_{t'} = \dots = 0)$ to $v'(v_t = v_{t'} = 1, |\ell_t + \ell_{t'}| = 2; v_{t''} = \dots = 0)$; $J \simeq K \simeq 1$

	$Q_{R^{\pm}_{J,\|K\|}}$ $Q_{P^{\pm}_{J,\|K\|}}$		k: a	b	c
	$\frac{E_{v'} - E_{v''}}{hc}$		0	2	4
	$+ m(B^{xx}_{v'} + B^{xx}_{v''})$		2	4	
	$+ m^2(B^{xx}_{v'} - B^{xx}_{v''})$		–	4	
	$+ K^2[(B^{zz}_{v'} - B^{zz}_{v''}) - (B^{xx}_{v'} - B^{xx}_{v''})]$		–	4	
$\|K\| \neq 0$	$\pm\left\{4K^2(\zeta^z_t + \zeta^z_{t'})^2(B^{zz}_e)^2 + (S^{tt'})^2\right\}^{\frac{1}{2}}$		2		
	$\mp 2\left\{\sum_s\left[K^2(\zeta^z_t + \zeta^z_{t'})B^{zz}_e(\eta_{ts} + \eta_{t's}) - 2S^{tt'} g^{tt',s}_{22+}\right](v_s + \frac{d_s}{2})\right\}\left[4K^2(\zeta^z_t + \zeta^z_{t'})^2(B^{zz}_e)^2 + (S^{tt'})^2\right]^{-\frac{1}{2}}$		4		
$K = 0$	$\Sigma\Delta v^{(2)}_n$ even (C_{4v} group) or $\Sigma\Delta v^{(2)}_n$ odd (D_{2d} group) ($\psi^+ \longrightarrow \psi^+$ transition)	$+S^{tt'}_{eff}$	2	4	
	$\Sigma\Delta v^{(2)}_n$ odd (C_{4v} group) or $\Sigma\Delta v^{(2)}_n$ even (D_{2d} group) ($\psi^+ \longrightarrow \psi^-$ transition)	$-S^{tt'}_{eff}$	2	4	

TABLE LIX, 6 Bα. Transition from $v''(v_t = v_{t'} = \ldots = 0)$ to $v'(v_t = v_{t'} = 1$, $|\ell_t + \ell_{t'}| = 2$; $v_{t''} = \ldots = 0)$; $J \simeq K \simeq 30$

${}^{Q}Q_{J,\,K\ell_t}$	k: a	b	c
$\frac{E_{v'} - E_{v''}}{hc}$	0	2	4
$+ J(J+1)(B^{xx}_{v'} - B^{xx}_{v''})$	–	2	4
$+ (K\ell_t)^2[(B^{zz}_{v'} - B^{zz}_{v''}) - (B^{xx}_{v'} - B^{xx}_{v''})]$	–	2	4
$- 2[(\zeta_t B^{zz})_{v'} + (\zeta_{t'} B^{zz})_{v'}]K\ell_t$	1	3	
$- J^2(J+1)^2(D^J_{v'} - D^J_{v''})$	–	4	
$- J(J+1)(K\ell_t)^2(D^{JK}_{v'} - D^{JK}_{v''})$	–	4	
$- (K\ell_t)^4(D^K_{v'} - D^K_{v''})$	–	4	
$-\{(s^{tt'})^2/4B^{zz}_e(\zeta^z_t + \zeta^z_{t'})K\ell_t\}$	3		
$+ J(J+1)K\ell_t(\eta^J_t + \eta^J_{t'})$	3		
$+ (K\ell_t)^3(\eta^K_t + \eta^K_{t'})$	3		

The contributions of this table take into account the relation $\ell_{t'} = \ell_t$

TABLE LIX, 6 Bβ. Transition from $v''(v_t = v_{t'} = \dots = 0)$ to $v'(v_t = v_{t'} = 1$, $|\ell_t + \ell_{t'}| = 2$; $v_{t''} = \dots = 0)$; $J \simeq K \simeq 30$

$^QR_{J,K\ell_t}$ $^QP_{J,K\ell_t}$	k a	k b	k c
$\frac{E_{v'} - E_{v''}}{hc}$	0	2	4
$+ m(B^{xx}_{v'} + B^{xx}_{v''})$	1	3	
$+ m^2(B^{xx}_{v'} - B^{xx}_{v''})$	-	2	4
$+(K\ell_t)^2[(B^{zz}_{v'} - B^{zz}_{v''}) - (B^{xx}_{v'} - B^{xx}_{v''})]$	-	2	4
$- 2[(\zeta_t B^{zz})_{v'} + (\zeta_{t'} B^{zz})_{v'}] K\ell_t$	1	3	
$- 4 m^3 D^J_e$	3		
$- 2m(K\ell_t)^2 D^{JK}_e$	3		
$+ m^2(\eta^J_t + \eta^J_{t'})K\ell_t$	3		
$+(\eta^K_t + \eta^K_{t'})(K\ell_t)^3$	3		
$- \{(s^{tt'})^2/4B^{zz}_e(\zeta^z_t + \zeta^z_{t'})K\ell_t\}$	3		
$+ mK\ell_t(\eta^J_t + \eta^J_{t'})$	4		
$- m^4(D^J_{v'} - D^J_{v''})$	4		
$- m^2(K\ell_t)^2(D^{JK}_{v'} - D^{JK}_{v''})$	4		
$-(K\ell_t)^4(D^K_{v'} - D^K_{v''})$	4		

The contributions of this table take into account the relation $\ell_{t'} = \ell_t$

CHAPTER XII

LINEAR MOLECULES

INTRODUCTION

The study of linear molecules is far from being an original subject Many authors have carried out high order calculations on individual molecules and given results with a great accuracy. However, no systematic work, applicable to any linear molecule has been published at this time ; it has moreover seemed convenient in this Monograph to give, for linear molecules, general formulae for frequencies of transitions, as done precedently for symmetric tops. This is all the more justified in that linear molecules may be considered as a limiting case of axially symmetric molecules, using the Sayvetz relation [27]

$$K = \sum_t \ell_t$$

Consequently, we have used the general method described in the preceding chapters to compute energy levels, then selection rules, and finally line frequencies for transitions in the rotation and rotation-vibration spectra, for $C_{\infty v}$ and $D_{\infty h}$ molecules. The formulae obtained hold for transitions between levels which are not perturbed by accidental resonances.

ROTATION-VIBRATION ENERGIES FOR MOLECULES BELONGING TO SYMMETRY GROUPS $C_{\infty v}$ AND $D_{\infty h}$

We have calculated the energies of rotation-vibration in the following vibrational states :

1 $v_t = v_{t'} = \dots = 0$

2 $v_t = 1$; $\ell_t = \pm 1$

3 $v_t = 2$; $\ell_t = 0 , \pm 2$

4 $v_t = 3$; $\ell_t = \pm 1 , \pm 3$

(cases 2–4: $v_{t'} = \dots = 0$)

5 $v_t = v_{t'} = 1$ $\begin{cases} \ell_t = -\ell_{t'} = \pm 1 \\ \ell_t = \ell_{t'} = \pm 1 \end{cases}$

(cases 1–5: $v_n , v_{n'} \dots$ arbitrary)

For linear molecules, only two cases are to be considered for values of rotational quantum numbers :

$$A \qquad J \simeq 1$$

$$C \qquad J \simeq 30$$

because the condition $K = \sum_t \ell_t$ implies $K \simeq 1$ in vibrational states generally observed. In the case C , we have included in energy contributions up to fourth order ; in case A , we have included up to sixth order contributions originating from rotational operators. The diagonal matrix elements from $h_0^\dagger + h_1^\dagger + h_2^\dagger + h_3^\dagger$ are well known and are recalled in Table XXXVIII of Chapter VIII. The diagonal contributions of $h_4^\dagger$ for linear molecules have been derived from contributions of $h_4^\dagger$ for symmetric tops.

For linear molecules, only two kinds of nondiagonal matrix elements exist. With all contributions coming from $h_0^\dagger + \dots + h_4^\dagger$, we have written them[1] :

$$(\ell_t \dots | h^\dagger | \ \ell_t \pm 2 \dots) = hc\left[q_t^0 + f_{22}^{tJ} J(J+1) + \sum_s f_{22}^{t,s} \left(v_s + \frac{d_s}{2}\right)\right]$$

$$\times \left\{\left[J(J+1) - \ell_t(\ell_t \pm 1)\right]\left[J(J+1) - (\ell_t \pm 1)(\ell_t \pm 2)\right](v_t \pm \ell_t + 2)(v_t \mp \ell_t)\right\}^{\frac{1}{2}}$$

and

$$(\ell_t \ell_{t'} \ldots | h^{\dagger} | \ell_t \pm 2, \ell_{t'} \mp 2 \ldots) =$$

$$hc\left[\frac{r^{0}_{tt'}}{4} + g_{22}^{tt'J} J(J+1) + \sum_{s} g_{22}^{tt',s}\left(v_s + \frac{d_s}{2}\right)\right]$$

$$\times\left[(v_t + \ell_t + 1 \pm 1)(v_t - \ell_t + 1 \mp 1)(v_{t'} + \ell_{t'} + 1 \mp 1)(v_{t'} - \ell_{t'} + 1 \pm 1)\right]^{\frac{1}{2}}$$

In Tables LX, the rotation-vibration energies are given in terms of quantum numbers J and $\ell = \sum_t \ell_t$, and coefficients depending on molecular parameters and vibrational quantum numbers. Some of these coefficients are expanded in a power series with respect to vibrational quantum numbers in Table LXI. The presentation of Tables LX in this chapter is very similar to that of Tables XLVIII (Chapter X) and LIV (Chapter XI). The energies are obtained by adding up the various contributions written one under the other. When a splitting occurs for a given value of $|\ell|$ between the levels corresponding to $\psi^{+}_{|\ell|}$ and $\psi^{-}_{|\ell|}$ function[2], we indicate the function corresponding to each level.

Moreover in vibrational states $v_t = 2, 3$ and $v_t = v_{t'} = 1$, for $J \simeq 30$, there is a strong rotational ℓ-type resonance. Then, the diagonalization of the hamiltonian matrix requires the mixing of ψ^{+} functions corresponding to $|\ell|$ and $|\ell \pm 2|$; the same mixing happens for ψ^{-} functions. This situation occurs in Tables LX, 3 C, 4 C and 5C, summarized hereafter :

TABLE LX, 3 C

<table>
<tr><td colspan="2">- - - - - - - - - -</td><td>(I)</td></tr>
<tr><td>$\ell = 0$</td><td>- - - - - - -</td><td>(II)</td></tr>
<tr><td rowspan="2">$|\ell| = 2$</td><td>- - - - - - -</td><td>(III)</td></tr>
<tr><td>0</td><td>(IV)</td></tr>
</table>

The energies are obtained by adding up two of the contributions numbered from (I) to (IV):

If $\ell = 0$: (I) + (II)

If $|\ell| = 2$: $\begin{cases} \text{(I) + (III)} \\ \text{(I) + (IV)} \end{cases}$

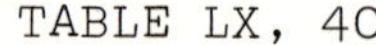
TABLE LX, 4C

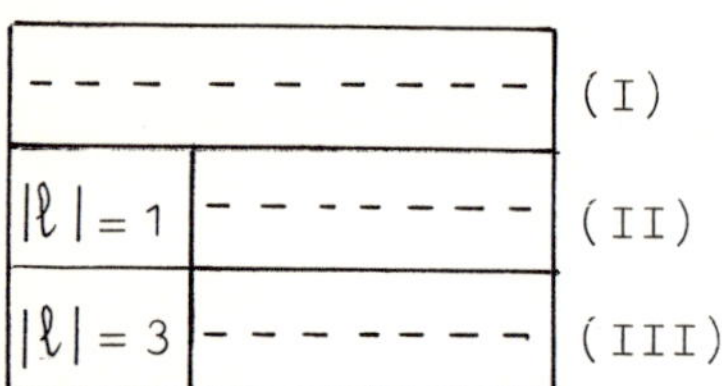

- - - - - - - - - -		(I)
$\|\ell\| = 1$	- - - - - - - -	(II)
$\|\ell\| = 3$	- - - - - - - -	(III)

The energies are :

If $|\ell| = 1$: (I) + (II) with the upper signs; (I) + (II) with the lower signs

If $|\ell| = 3$: (I) + (III) with the upper signs; (I) + (III) with the lower signs

TABLE LX, 5 C

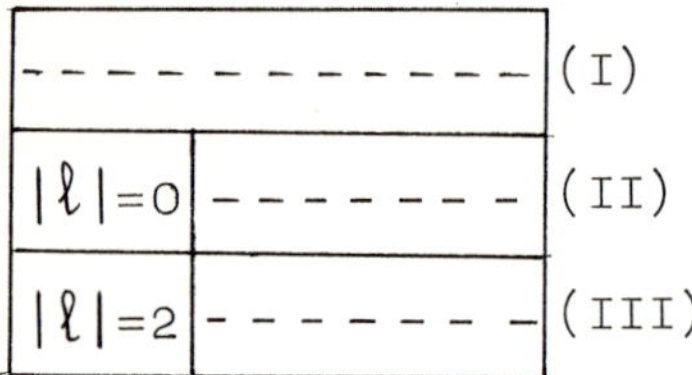

- - - - - - - - - - - -		(I)
$\|\ell\| = 0$	- - - - - - - -	(II)
$\|\ell\| = 2$	- - - - - - - -	(III)

The energies are :

If $|\ell| = 0$: (I) + (II) with the upper signs; (I) + (II) with the lower signs

If $|\ell| = 2$: (I) + (III) with the upper signs; (I) + (III) with the lower signs

FREQUENCIES OF ROTATION AND ROTATION-VIBRATION LINES FOR LINEAR MOLECULES

A . Selection rules

The rotation and rotation-vibration selection rules for linear molecules are given in Table LXII. These rules take into account all possible splittings for the levels[3].

B . Frequencies of Rotation Lines

Only $C_{\infty v}$ molecules have a pure rotation spectrum. The line frequencies have been computed for low values of J in the vibrational states previously studied, that is to say :

- in a vibrational state where only non-degenerate vibrations are excited ; the results are given in Table LXIII,1
- in a vibrational state where, besides non-degenerate vibrations, one twofold degenerate vibration is excited by 1, 2, or 3 quanta ; the results are given in Tables LXIII,2, 3 and 4 respectively
- in a vibrational state where, besides non-degenerate vibrations, two twofold degenerate vibrations are excited simultaneously, each by

one quantum : Table LXIII,5.

For this calculation, we have used expressions of the energies to sixth order.

Frequencies are expressed in terms of the rotational quantum number J associated with the lower level of the transition. They are called ν_J in the vibrational states $v_t = 0$ and $v_t = 1$ (Tables LXIII,1 and 2). When $v_t = 2$ or 3 and $v_t = v_{t'} = 1$ we give the frequencies of transitions, called $\nu_{J,|\ell|}$, in the two states defined by the two possible values of $|\ell|$. The frequencies are obtained from Tables LXIII by adding up the various contributions written one under the other, taking into account the value of $|\ell|$ when it is necessary. Let us give an example, for instance Table LXIII,4, summarized hereafter :

TABLE LXIII,4

– – – – – – – – – – – – – –		(I)
$\|\ell\| = 1$	$\psi^+ \rightarrow \psi^+$ transition	(II)
	$\psi^- \rightarrow \psi^-$ transition	(III)
$\|\ell\| = 3$	– – – – – – – – – –	(IV)

When $|\ell| = 1$, the frequencies of the transitions are :

$$\psi^+ \rightarrow \psi^+ \text{ transition} : \nu_{J,1++} = (I) + (II)$$

$$\psi^- \rightarrow \psi^- \text{ transition} : \nu_{J,1--} = (I) + (III)$$

When $|\ell| = 3$, the frequency is given by :

$$\nu_{J,3} = (I) + (IV)$$

C . Frequencies of Rotation-Vibration Lines

Tables LXIV, from 1 to 5, give frequencies of transitions between a state where only non-degenerate vibrations are excited and various vibrational states, for $C_{\infty v}$ and $D_{\infty h}$ molecules. The number from 1 to 5 characterizes the upper vibrational state, as indicated before.

The Tables LXIV, 6 and 7 give frequencies of the lines for hot bands between a state where, besides non-degenerate vibrations, one two-fold degenerate vibration is excited by one quantum and a state where besides non-degenerate vibrations, the same twofold degenerate vibration is excited by two quanta (for example, hot bands $2\nu_t - \nu_t$). In Tables LXIV, 6 are the transitions ending at $\ell = 0$ levels, in Tables LXIV, 7 , the transitions ending at $|\ell| = 2$ levels.

The allowed transitions for the hot band $v_t = 1, |\ell| = 1 \rightarrow v_t = 2, |\ell| = 2$ are summarized in figure (2) where the splitting of the $|\ell| = 2$ state is taken into account. In this figure, it may be seen that two kinds of transitions are possible, due to the ℓ-type splitting in the lower state $v_t = 1, |\ell| = 1$.

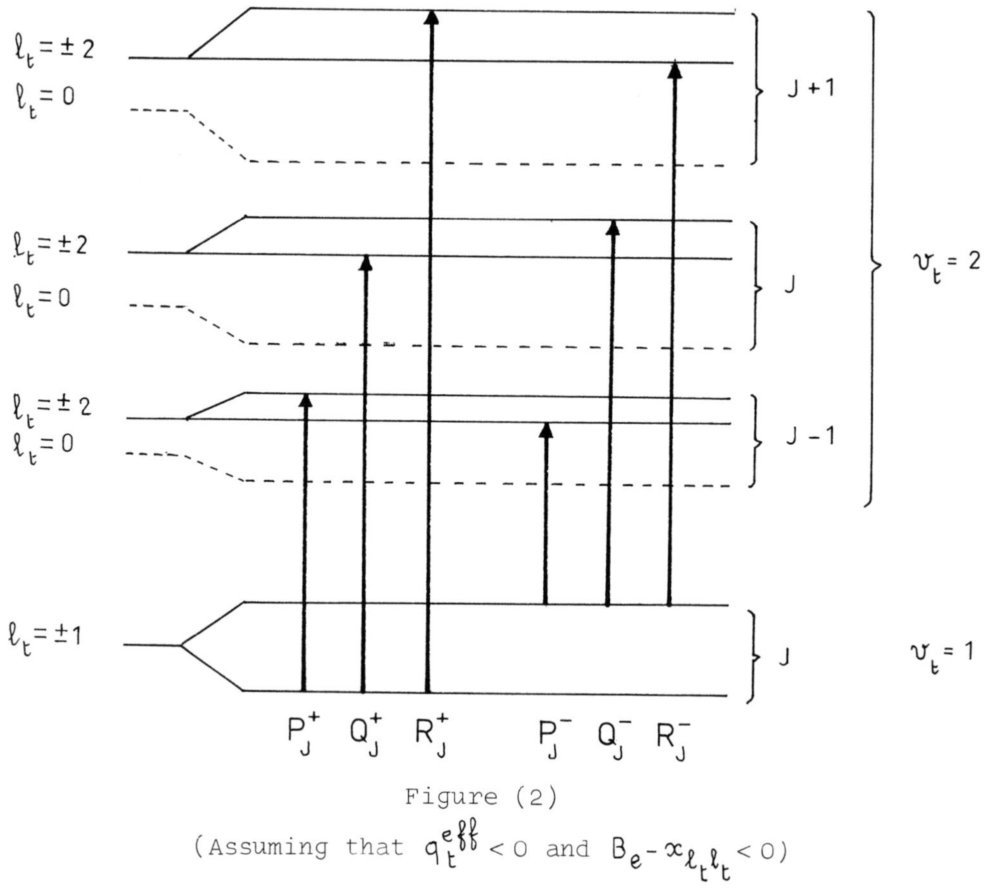

Figure (2)

(Assuming that $q_t^{eff} < 0$ and $B_e - x_{\ell_t \ell_t} < 0$)

In all the Tables LXIV the capital letters A and C refer respectively to the cases $J \simeq 1$ and $J \simeq 30$; the greek letters α and β refer to lines belonging to Q branches and R or P branches respectively. The symbols v' and v'' characterize respectively the upper and the lower vibrational states of the transitions. Quantum number J is that associated with the lower state of the transition. The following conventions have been used in Tables β :

$$m = J + 1 \text{ for } R_J \text{ lines}$$

$$m = -J \text{ for } P_J \text{ lines}$$

The frequencies are obtained from Tables LXIV by adding up the various contributions written one under the other. This process is straightforward for Tables LXIV, 1 A to 3 A, and Tables LXIV, 4 A, 5 A, 6 A and 6 C, but the other Tables present some difficulties.

Table LXIV, 3 Cβ : in this case, the upper state of the transition is $v_t = 2$, $\ell = 0$. However, there is a strong resonance between the $\ell = 0$ and $|\ell| = 2$ levels, the later being split in two components of which only one is displaced. Because of this strong resonance, the forbidden transition between the ground state and the displaced state $v_t = 2$, $|\ell| = 2$ may occur with a low intensity. We give in Table LXIV, 3 C the frequencies for the two transitions. The upper signs refer to the $v_t = 0 \longrightarrow v_t = 2$, $\ell = 0$ transition, the lower signs to the $v_t = 0 \longrightarrow v_t = 2$, $|\ell| = 2$ transition, this last transition being considered as a secondary series accompanying the main series.

The same situation arises in Tables LXIV, 4 Cα and 4 Cβ where the upper state is $v_t = 3$, $|\ell| = 1$. Due to a strong resonance, the transitions ending at the $v_t = 3$, $|\ell| = 3$ levels may occur. Their frequencies are obtained by using the lower signs in the tables.

Similarly, in Table LXIV, 5 Cβ the coupling between the state

$v_t = v_{t'} = 1$, $\ell = 0$ and the state $v_t = v_{t'} = 1$, $|\ell| = 2$ may induce a transition ending at the $|\ell| = 2$ state. The main series is obtained by taking the upper signs, the secondary series by taking the lower signs.

At last, in Tables LXIV, 7 where two kinds of transitions are possible, due to the ℓ-type splitting in the lower state $v_t = 1, |\ell| = 1$, the transitions starting from the ψ^+ component are called R_J^+, Q_J^+ and P_J^+ those starting from the ψ^- component are called R_J^-, Q_J^- and P_J^-

NOTES TO CHAPTER XII

1. In Table XXXVIII of Chapter VIII, $F_\pm^t$ is in place of q_t^0 and $R_\pm^{tt'}$ in place of $r_{tt'}^0/4$

2. For linear molecules the functions ψ^+ and ψ^- are defined by :

$$\psi^{\pm}(|\ell| \ell_t \ell_{t'} \ldots) = \frac{1}{\sqrt{2}}\left[\psi_0(\ell = |\ell|, \ell_t, \ell_{t'} \ldots) \pm \psi_0(\ell = -|\ell|, -\ell_t, -\ell_{t'})\right]$$

3. In addition to these rules, there is of course a vibrational selection rule $g \longleftrightarrow u$ for rotation-vibration transitions of $D_{\infty h}$ molecules.

4. Throughout the Tables LXIV the values of the quantum numbers $v_n, v_{n'} \ldots$ defining the upper vibrational state v' are said to be arbitrary. This statement holds actually only for linear molecules belonging to the symmetry group $C_{\infty v}$. In the case of symmetric linear molecules belonging to the group $D_{\infty h}$, the selection rule

$$\sum \Delta v_s^{(u)} = \pm 1, \pm 3, \pm 5 \ldots$$

must be taken into account (see above Note 3).

TABLES LX . ROTATION-VIBRATION ENERGIES FOR LINEAR MOLECULES

TABLE LX, 1 A . $v_t = v_{t'} = \ldots = 0$; $v_n , v_{n'} \ldots$ arbitrary ; $J \simeq 1$

Contributions	k: a	b	c
$\frac{E_v}{hc}$	0	2	4
$+J(J+1)B_v$	2	4	6
$-J^2(J+1)^2 D_e$	6		

TABLE LX, 1 C . $v_t = v_{t'} = \ldots = 0$; $v_n , v_{n'} \ldots$ arbitrary ; $J \simeq 30$

Contributions	k: a	b	c
$\frac{E_v}{hc}$	0	2	4
$+J(J+1)B_v$	0	2	4
$-J^2(J+1)^2 D_v$	2	4	
$+J^3(J+1)^3 H$	4		

TABLE LX, 2 A . $v_t = 1, |\ell_t| = 1$; $v_{t'} = \ldots = 0$; v_n , $v_{n'} \ldots$ arbitrary ; $J \simeq 1$

Contributions		k: a	b	c
$\frac{E_v}{hc}$		0	2	4
$+[J(J+1)-1]B_v$		2	4	6
$-[J(J+1)-1]^2 D_e$		6		
ψ_1^+ state	$+2q_t^{eff} J(J+1)$	4	6	
ψ_1^- state	$-2q_t^{eff} J(J+1)$	4	6	

TABLE LX, 2 C . $v_t = 1, |\ell_t| = 1$; $v_{t'} = \ldots = 0$; v_n , $v_{n'} \ldots$ arbitrary ; $J \simeq 30$

Contributions		k: a	b	c
$\frac{E_v}{hc}$		0	2	4
$+[J(J+1)-1]B_v$		0	2	4
$-[J(J+1)-1]^2 D_v$		2	4	
$+[J(J+1)-1]^3 H$		4		
ψ_1^+ state	$+2q_t^{eff} J(J+1)$	2	4	
	$+2f_{22}^{t,J} J^2(J+1)^2$	4		
ψ_1^- state	$-2q_t^{eff} J(J+1)$	2	4	
	$-2f_{22}^{t,J} J^2(J+1)^2$	4		

TABLE LX, 3 A . $v_t = 2, |\ell_t| = 0, 2 ; v_{t'} = \ldots = 0 ; v_n, v_{n'} \ldots$ arbitrary; $J \simeq 1$

Contributions			k: a	b	c
$\frac{E_v}{hc}$			0	2	4
$+[J(J+1)-\ell^2]B_v$			2	4	6
$-[J(J+1)-\ell^2]^2 D_e$			6		
$\ell = 0$	$+\frac{4(q_t^0)^2 J(J+1)[J(J+1)-2]}{B_e - x_{\ell_t \ell_t}}$		6		
$\lvert\ell\rvert = 2$	ψ_2^- state	0			
	ψ_2^+ state	$-\frac{4(q_t^0)^2 J(J+1)[J(J+1)-2]}{B_e - x_{\ell_t \ell_t}}$	6		

TABLE LX, 3 C . $v_t = 2, |\ell_t| = 0, 2; v_{t'} = \dots = 0; v_n, v_{n'} \dots$ arbitrary; $J \simeq 30$

	Contributions	k: a	b	c
	$\frac{E_v}{hc}$	0	2	4
	$+[J(J+1)-\ell^2]B_v$	0	2	4
	$-[J(J+1)-\ell^2]^2 D_v$	2	4	
	$+[J(J+1)-\ell^2]^3 H$	4		
$\ell = 0$	$+2\frac{\rho}{\lvert\rho\rvert}\sqrt{\rho^2+4J^2(J+1)^2(q_t^0)^2} - 2\rho$	2		
	$+2\sum_s \tau_s\left(v_s+\frac{d_s}{2}\right)+4J(J+1)D_e$	4		
	$-\frac{\rho}{\lvert\rho\rvert}\left\{2\rho\sum_s \tau_s\left(v_s+\frac{d_s}{2}\right)+4J(J+1)\left[\rho D_e+2(q_t^0)^2\right]-8J^2(J+1)^2 q_t^0 \sum_s f_{22}^{t,s}\left(v_s+\frac{d_s}{2}\right)\right.$ $\left.-8J^3(J+1)^3 q_t^0 f_{22}^{t,J}\right\}\left[\rho^2+4J^2(J+1)^2(q_t^0)^2\right]^{-\frac{1}{2}}$	4		

$\lvert \ell \rvert = 2$	$-2\frac{\rho}{\lvert\rho\rvert}\sqrt{\rho^2+4J^2(J+1)^2(q_t^o)^2}+2\rho$	2		
	$-2\sum_s \tau_s\left(v_s+\frac{d_s}{2}\right)-4J(J+1)D_e$	4		
	$+\frac{\rho}{\lvert\rho\rvert}\Big\{2\rho\sum_s \tau_s\left(v_s+\frac{d_s}{2}\right)+4J(J+1)\left[\rho D_e+2(q_t^o)^2\right]-8J^2(J+1)^2 q_t^o \sum_s f_{22}^{t,s}\left(v_s+\frac{d_s}{2}\right)$			
	$-8J^3(J+1)^3 q_t^o f_{22}^{t,J}\Big\}\left[\rho^2+4J^2(J+1)^2(q_t^o)^2\right]^{-\frac{1}{2}}$	4		
	0			

$$\rho = B_e - x_{\ell_t \ell_t}$$

$$\tau_s = y_{s\ell_t\ell_t} + \eta_{ts} + \alpha_s \quad (\eta_{ts} = 0 \text{ when } s \neq n)$$

TABLE LX, 4 A . $v_t = 3, |\ell_t| = 1, 3$; $v_{t'} = \ldots = 0$; $v_n, v_{n'} \ldots$ arbitrary ; $J \simeq 1$

Contributions			k: a	b	c
$\frac{E_v}{hc}$			0	2	4
$+\left[J(J+1)-\ell^2\right]B_v$			2	4	6
$-\left[J(J+1)-\ell^2\right]^2 D_e$			6		
$\lvert\ell\rvert = 1$	ψ_1^+ state	$+4\, q_t^{eff}\, J(J+1)$	4	6	
		$+\frac{3\left[J(J+1)-2\right]\left[J(J+1)-6\right](q_t^0)^2}{2(B_e - x_{\ell_t\ell_t})}$	6		
	ψ_1^- state	$-4\, q_t^{eff}\, J(J+1)$	4	6	
		$+\frac{3\left[J(J+1)-2\right]\left[J(J+1)-6\right](q_t^0)^2}{2(B_e - x_{\ell_t\ell_t})}$	6		

$\lvert \ell \rvert = 3$	$-\dfrac{3[J(J+1)-2][J(J+1)-6](q_t^0)^2}{2(B_e - x_{\ell_t \ell_t})}$	6		

TABLE LX, 4 C . $v_t = 3, |\ell_t| = 1, 3 ; v_{t'} = \ldots = 0 ; v_n, v_{n'} \ldots$ arbitrary ; $J \simeq 30$

Contributions	k: a	b	c
$\frac{E_v}{hc}$	0	2	4
$+\left[J(J+1)-\ell^2\right]B_v$	0	2	4
$-\left[J(J+1)-\ell^2\right]^2 D_v$	2	4	
$+\left[J(J+1)-\ell^2\right]^3 H$	4		
$\pm 2J(J+1)q_t^{eff}$	2	4	
$\pm 2J^2(J+1)^2 f_{22}^{t,J}$	4		
$+4\frac{\rho}{\lvert\rho\rvert}\sqrt{\rho^2 \pm \rho J(J+1)q_t^o + J^2(J+1)^2(q_t^o)^2} - 4\rho$	2		
$+4\sum_s \tau_s\left(v_s + \frac{d_s}{2}\right) + 8D_e J(J+1)$	4		

$\lvert l \rvert = 1$	$+\frac{\rho}{\lvert\rho\rvert}\Big\{-4\rho\sum_s \tau_s\left(v_s+\frac{d_s}{2}\right)-2J(J+1)\Big[4\rho D_e+6(q_t^0)^2\pm\sum_s\left(q_t^0\tau_s-\rho f_{22}^{t,s}\right)\left(v_s+\frac{d_s}{2}\right)\Big]$			
	$\pm 2J^2(J+1)^2\Big[\rho f_{22}^{t,J}-2D_e q_t^0\pm 2q_t^0\sum_s f_{22}^{t,s}\left(v_s+\frac{d_s}{2}\right)\Big]+4J^3(J+1)^3 q_t^0 f_{22}^{t,J}\Big\}$			
	$\times\left[\rho^2\pm\rho J(J+1)q_t^0+J^2(J+1)^2(q_t^0)^2\right]^{-\frac{1}{2}}$	4		
	$-4\frac{\rho}{\lvert\rho\rvert}\sqrt{\rho^2\pm\rho J(J+1)q_t^0+J^2(J+1)^2(q_t^0)^2}\quad+4\rho$	2		
	$-4\sum_s \tau_s\left(v_s+\frac{d_s}{2}\right)-8D_e J(J+1)$	4		
$\lvert l \rvert = 3$	$-\frac{\rho}{\lvert\rho\rvert}\Big\{-4\rho\sum_s \tau_s\left(v_s+\frac{d_s}{2}\right)-2J(J+1)\Big[4\rho D_e+6(q_t^0)^2\pm\sum_s\left(q_t^0\tau_s-\rho f_{22}^{t,s}\right)\left(v_s+\frac{d_s}{2}\right)\Big]$			
	$\pm 2J^2(J+1)^2\Big[\rho f_{22}^{t,J}-2D_e q_t^0\pm 2q_t^0\sum_s f_{22}^{t,s}\left(v_s+\frac{d_s}{2}\right)\Big]+4J^3(J+1)^3 q_t^0 f_{22}^{t,J}\Big\}$			
	$\times\left[\rho^2\pm\rho J(J+1)q_t^0+J^2(J+1)^2(q_t^0)^2\right]^{-\frac{1}{2}}$	4		

$$\rho = B_e - x_{l_t l_t}$$

$$\tau_s = y_{s l_t l_t} + \eta_{ts} + \alpha_s \qquad (\eta_{ts}=0 \text{ when } s \neq n)$$

TABLE LX, 5 A . $v_t = v_{t'} = 1$, $|\ell_t + \ell_{t'}| = 0, 2$; $v_{t''} = \ldots = 0$; $v_n, v_{n'} \ldots$ arbitrary ; $J \simeq 1$

		Contributions	k: a	k: b	k: c
		$\frac{E_v}{hc}$	0	2	4
		$+[J(J+1)-\ell^2]B_v$	2	4	6
		$-[J(J+1)-\ell^2]^2 D_e$	6		
$\ell = 0$	$\psi^+_{1,-1}$ state	$+\, r^{eff}_{tt'}$	2	4	6
		$+4 g^{tt';J}_{22} J(J+1) + \frac{4J(J+1)[J(J+1)-2][q^0_{t'}+q^0_t]^2}{4B_e - 2x_{\ell_t\ell_{t'}} + r^0_{tt'}}$	6		
	$\psi^-_{1,-1}$ state	$-\, r^{eff}_{tt'}$	2	4	6
		$-4 g^{tt';J}_{22} J(J+1) + \frac{4J(J+1)[J(J+1)-2][q^0_{t'}-q^0_t]^2}{4B_e - 2x_{\ell_t\ell_{t'}} - r^0_{tt'}}$	6		

$\lvert \ell \rvert = 2$	$\psi^{+}_{1,1}$ state	$-\dfrac{4J(J+1)\left[J(J+1)-2\right]\left[q^{o}_{t'}+q^{o}_{t}\right]^{2}}{4B_e-2x_{\ell_t\ell_{t'}}+r^{o}_{tt'}}$	6		
	$\psi^{-}_{1,1}$ state	$-\dfrac{4J(J+1)\left[J(J+1)-2\right]\left[q^{o}_{t'}-q^{o}_{t}\right]^{2}}{4B_e-2x_{\ell_t\ell_{t'}}-r^{o}_{tt'}}$	6		

TABLE LX , 5 C . $v_t = v_{t'} = 1, |\ell_t + \ell_{t'}| = 0, 2 ; v_{t''} = \ldots = 0 ; v_n, v_{n'} \ldots$ arbitrary ; $J \simeq 30$

Contributions	k: a	b	c
$\frac{E_v}{hc}$	0	2	4
$+\left[J(J+1)-\ell^2\right] B_v$	0	2	4
$-\left[J(J+1)-\ell^2\right]^2 D_v$	2	4	
$+\left[J(J+1)-\ell^2\right]^3 H$	4		
$\pm \frac{1}{2} r^{eff}_{tt'}$	2	4	
$\pm 2 g_{22}^{tt';J} J(J+1)$	4	.	
$+ \frac{\gamma^{\pm}}{\lvert\gamma^{\pm}\rvert} \left[(\gamma^{\pm})^2 + 4J^2(J+1)^2(q^o_{t'} \pm q^o_t)^2\right]^{\frac{1}{2}} - \rho'$	2		
$+\sum_s \tau_s (v_s + \frac{d_s}{2}) + 4J(J+1)D_e$	4		

$\ell = 0$	$+ \frac{\gamma^{\pm}}{\vert\gamma^{\pm}\vert}\Big\{-\gamma^{\pm}\sum_{s}(\tau_s \mp 2g_{22}^{tt';s})(v_s + \frac{d_s}{2}) + 2J(J+1)\big[\gamma^{\pm}(-2D_e \pm g_{22}^{tt';J}) - 2(q_{t'}^{o} \pm q_{t}^{o})^2\big]$ $+ 4J^2(J+1)^2(q_{t'}^{o} \pm q_{t}^{o})\sum_{s}(f_{22}^{t';s} \pm f_{22}^{t,s})(v_s + \frac{d_s}{2}) + 4J^3(J+1)^3(q_{t'}^{o} \pm q_{t}^{o})(f_{22}^{t';J} \pm f_{22}^{t,J})\Big\}$ $\times\big[(\gamma^{\pm})^2 + 4J^2(J+1)^2(q_{t'}^{o} \pm q_{t}^{o})^2\big]^{-\frac{1}{2}}$	4		
$\vert\ell\vert = 2$	$-\frac{\gamma^{\pm}}{\vert\gamma^{\pm}\vert}\big[(\gamma^{\pm})^2 + 4J^2(J+1)^2(q_{t'}^{o} \pm q_{t}^{o})^2\big]^{\frac{1}{2}} + \rho'$	2		
	$-\sum_{s}\tau_s(v_s + \frac{d_s}{2}) - 4J(J+1)D_e$	4		
	$-\frac{\gamma^{\pm}}{\vert\gamma^{\pm}\vert}\Big\{-\gamma^{\pm}\sum_{s}(\tau_s \mp 2g_{22}^{tt';s})(v_s + \frac{d_s}{2}) + 2J(J+1)\big[\gamma^{\pm}(-2D_e \pm g_{22}^{tt';J}) - 2(q_{t'}^{o} \pm q_{t}^{o})^2\big]$ $+ 4J^2(J+1)^2(q_{t'}^{o} \pm q_{t}^{o})\sum_{s}(f_{22}^{t';s} \pm f_{22}^{t,s})(v_s + \frac{d_s}{2}) + 4J^3(J+1)^3(q_{t'}^{o} \pm q_{t}^{o})(f_{22}^{t';J} \pm f_{22}^{t,J})\Big\}$ $\times\big[(\gamma^{\pm})^2 + 4J^2(J+1)^2(q_{t'}^{o} \pm q_{t}^{o})^2\big]^{-\frac{1}{2}}$	4		

$$\rho' = 2B_e - x_{\ell_t \ell_{t'}}$$

$$\gamma^{\pm} = \rho' \pm r_{tt'}^{o}/2$$

TABLE LXI . Coefficients appearing in the expressions of energies or frequencies

$\frac{E_v}{hc}$	(a)	$\sum_{s} \omega_s (v_s + \frac{d_s}{2})$
	(b)	$\sum_{\substack{ss' \\ s \leq s'}} x_{ss'} (v_s + \frac{d_s}{2})(v_{s'} + \frac{d_{s'}}{2}) + \sum_{\substack{tt' \\ t \leq t'}} x_{l_t l_{t'}} l_t l_{t'}$
	(c)	$\sum_{\substack{ss's'' \\ s \leq s' \leq s''}} y_{ss's''} (v_s + \frac{d_s}{2})(v'_s + \frac{d_{s'}}{2})(v_{s''} + \frac{d_{s''}}{2}) + \sum_{\substack{tt's \\ t \leq t'}} y_{s l_t l_{t'}} l_t l_{t'} (v_s + \frac{d_s}{2}) + \sum_{s} \Delta \omega_s (v_s + \frac{d_s}{2})$
B_v	(a)	B_e
	(b)	$- \sum_{s} \alpha_s (v_s + \frac{d_s}{2})$
	(c)	$\sum_{\substack{ss' \\ s \leq s'}} \gamma_{ss'} (v_s + \frac{d_s}{2})(v_{s'} + \frac{d_{s'}}{2}) + \sum_{\substack{tt' \\ t \leq t'}} \gamma_{l_t l_{t'}} l_t l_{t'} + \Delta B_e$

D_v	(a)	D_e
	(b)	$\sum_s \beta_s (v_s + \frac{d_s}{2})$
q_t^{eff}	(a)	q_t^0
	(b)	$\sum_s f_{22}^{t,s} (v_s + \frac{d_s}{2})$
$r_{tt'}^{eff}$	(a)	$r_{tt'}^0$
	(b)	$4 \sum_s g_{22}^{tt',s} (v_s + \frac{d_s}{2})$

An index s designates any vibration(non degenerate or twofold degenerate).

An index n designates a non degenerate vibration, an index t a twofold degenerate vibration.

The coefficient $y_{s l_t l_{t'}}$ partly arises from the operator r^6 of h_4^+, partly from the operator $r^4 P$ of h_3^+

TABLE LXII . Selection rules for dipole transitions of molecules $C_{\infty v}$ (Rotation and Rotation-Vibration) and $D_{\infty h}$ (Rotation-Vibration)

Parallel bands $\Delta \ell = 0$	R and P branches $\Delta J = \pm 1$	$\psi^+ \longleftrightarrow \psi^+$ $\psi^- \longleftrightarrow \psi^-$
Perpendicular bands $\Delta \ell = \pm 1$	Q branches $\Delta J = 0$	$\psi^+ \longleftrightarrow \psi^-$
	R and P branches $\Delta J = \pm 1$	$\psi^+ \longleftrightarrow \psi^+$ $\psi^- \longleftrightarrow \psi^-$
Rotational transitions $\Delta \ell = 0$	$\Delta J = +1$	$\psi^+ \longleftrightarrow \psi^+$ $\psi^- \longleftrightarrow \psi^-$

TABLES LXIII . FREQUENCIES OF LINES IN THE ROTATION SPECTRUM OF MOLECULES $C_{\infty v}$

TABLE LXIII, 1 . Rotational frequencies for $v_t = v_{t'} = \dots = 0$; $v_n, v_{n'} \dots$ arbitrary

ν_J	k: a	b	c
$2(J+1)B_v$	2	4	6
$-4(J+1)^3 D_e$	6		

TABLE LXIII, 2 . Rotational frequencies for $v_t = 1$, $\lvert l_t \rvert = 1$; $v_{t'} = \dots = 0$; $v_n, v_{n'} \dots$ arbitrary

ν_J		k: a	b	c
$2(J+1)[B_v$		2	4	6
$+2D_e]$		6		
$-4(J+1)^3 D_e$		6		
$\psi^+ \longrightarrow \psi^+$ transition	$+4q_t^{eff}(J+1)$	4	6	
$\psi^- \longrightarrow \psi^-$ transition	$-4q_t^{eff}(J+1)$	4	6	

TABLE LXIII, 3 . Rotational frequencies for $v_t = 2$; $|\ell_t| = 0, 2$; $v_{t'} = \ldots = 0$; $v_n, v_{n'} \ldots$ arbitrary

$\nu_{J,\lvert\ell\rvert}$			k: a	k: b	k: c
$2B_v(J+1)$			2	4	6
$-4(J+1)^3 D_e$			6		
$+4(J+1)\ell^2 D_e$			6		
$\ell = 0$	$+16(J+1)[(J+1)^2-1]\{(q_t^o)^2/(B_e - x_{\ell_t\ell_t})\}$		6		
$\lvert\ell\rvert = 2$	$\psi^+ \longrightarrow \psi^+$ transition	$-16(J+1)[(J+1)^2-1]\{(q_t^o)^2/(B_e - x_{\ell_t\ell_t})\}$			
	$\psi^- \longrightarrow \psi^-$ transition	0	6		

TABLE LXIII, 4 . Rotational frequencies for $v_t = 3, |\ell_t| = 1,3 ; v_{t'} = \ldots = 0 ; v_n, v_{n'} \ldots$ arbitrary

$\nu_{J,\lvert\ell\rvert}$			k: a	k: b	k: c
$2(J+1)B_v$			2	4	6
$-4(J+1)^3 D_e$			6		
$+4(J+1)\ell^2 D_e$			6		
$\lvert\ell\rvert = 1$	$\psi^+ \longrightarrow \psi^+$ transition	$+8 q_t^{eff}(J+1)$	4	6	
		$+6(J+1)[(J+1)^2-4]\{(q_t^0)^2/(B_e - x_{\ell_t\ell_t})\}$	6		
	$\psi^- \longrightarrow \psi^-$ transition	$-8 q_t^{eff}(J+1)$	4	6	
		$+6(J+1)[(J+1)^2-4]\{(q_t^0)^2/(B_e - x_{\ell_t\ell_t})\}$	6		
$\lvert\ell\rvert = 3$	$-6(J+1)[(J+1)^2-4]\{(q_t^0)^2/(B_e - x_{\ell_t\ell_t})\}$		6		

TABLE LXIII, 5 . Rotational frequencies for $v_t = v_{t'} = 1, |\ell_t + \ell_{t'}| = 0, 2; v_{t''} = \ldots = 0; v_n, v_{n'} \ldots$ arbitrary

$\nu_{J,\|\ell\|}$			k a	k b	k c
$2(J+1)B_v$			2	4	6
$-4(J+1)^3 D_e + 4(J+1)\ell^2 D_e$			6		
$\ell = 0$	$\psi^+ \longrightarrow \psi^+$ transition	$+8(J+1) g_{22}^{tt';J}$	6		
		$+16(J+1)\left[(J+1)^2-1\right]\left\{(q_{t'}^0 + q_t^0)^2/(4B_e - 2x_{\ell_t\ell_t} + r_{tt'}^0)\right\}$	6		
	$\psi^- \longrightarrow \psi^-$ transition	$-8(J+1) g_{22}^{tt;J}$	6		
		$+16(J+1)\left[(J+1)^2-1\right]\left\{(q_{t'}^0 - q_t^0)^2/(4B_e - 2x_{\ell_t\ell_t} - r_{tt'}^0)\right\}$	6		
$\|\ell\| = 2$	$\psi^+ \longrightarrow \psi^+$ transition	$-16(J+1)\left[(J+1)^2-1\right]\left\{(q_{t'}^0 + q_t^0)^2/(4B_e - 2x_{\ell_t\ell_t} + r_{tt'}^0)\right\}$	6		
	$\psi^- \longrightarrow \psi^-$ transition	$-16(J+1)\left[(J+1)^2-1\right]\left\{(q_{t'}^0 - q_t^0)^2/(4B_e - 2x_{\ell_t\ell_t} - r_{tt'}^0)\right\}$	6		

TABLES LXIV. FREQUENCIES OF LINES IN THE ROTATION-VIBRATION SPECTRUM OF LINEAR MOLECULES(*)

TABLE LXIV, 1 Aβ. Transition from $v''(v_t = v_{t'} = \ldots = 0\,;\, v_n, v_{n'} \ldots$ arbitrary) to $v'(v_t = v_{t'} = \ldots = 0\,;\, v_n, v_{n'} \ldots$ arbitrary) ; $J \simeq 1$

R_J P_J	k: a	b	c
$\dfrac{E_{v'} - E_{v''}}{hc}$	0	2	4
$+ m(B_{v'} + B_{v''})$	2	4	
$+ m^2(B_{v'} - B_{v''})$	-	4	

TABLE LXIV, 1 Cβ. Transition from $v''(v_t = v_{t'} = \ldots = 0\,;\, v_n, v_{n'} \ldots$ arbitrary) to $v'(v_t = v_{t'} = \ldots = 0\,;\, v_n, v_{n'} \ldots$ arbitrary) ; $J \simeq 30$

R_J P_J	k: a	b	c
$\dfrac{E_{v'} - E_{v''}}{hc}$	0	2	4
$+ m(B_{v'} + B_{v''})$	1	3	
$+ m^2(B_{v'} - B_{v''})$	-	2	4
$- 4m^3 D_e$	3		
$- m^4(D_{v'} - D_{v''})$	-	4	

(*) See note 4, page 398

TABLE LXIV, 2 A α . Transition from $v''(v_t = v_{t'} = \ldots = 0\,;\, v_n, v_{n'} \ldots$ arbitrary) to $v'(v_t = 1, |\ell_t| = 1\,;\, v_{t'} = \ldots = 0\,;\, v_n, v_{n'} \ldots$ arbitrary); $J \simeq 1$

Q_J	k: a	b	c
$\dfrac{E_{v'} - E_{v''}}{hc}$	0	2	4
$- B_{v'}$	2	4	
$+ J(J+1)\,[B_{v'} - B_{v''}$	–	4	
$- 2q_t^0]$	4		

TABLE LXIV, 2 A β . Transition from $v''(v_t = v_{t'} = \ldots = 0\,;\, v_n, v_{n'} \ldots$ arbitrary) to $v'(v_t = 1, |\ell_t| = 1\,;\, v_{t'} = \ldots = 0\,;\, v_n, v_{n'} \ldots$ arbitrary); $J \simeq 1$

R_J P_J	k: a	b	c
$\dfrac{E_{v'} - E_{v''}}{hc}$	0	2	4
$- B_{v'}$	2	4	
$+ m\,[B_{v'} + B_{v''}$	2	4	
$+ 2q_0^t]$	4		
$+ m^2\,[B_{v'} - B_{v''}$	–	4	
$+ 2q_t^0]$	4		

TABLE LXIV, 2 C α . Transition from v'' ($v_t = v_{t'} = \ldots = 0$; v_n , $v_{n'} \ldots$ arbitrary) to v' ($v_t = 1$, $|\ell_t| = 1$; $v_{t'} = \ldots = 0$; v_n , $v_{n'} \ldots$ arbitrary) ; $J \simeq 30$

a_J	k: a	b	c
$\dfrac{E_{v'} - E_{v''}}{hc}$	0	2	4
$- B_{v'}$	2	4	
$+ J(J+1)\left[B_{v'} - B_{v''}\right.$	–	2	4
$- 2q_t^{eff}$	2	4	
$\left. + 2D_e\right]$	4		
$- J^2(J+1)^2\left[D_{v'} - D_{v''}\right.$	–	4	
$\left. + 2f_{22}^{t,J}\right]$	4		

TABLE LXIV, 2 Cβ. Transition from $v''(v_t = v_{t'} = \ldots = 0\,;\, v_n, v_{n'} \ldots$ arbitrary) to $v'(v_t = 1, |\ell_t| = 1\,;\, v_{t'} = \ldots = 0\,;\, v_n, v_{n'} \ldots$ arbitrary); $J \simeq 30$

R_J P_J	k a	k b	k c
$\dfrac{E_{v'} - E_{v''}}{hc}$	0	2	4
$- B_{v'}$	2	4	
$+ m\,[B_{v'} + B_{v''}$	1	3	
$+ 2 q_t^0\,]$	3		
$+ m^2\,[B_{v'} - B_{v''}$	–	2	4
$+ 2 q_t^{eff}$	2	4	
$+ 2 D_e\,]$	4		
$- 4 m^3 D_e$	3		
$- m^4\,[D_{v'} - D_{v''}$	–	4	
$- 2 f_{22}^{t,J}\,]$	4		

TABLE LXIV, 3 Aβ. Transition from $v''(v_t = v_{t'} = \ldots = 0\,;\, v_n, v_{n'} \ldots$ arbitrary) to $v'(v_t = 2, |\ell_t| = 0,2\,;\, v_{t'} = \ldots = 0\,;\, v_n, v_{n'} \ldots$ arbitrary); $J \simeq 1$

R_J P_J	k a	k b	k c
$\dfrac{E_{v'}(\ell = 0) - E_{v''}}{hc}$	0	2	4
$+ m\,(B_{v'} + B_{v''})$	2	4	
$+ m^2\,(B_{v'} - B_{v''})$	–	4	

TABLE LXIV, 3 C β . Transition from v'' ($v_t = v_{t'} = \ldots = 0$; v_n , $v_{n'} \ldots$ arbitrary) to v' ($v_t = 2$, $|\ell_t| = 0, 2$; $v_{t'} = \ldots = 0$; v_n , $v_{n'} \ldots$ arbitrary); $J \simeq 30$

R_J P_J	k: a	b	c
$\dfrac{E_{v'}(\ell=0) - E_{v''}}{hc}$	0	2	4
$-2B_{v'}$	2	4	
$+2x_{\ell_t \ell_t}$	2		
$+2\sum_s (y_{s\ell_t\ell_t} + \eta_{ts})(v'_s + \frac{d_s}{2})$	4		
$+m(B_{v'} + B_{v''})$	1	3	
$+m^2[B_{v'} - B_{v''}$	–	2	4
$+4D_e]$	4		
$-4m^3 D_e$	3		
$-m^4(D_{v'} - D_{v''})$	–	4	
$\pm 2\frac{\rho}{\lvert\rho\rvert}\left[\rho^2 + 4m^4(q_t^0)^2\right]^{\frac{1}{2}}$	2		
$\pm 8\frac{\rho}{\lvert\rho\rvert} m^3 (q_t^0)^2 \left[\rho^2 + 4m^4(q_t^0)^2\right]^{-\frac{1}{2}}$	3		
$\mp \frac{\rho}{\lvert\rho\rvert}\left[\rho^2 + 4m^4(q_t^0)^2\right]^{-\frac{1}{2}} \Big\{ 2\rho\sum_s \tau_s (v'_s + \frac{d_s}{2}) + 4m^2[\rho D_e$ $+ (q_t^0)^2] - 8m^4 q_t^0 \sum_s f_{22}^{t,s}(v'_s + \frac{d_s}{2}) - 8m^6 q_t^0 f_{22}^{t,J} \Big\}$	4		

$\rho = B_e - x_{\ell_t \ell_t}$; $\tau_s = y_{s\ell_t\ell_t} + \eta_{ts} + \alpha_s$ ($\eta_{ts} = 0$ when $s \neq n$)

TABLE LXIV, 4 Aα. Transition from $v''(v_t = v_{t'} = \ldots = 0\,; v_n, v_{n'} \ldots$ arbitrary) to v' $(v_t = 3, |\ell_t| = 1,3\,; v_{t'} = \ldots = 0\,; v_n, v_{n'} \ldots$arbitrary) ; $J \simeq 1$

Q_J	k		
	a	b	c
$\frac{E_{v'}(\lvert\ell\rvert = 1) - E_{v''}}{hc}$	0	2	4
$-B_{v'}$	2	4	
$+ J(J+1)\,[B_{v'} - B_{v''}$	–	4	
$-4q_t^0\,]$	4		

TABLE LXIV, 4 Aβ . Transition from v'' $(v_t = v_{t'} = \ldots = 0\,; v_n, v_{n'} \ldots$ arbitrary) to v' $(v_t = 3, |\ell_t| = 1,3\,; v_{t'} = \ldots = 0\,; v_n, v_{n'} \ldots$arbitrary) ; $J \simeq 1$

R_J P_J	k		
	a	b	c
$\frac{E_{v'}(\lvert\ell\rvert = 1) - E_{v''}}{hc}$	0	2	4
$-B_{v'}$	2	4	
$+m\,[B_{v'} + B_{v''}$	2	4	
$-4q_t^0\,]$	4		
$+m^2\,[B_{v'} - B_{v''}$	–	4	
$+4q_t^0\,]$	4		

TABLE LXIV, 4 Cα . Transition from $v''(v_t = v_{t'} = \ldots = 0\ ; v_n, v_{n'} \ldots$ arbitrary) to $v'(v_t = 3, |\ell_t| = 1,3\ ; v_{t'} = \ldots = 0\ ; v_n, v_{n'} \ldots$ arbitrary); $J \simeq 30$

Q_J	k: a	b	c
$\dfrac{E_{v'}(\lvert\ell\rvert=1) - E_{v''}}{hc}$	0	2	4
$- 5 B_{v'}$	2	4	
$+ 4 x_{\ell_t \ell_t}$	2		
$+ 4 \sum_s (y_{s\ell_t\ell_t} + \eta_{ts})(v'_s + \frac{d_s}{2})$	4		
$+ J(J+1)[B_{v'} - B_{v''}$	–	2	4
$- 2 q_t^{\text{eff}}$	2	4	
$+ 10 D_e]$	4		
$- J^2(J+1)^2 [D_{v'} - D_{v''}$	–	4	
$+ 2 f_{22}^{t,J}]$	4		
$\pm 4 \frac{\rho}{\lvert\rho\rvert} [\rho^2 - \rho J(J+1) q_t^0 + J^2(J+1)^2 (q_t^0)^2]^{\frac{1}{2}}$	2		
$\pm \frac{\rho}{\lvert\rho\rvert} \{-4\rho \sum_s \tau_s (v'_s + \frac{d_s}{2}) - 2J(J+1)[4\rho D_e + 6(q_t^0)^2 - \sum_s (q_t^0 \tau_s - \rho f_{22}^{t,s})(v'_s + \frac{d_s}{2})] - 2J^2(J+1)^2[\rho f_{22}^{t,J} - 2D_e q_t^0 - 2q_t^0 \sum_s f_{22}^{t,s}(v'_s + \frac{d_s}{2})] + 4J^3(J+1)^3 q_t^0 f_{22}^{t,J}\} [\rho^2 - \rho J(J+1) q_t^0 + J^2(J+1)^2 (q_t^0)^2]^{-\frac{1}{2}}$	4		

$\rho = B_e - x_{\ell_t \ell_t}$; $\tau_s = y_{s\ell_t\ell_t} + \eta_{ts} + \alpha_s$ ($\eta_{ts} = 0$ when $s \neq n$)

TABLE LXIV, $4C\beta$. Transition from v'' ($v_t = v_{t'} = \dots = 0$; $v_n, v_{n'} \dots$ arbitrary) to v' ($v_t = 3, |\ell_t| = 1,3$; $v_{t'} = \dots = 0$; $v_n, v_{n'} \dots$ arbitrary); $J \simeq 30$

R_J P_J	k: a	b	c
$\frac{E_{v'}(\lvert\ell\rvert=1) - E_{v''}}{hc}$	0	2	4
$-5B_{v'}$	2	4	
$+4x_{\ell_t\ell_t}$	2		
$+4\sum_s (y_{s\ell_t\ell_t} + \eta_{ts})(v'_s + \frac{d_s}{2})$	4		
$+m[B_{v'} + B_{v''}$	1	3	
$+2q_t^0]$	3		
$+m^2[B_{v'} - B_{v''}$	–	2	4
$+2q_t^{eff}$	2	4	
$+10D_e]$	4		
$-4m^3 D_e$	3		

$-m^4\left[D_{v'} - D_{v''}\right.$	–	4	
$\left. - 2 f_{22}^{t,J}\right]$	4		
$\pm 4 \frac{\rho}{\lvert\rho\rvert}\left[\rho^2 + \rho m^2 q_t^0 + m^4 (q_t^0)^2\right]^{\frac{1}{2}}$	2		
$\pm 2 \frac{\rho}{\lvert\rho\rvert}\, m \left[\rho q_t^0 + 2m^2 (q_t^0)^2\right]\left[\rho^2 + \rho m^2 q_t^0 + m^4 (q_t^0)^2\right]^{-\frac{1}{2}}$	3		
$\pm \frac{\rho}{\lvert\rho\rvert}\left\{-4\rho \sum_s \tau_s (v'_s + \frac{d_s}{2}) - 2m^2\left[4\rho D_e + 5(q_t^0)^2 + \sum_s (q_t^0 \tau_s - \rho f_{22}^{t,s})(v'_s + \frac{d_s}{2})\right] + 2m^4\left[\rho f_{22}^{t,J}\right.\right.$			
$\left.\left. - 2D_e q_t^0 + 2 q_t^0 \sum_s f_{22}^{t,s}(v'_s + \frac{d_s}{2})\right] + 4m^6 q_t^0 f_{22}^{t,J}\right\}\left[\rho^2 + \rho m^2 q_t^0 + m^4 (q_t^0)^2\right]^{-\frac{1}{2}}$	4		

$$\rho = B_e - x_{l_t l_t}$$

$$\tau_s = y_{s l_t l_t} + \eta_{ts} + \alpha_s \qquad (\eta_{ts} = 0 \quad \text{when} \quad s \neq n)$$

TABLE LXIV, 5 Aβ . Transition from $v''(v_t = v_{t'} = \dots = 0 ; v_n , v_{n'} \dots$ arbitrary) to v' $(v_t = v_{t'} = 1$, $|\ell_t + \ell_{t'}| = 0$; $v_{t''} = \dots = 0 ; v_n , v_{n'} \dots$ arbitrary) ; $J \simeq 1$

R_J P_J	k: a	b	c
$\frac{E_{v'}(\ell = 0) - E_{v''}}{hc}$	0	2	4
$+ r^{eff}_{tt'}$	2	4	
$+ m(B_{v'} + B_{v''})$	2	4	
$+ m^2(B_{v'} - B_{v''})$	–	4	

TABLE LXIV, 5 Cβ . Transition from $v''(v_t = v_{t'} = \dots = 0 ; v_n , v_{n'} \dots$ arbitrary) to v' $(v_t = v_{t'} = 1$, $|\ell_t + \ell_{t'}| = 0$; $v_{t''} = \dots = 0 ; v_n , v_{n'} \dots$ arbitrary) ; $J \simeq 30$

R_J P_J	k: a	b	c
$\frac{E_{v'}(\lvert\ell\rvert = 0) - E_{v''}}{hc}$	0	2	4
$- 2B_{v'} + r^{eff}_{tt'}/2$	2	4	

$+x_{l_t l_{t'}}$	2		
$+\sum_s (y_s l_t l_{t'} + \eta_{ts} + \eta_{t's})(v'_s + \frac{d_s}{2})$	4		
$+m(B_{v'} + B_{v''})$	1	3	
$+m^2 [B_{v'} - B_{v''}$	–	2	4
$+2g_{22}^{tt';J} + 4D_e]$	4		
$\pm \frac{\gamma^+}{\lvert\gamma^+\rvert} [(\gamma^+)^2 + 4m^4 (q_{t'}^o + q_t^o)^2]^{\frac{1}{2}}$	2		
$\pm 4 \frac{\gamma^+}{\lvert\gamma^+\rvert} m^3 (q_{t'}^o + q_t^o)^2 [(\gamma^+)^2 + 4m^4 (q_{t'}^o + q_t^o)^2]^{-\frac{1}{2}} \quad - 4m^3 D_e$	3		
$-m^4 (D_{v'} - D_{v''}$	–	4	
$\pm \frac{\gamma^+}{\lvert\gamma^+\rvert} \{ -\gamma^+ \sum_s (\tau_s - 2g_{22}^{tt';s})(v'_s + \frac{d_s}{2}) + 2m^2 [\gamma^+(-2D_e + g_{22}^{tt';J}) - (q_{t'}^o + q_t^o)^2]$			
$+4m^4 (q_{t'}^o + q_t^o) \sum_s (f_{22}^{t';s} + f_{22}^{t,s})(v'_s + \frac{d_s}{2}) + 4m^6 (q_{t'}^o + q_t^o)(f_{22}^{t';J} + f_{22}^{t,J}) \} [(\gamma^+)^2$			
$+4m^4 (q_{t'}^o + q_t^o)^2]^{-\frac{1}{2}}$	4		

$$\rho' = 2B_e - x_{l_t l_{t'}} \qquad ; \gamma^+ = \rho' + r_{tt'}^o / 2$$

TABLE LXIV, 6 A α. Transition from v'' ($v_t = 1, |\ell_t| = 1, v_{t'} = \ldots = 0$; $v_n, v_{n'} \ldots$ arbitrary) to v' ($v_t = 2, \ell_t = 0$; $v_{t'} = \ldots = 0$; $v_n, v_{n'} \ldots$ arbitrary); $J \simeq 1$

Q_J ($J = 1,2,3\ldots$)	k: a	b	c
$\dfrac{E_{v'}(\ell = 0) - E_{v''}(\lvert\ell\rvert = 1)}{hc}$	0	2	4
$+ B_{v'}$	2	4	
$+ [J(J+1) - 1](B_{v'} - B_{v''})$	-	4	
$+ 2 q_t^{eff} J(J+1)$	4		

TABLE LXIV, 6 A β. Transition from v'' ($v_t = 1, |\ell_t| = 1$; $v_{t'} = \ldots = 0$; $v_n, v_{n'} \ldots$ arbitrary) to v' ($v_t = 2, \ell_t = 0$; $v_{t'} = \ldots = 0$; $v_n, v_{n'} \ldots$ arbitrary); $J \simeq 1$

R_J ($J = 1,2,3\ldots$) P_J ($J = 1,2,3\ldots$)	k: a	b	c
$\dfrac{E_{v'}(\ell = 0) - E_{v''}(\lvert\ell\rvert = 1)}{hc}$	0	2	4
$+ B_{v'}$	2	4	
$+ m(B_{v'} + B_{v''})$	2	4	
$+ (m^2 - 1)(B_{v'} - B_{v''})$	-	4	
$- 2m(m-1) q_t^{eff}$	4		

TABLE LXIV, 6 C α . Transition from v'' ($v_t = 1, |\ell_t| = 1$; $v_{t'} = \ldots = 0$; $v_n, v_{n'} \ldots$ arbitrary) to v' ($v_t = 2$, $\ell_t = 0$; $v_{t'} = \ldots = 0$; $v_n, v_{n'} \ldots$ arbitrary) ; $J \simeq 30$

Q_J ($J = 1,2,3\ldots$)	k: a	b	c
$\dfrac{E_{v'}(\ell=0) - E_{v''}(\|\ell\|=1)}{hc}$	0	2	4
$- B_{v'}$	2	4	
$-(B_{v'} - B_{v''})$	–	4	
$+ J(J+1)\left[B_{v'} - B_{v''}\right.$	–	2	4
$+ 2q_t^{eff}$	2	4	
$\left.+ 2D_e\right]$	4		
$- J^2(J+1)^2\left[D_{v'} - D_{v''}\right.$	–	4	
$\left.- 2f_{22}^{t,J}\right]$	4		
$+ 2x_{\ell_t\ell_t}$	2		
$+ 2\sum_s (y_{s\ell_t\ell_t} + \eta_{ts})(v'_s + \frac{d_s}{2})$	4		
$+ 2\frac{\rho}{\|\rho\|}\left[\rho^2 + 4(q_t^0)^2 J^2(J+1)^2\right]^{\frac{1}{2}}$	2		
$- \frac{\rho}{\|\rho\|}\left[\rho^2 + 4(q_t^0)^2 J^2(J+1)^2\right]^{-\frac{1}{2}}\left\{2\rho\sum_s \tau_s(v'_s + \frac{d_s}{2}) + 4J(J+1)\left[\rho D_e + 2(q_t^0)^2\right] - 8J^2(J+1)^2 q_t^0 \sum_s f_{22}^{t,s}(v'_s + \frac{d_s}{2}) - 8J^3(J+1)^3 q_t^0 f_{22}^{t,J}\right\}$	4		

$$\rho = B_e - x_{\ell_t\ell_t}$$

$$\tau_s = y_{s\ell_t\ell_t} + \eta_{ts} + \alpha_s \quad (\eta_{ts} = 0 \text{ when } s \neq n)$$

TABLE LXIV, 6 Cβ. Transition from v'' ($v_t = 1, |\ell_t| = 1$; $v_{t'} = \ldots = 0$; $v_n, v_{n'} \ldots$ arbitrary) to v' ($v_t = 2$, $\ell_t = 0$; $v_{t'} = \ldots = 0$; $v_n, v_{n'} \ldots$ arbitrary) ; $J \simeq 30$

R_J (J = 1,2,3...) P_J (J = 2,3,4...)	k a	k b	k c
$\frac{E_{v'}(\ell = 0) - E_{v''}(\lvert\ell\rvert = 1)}{hc}$	0	2	4
$- B_{v'}$	2	4	
$- (B_{v'} - B_{v''})$	–	4	
$+ m\,[B_{v'} + B_{v''}$	1	3	
$+ 2 q_t^{eff}]$	3		
$+ m^2\,[B_{v'} - B_{v''}$	–	2	4
$+ 2 q_t^{eff}$	2	4	
$+ 2 D_e]$	4		
$- 4 m^3 D_e$	3		

$-m^4\left[D_{v'}-D_{v''}\right.$	–	4	
$\left.+2f_{22}^{tJ}\right]$	4		
$+2x_{l_t l_t}$	2		
$+2\sum_s (y_s l_t l_t+\eta_{ts})(v'_s+\frac{d_s}{2})$	4		
$+2\frac{\rho}{\lvert\rho\rvert}\left[\rho^2+4(q_t^0)^2 m^4\right]^{\frac{1}{2}}$	2		
$-8\frac{\rho}{\lvert\rho\rvert} m^3 (q_t^0)^2\left[\rho^2+4(q_t^0)^2 m^4\right]^{-\frac{1}{2}}$	3		
$-\frac{\rho}{\lvert\rho\rvert}\left\{2\rho\sum_s \tau_s (v'_s+\frac{d_s}{2})+4m^2\left[\rho D_e+(q_t^0)^2\right]\right.$			
$\left.-8m^4 q_t^0\sum_s f_{22}^{t,s}(v'_s+\frac{d_s}{2})-8m^6 q_t^0 f_{22}^{t,J}\right\}\left[\rho^2+4m^4(q_t^0)^2\right]^{-\frac{1}{2}}$	4		

$$\rho = B_e - x_{l_t l_t}$$

$$\tau_s = y_s l_t l_t + \eta_{ts} + \alpha_s \qquad (\eta_{ts}=0 \text{ when } s \neq n)$$

TABLE LXIV, 7 A α . Transition from $v''(v_t = 1, |\ell_t| = 1\ ;\ v_{t'} = \ldots = 0\ ;\ v_n, v_{n'} \ldots$ arbitrary) to $v'(v_t = 2, |\ell_t| = 2\ ;\ v_{t'} = \ldots = 0\ ; v_n, v_{n'} \ldots$ arbitrary) ; $J \simeq 1$

$Q_J^{\pm}$ $(J = 2,3,4\ldots)$	k: a	b	c
$\frac{E_{v'}(\lvert\ell\rvert = 2) - E_{v''}(\lvert\ell\rvert = 1)}{hc}$	0	2	4
$-3B_{v'}$	2	4	
$+[J(J+1)-1](B_{v'} - B_{v''})$	–	4	
$\mp 2 q_t^{eff} J(J+1)$	4		

TABLE LXIV, 7 A β . Transition from $v''(v_t = 1, |\ell_t| = 1,\ v_{t'} = \ldots = 0\ ;\ v_n, v_{n'} \ldots$ arbitrary) to $v'(v_t = 2, |\ell_t| = 2\ ;\ v_{t'} = \ldots = 0\ ; v_n, v_{n'} \ldots$ arbitrary) ; $J \simeq 1$

$R_J^{\pm}$ $(J = 1,2,3\ldots)$ $P_J^{\pm}$ $(J = 3,4,5\ldots)$	k: a	b	c
$\frac{E_{v'}(\lvert\ell\rvert = 2) - E_{v''}(\lvert\ell\rvert = 1)}{hc}$	0	2	4
$-3B_{v'}$	2	4	
$+m(B_{v'} + B_{v''})$	2	4	
$+(m^2-1)(B_{v'} - B_{v''})$	–	4	
$\mp 2m(m-1) q_t^{eff}$	4		

TABLE LXIV, 7 C α . Transition from $v''(v_t=1, |\ell_t|=1\ ;\ v_{t'}=\ldots=0\ ;\ v_n, v_{n'}\ldots$ arbitrary) to $v'(v_t=2, |\ell_t|=2\ ;\ v_{t'}=\ldots=0\ ;\ v_n, v_{n'}\ldots$ arbitrary) ; $J \simeq 30$

Q_J^+ and Q_J^- $(J = 2,3,4\ldots)$		k: a	b	c
$\frac{E_{v'}(\vert\ell\vert=2) - E_{v''}(\vert\ell\vert=1)}{hc}$		0	2	4
$-3B_{v'}$		2	4	
$-(B_{v'} - B_{v''})$		–	4	
$+J(J+1)\left[B_{v'} - B_{v''}\right.$		–	2	4
$\left. + 6D_e\right]$		4		
$-J^2(J+1)^2(D_{v'} - D_{v''})$		–	4	
Q_J^+	$-2q_t^{eff}J(J+1)$	2	4	
	$-2f_{22}^{t,J}J^2(J+1)^2$	4		
	$+J(J+1)\left[q_t^{eff}\right.$	2	4	
	$\left. -4D_e\right]$	4		

Q_J^-	$+f_{22}^{tJ} J^2(J+1)^2$	4		
	$+2B_{v'}$	2	4	
	$-2x_{l_t l_t}$	2		
	$-2\sum_s (y_{s l_t l_t} + \eta_{ts})(v'_s + \frac{d_s}{2})$	4		
	$-2\frac{\rho}{\lvert\rho\rvert}\left[\rho^2 + 4(q_t^0)^2 J^2(J+1)^2\right]^{\frac{1}{2}}$	2		
	$+\frac{\rho}{\lvert\rho\rvert}\Big\{2\rho\sum_s \tau_s(v'_s + \frac{d_s}{2}) + 4J(J+1)\left[\rho D_e + 2(q_t^0)^2\right] - 8J^2(J+1)^2 q_t^0 \sum_s f_{22}^{t,s}(v'_s + \frac{d_s}{2})$			
	$-8J^3(J+1)^3 q_t^0 f_{22}^{t,s}\Big\}\left[\rho^2 + 4(q_t^0)^2 J^2(J+1)^2\right]^{-\frac{1}{2}}$	4		

$$\rho = B_e - x_{l_t l_t}$$

$$\tau_s = y_{s l_t l_t} + \eta_{ts} + \alpha_s \qquad (\eta_{ts} = 0 \text{ when } s \neq n)$$

TABLE LXIV, 7 Cβ . Transition from v'' ($v_t = 1, |\ell_t| = 1$; $v_{t'} = \ldots = 0$; $v_n, v_{n'} \ldots$ arbitrary) to v' ($v_t = 2$, $|\ell_t| = 2$; $v_{t'} = \ldots = 0$; $v_n, v_{n'} \ldots$ arbitrary) ; $J \simeq 30$

R_J^+ and R_J^- ($J = 1,2,3\ldots$) ; P_J^+ and P_J^- ($J = 3,4,5\ldots$)	k: a	b	c
$\frac{E_{v'}(\lvert\ell\rvert = 2) - E_{v''}(\lvert\ell\rvert = 1)}{hc}$	0	2	4
$- 3 B_{v'}$	2	4	
$- (B_{v'} - B_{v''})$	–	4	
$+ m (B_{v'} + B_{v''})$	1	3	
$+ m^2 [B_{v'} - B_{v''}$	–	2	4
$+ D_e]$	4		
$- 4 m^3 D_e$	3		
$- m^2 [2 q_t^{eff}$	2	4	
$+ 4 D_e]$	4		
$+ 2 q_t^{eff} m$	3		

	$-2 f_{22}^{tJ} m^4$	4		
	$+2 B_{v'}$	2	4	
$R_J^+ \quad P_J^+$	$-2 x_{l_t l_t}$	2		
	$-2 \sum_s (y_{s l_t l_t} + \eta_{ts})(v'_s + \frac{d_s}{2})$	4		
	$-2 \frac{\rho}{\lvert\rho\rvert} [\rho^2 + 4(q_t^o)^2 m^4]^{\frac{1}{2}}$	2		
	$+8 \frac{\rho}{\lvert\rho\rvert} m^3 (q_t^o)^2 [\rho^2 + 4(q_t^o)^2 m^4]^{-\frac{1}{2}}$	3		
	$+\frac{\rho}{\lvert\rho\rvert} \{ 2\rho \sum_s \tau_s (v'_s + \frac{d_s}{2}) + 4 m^2 [\rho D_e + (q_t^o)^2] - 8 m^4 q_t^o \sum_s f_{22}^{t,s} (v'_s + \frac{d_s}{2})$ $- 8 m^6 q_t^o f_{22}^{t,J} \} [\rho^2 + 4 m^4 (q_t^o)^2]^{-\frac{1}{2}}$	4		
	$+2 q_t^{eff} m^2$	2		
$R_J^- \quad P_J^-$	$-2 q_t^{eff} m$	3		
	$+2 f_{22}^{tJ} m^4$	4		

REFERENCES

[1] M.Goldsmith, G.Amat and H.H.Nielsen, J. Chem. Phys., 24, 1178 (1956)

[2] G.Amat, M.Goldsmith and H.H.Nielsen, J. Chem. Phys., 27, 838 (1957)

[3] G.Amat and H.H.Nielsen, J. Chem. Phys., 27, 845 (1957)

[4] G.Amat and H.H.Nielsen, J. Chem. Phys., 29, 665 (1958)

[5] G.Amat and H.H.Nielsen, J. Chem. Phys., 36, 1859 (1962)

[6] M.L.Grenier-Besson, G.Amat and H.H.Nielsen, J. Chem. Phys., 36, 3454 (1962)

[7] S.K.Kurtz, Thesis, The Ohio State University, Columbus, Ohio, 1960

[8] G.Tarrago, Cahiers Phys., 19, 149 (1965)

[9] P.Kupecek, These, Faculté des Sciences, Paris, 1970

[10] H.H.Nielsen, Rev. Mod. Phys., 23, 90 (1951)

[11] E.B.Wilson Jr. and J.B.Howard, J.Chem. Phys., 4, 262 (1936)

[12] B.T.Darling and D.M.Dennison, Phys. Rev., 57, 128 (1940)

[13] G.Amat and L.Henry, Cahiers Phys., 12, 273 (1958)

[14] J.K.G.Watson, Mol. Phys., 15, 479 (1968)

[15] W.H.Shaffer, H.H.Nielsen and L.H.Thomas, Phys. Rev., 56, 895 (1939)

[16] R.C.Herman and W.H.Shaffer, J. Chem. Phys., 16, 453 (1948)

[17] S.Maes and G.Amat, Cahiers Phys., 11, 277 (1957)

[18] G.Amat, M.Goldsmith and H.H.Nielsen, J. Phys. Rad., 16, 854 (1955)

[19] H.H.Nielsen, Phys. Rev., 68, 181 (1945)

[20] S.Maes, Cahiers Phys., 14, 125 (1960)

[21] M.H.Andrade e Silva and G.Amat, J. Mol. Spectrosc., 9, 354 (1962)

[22] M.H.Andrade e Silva and J.Ramadier, J. Phys. Rad., 26, 246 (1965)

[23] M.L.Grenier-Besson and G.Amat, Technical Note NR 3, March 1962, AF Contract NR 61 (052)-369

[24] S.Maes, J. Phys., 27, 37 (1966)

[25] G.Amat, Compt. Rend. Acad. Sci., 250, 1439 (1960)

[26] L.Henry and G.Amat, Cahiers Phys., 14, 230 (1960)

[27] A.Sayvetz, J. Chem. Phys., 7, 282 (1939)

[28] H.A.Jahn, Proc. Roy. Soc. A 168, 469 (1938)

[29] H.A.Jahn, Proc. Roy. Soc. A 168, 495 (1938)

[30] W.H.Childs and H.A.Jahn, Proc. Roy. Soc. A 169, 459 (1939)

[31] W.H.Shaffer, H.H.Nielsen and L.H.Thomas, Phys. Rev., 56, 1051 (1939)

[32] J.D.Louck, Thesis, The Ohio State University, Columbus, Ohio, 1958

[33] K.T.Hecht, J. Mol. Spectrosc., 5, 355 (1960)

[34] K.T.Hecht, J. Mol. Spectrosc., 5, 390 (1960)

[35] J.Herranz, J. Mol. Spectrosc., 6, 343 (1961)

[36] J.Moret-Bailly, Cahiers Phys., 15, 237 (1961)

[37] J.Moret-Bailly, J. Mol. Spectrosc., 15, 344 (1965)

[38] K.Fox, J. Mol. Spectrosc., 9, 381 (1962) and 16, 35 (1965) ; J. Chem. Phys., 43, 525 (1965)

[39] K.Fox, O.S.U. Symposium on Molecular Structure and Spectroscopy, The Ohio State University, Columbus, Ohio, June 1965

[40] C.F.Curtis, Symposium on Molecular Structure and Spectroscopy, The Ohio State University, Columbus, Ohio, June 1957

[41] C.C.Costain, J. Mol. Spectrosc. 9, 317 (1962)

[42] S.Maes and G.Amat, Canadian J. Phys., 43, 321 (1965)

[43] H.H.Nielsen, Handbuch der Physik, Vol. XXXVII/1, Springer Verlag (1959)

[44] M.L.Grenier-Besson, J. Phys., 25, 757 (1964)

[45] M.L.Grenier-Besson, J. Phys., 21, 555 (1960)

[46] G.Amat and L.Henry, J. Phys., 21, 728 (1960)

[47] G.Amat, M.L.Grenier-Besson and H.Z.Cummins, Compt. Rend. Acad. Sci., Paris, 244, 2380 (1957)

[48] G.Herzberg, Rev. Mod. Phys., 14, 219 (1942)

[49] G.Amat and H.H.Nielsen, J. Mol. Spectrosc., 2, 152 (1958)

[50] K.K.Yallabandi and P.M.Parker, J. Chem. Phys., 49, 420 (1968)

[51] B.R.Judd, Operator Techniques in Atomic Spectroscopy (McGraw-Hill, 1963)

[52] V.Fano and G.Racah, Irreductible Tensorial Sets (Academic Press Inc., New-York, 1959)

[53] J.Ramadier and G.Amat, J. Phys. Rad., 19, 915 (1958)

[54] M.H.Andrade e Silva, These, Faculté des Sciences, Paris, 1967

[55] H.Hönl and F.London, Zeits, Phys., 33, 803 (1925)

[56] H.Hönl and F.London, Ann. Phys . , 79, 273 (1926)

[57] D.M.Dennison, Phys. Rev., 28, 318 (1926)

[58] M.L.Grenier-Besson and G.Amat, J. Mol. Spectrosc., 8, 22 (1962)